Elements of General and Biological Chemistry

Elements of General and Biological Chemistry

Contributors

Muhammad Abdul Mojid Mondol, Hee Jae Shin et al.

www.aurisreference.com

Elements of General and Biological Chemistry

Contributors: Muhammad Abdul Mojid Mondol, Hee Jae Shin et al.

Published by Auris Reference Limited
www.aurisreference.com

United Kingdom

Copyright 2016
Printed in 2017 for Sale in the Indian Subcontinent

The information in this book has been obtained from highly regarded resources. The copyrights for individual articles remain with the authors, as indicated. All chapters are distributed under the terms of the Creative Commons Attribution License, which permit unrestricted use, distribution, and reproduction in any medium, provided the original author and source are credited.

Notice

Contributors, whose names have been given on the book cover, are not associated with the Publisher. The editors and the Publisher have attempted to trace the copyright holders of all material reproduced in this publication and apologise to copyright holders if permission has not been obtained. If any copyright holder has not been acknowledged, please write to us so we may rectify.

Reasonable efforts have been made to publish reliable data. The views articulated in the chapters are those of the individual contributors, and not necessarily those of the editors or the Publisher. Editors and/or the Publisher are not responsible for the accuracy of the information in the published chapters or consequences from their use. The Publisher accepts no responsibility for any damage or grievance to individual(s) or property arising out of the use of any material(s), instruction(s), methods or thoughts in the book.

Elements of General and Biological Chemistry

ISBN: 978-1-78154-851-6

British Library Cataloguing in Publication Data
A CIP record for this book is available from the British Library

Printed in the United Kingdom

Exclusively distributed by CBS Publishers & Distributors Pvt. Ltd.

Sales & Distribution Rights only for India, Pakistan, Bangladesh, Sri Lanka, Nepal and Bhutan. This book is not to be sold outside these territories.

Contents

List of Abbreviations .... *vii*
List of Contributors .... *ix*
Preface .... *xiii*

Chapter 1 **Diversity of Secondary Metabolites from Marine Bacillus Species: Chemistry and Biological Activity** 1
Muhammad Abdul Mojid Mondol, Hee Jae Shin and Mohammad Tofazzal Islam

Chapter 2 **Mentha suaveolens Ehrh. (Lamiaceae) Essential Oil and Its Main Constituent Piperitenone Oxide: Biological Activities and Chemistry** 33
Mijat Božović, Adele Pirolli and Rino Ragno

Chapter 3 **Novel Biological Activities of Allosamidins** 71
Shohei Sakuda, Hiromasa Inoue and Hiromichi Nagasawa

Chapter 4 **Biological Activities of Hydrazone Derivatives** 93
Sevim Rollas and Ş. Güniz Küçükgüzel

Chapter 5 **Synthesis of N-(6-Arylbenzo[d] thiazole-2-acetamide Derivatives and Their Biological Activities: An Experimental and Computational Approach** 133
Yasmeen Gull, Nasir Rasool, Mnaza Noreen, Ataf Ali Altaf, Syed Ghulam Musharraf, Muhammad Zubair, Faiz-Ul-Hassan Nasim, Asma Yaqoob, Vincenzo DeFeo and Muhammad Zia-Ul-Haq

Chapter 6 **A Survey of Chemical Compositions and Biological Activities of Yemeni Aromatic Medicinal Plants** 161
Bhuwan K. Chhetri, Nasser A. Awadh Ali and William N. Setzer

Chapter 7 **The Chemistry and Toxicology of Depleted Uranium** 193
Sidney A. Katz

Chapter 8 **Syntheses of Sulfo-Glycodendrimers Using Click Chemistry and Their Biological Evaluation** 231
Yoshiko Miura, Shunsuke Onogi and Tomohiro Fukuda

Chapter 9 Biological Chemistry of Reactive Oxygen and Nitrogen and
Radiation-Induced Signal Transduction Mechanisms 259
Ross B Mikkelsen and Peter Wardman

Citations 307

Index 309

List of Abbreviations

AChE	Acetylcholine esterase
ALT	Alanine amino transferase
ALP	Alkaline phosphatase
AST	Aspartate amino transferase
AFM	Atomic force microscopy
BAL	Bronchoalveolar lavage
BHT	Butylated hydroxytoluene
cLPs	CD Circular dichroism
DU	Depleted uranium
DMF	Dimethylformamide
DLS	Dynamic light scattering
EEG	Electroencephalographic
ESI-MS	Electrospray ionization mass spectroscopy
FAS	Fatty acid synthases
FITC	Fluorescein isothiocyanate containing sulfonated N-acetyl-D-glucosamine
GAGs	Glycosaminoglycans
HSV-1	Herpes simplex virus-1
HEQEP	Higher Education Quality Enhancement Project
HIV	Human immunodeficiency virus
ICRP	International Commission on Radiological Protection
ISNEs	Isonicotinoylhydrazones
MRSA	Methicillin-resistant Staphylococcus aureus
MRSA	Meticilin-resistant Staphylococcus aureus
MBC	Minimum bactericidal concentration
MIC	Minimum inhibitory concentrations
NNIs	Non-nucleoside inhibitors
NRPs	Nonribosomal peptides
PDF	Peptide deformylase
PO	Piperitenone oxide
PKS	Polyketide synthetase
PQQ	Pyrroloquinoline quinone
RNS	Reactive nitrogen species
ROS	Reactive oxygen species
RT	Reverse transcriptase
scPTZ	Subcutaneous pentylenetetrazole
SOD	Superoxide dismutase
TBARS	Thiobarbituric acid reactive substances
UNEP	United Nations Environment Programme
USFDA	US Food and Drug Administration

VRE	Vancomycin-resistant enterococcus
WGA	Wheat germ agglutinin
WHO	World Health Organization

List of Contributors

Muhammad Abdul Mojid Mondol
School of Science and Technology, Bangladesh Open University, Board Bazar, Gazipur 1705, Bangladesh

Hee Jae Shin
Marine Natural Products Chemistry Laboratory, Korea Institute of Ocean Science & Technology, Ansan, Seoul 425-600, Korea

Mohammad Tofazzal Islam
Department of Biotechnology, Bangabandhu Sheikh Mujibur Rahman Agricultural University, Gazipur 1706, Bangladesh

Mijat Božović
Rome Center for Molecular Design, Department of Drug Chemistry and Technology, Sapienza University, P.le Aldo Moro 5, 00185 Rome, Italy

Adele Pirolli
Rome Center for Molecular Design, Department of Drug Chemistry and Technology, Sapienza University, P.le Aldo Moro 5, 00185 Rome, Italy

Rino Ragno
Rome Center for Molecular Design, Department of Drug Chemistry and Technology, Sapienza University, P.le Aldo Moro 5, 00185 Rome, Italy

Shohei Sakuda
Department of Applied Biological Chemistry, the University of Tokyo, Bunkyo-ku, Tokyo 113-8657, Japan

Hiromasa Inoue
Department of Pulmonary Medicine, Kagoshima University, 8-35-1 Sakuragaoka, Kagoshima 890-8520, Japan

Hiromichi Nagasawa
Department of Applied Biological Chemistry, the University of Tokyo, Bunkyo-ku, Tokyo 113-8657, Japan

Sevim Rollas
Marmara University, Faculty of Pharmacy, Department of Pharmaceutical Chemistry, 34668, Turkey

Ş. Güniz Küçükgüzel
Marmara University, Faculty of Pharmacy, Department of Pharmaceutical Chemistry, 34668, Turkey

Yasmeen Gull
Department of Chemistry, Faculty of Science and Technology, Government College University Faisalabad, Faisalabad 38000, Pakistan
Department of Chemistry, Faculty of Science, University of Sargodhah, Bhakkar Campus, Bhakkar 30000, Pakistan

Nasir Rasool
Department of Chemistry, Faculty of Science and Technology, Government College University Faisalabad, Faisalabad 38000, Pakistan

Mnaza Noreen
Department of Chemistry, Faculty of Science and Technology, Government College University Faisalabad, Faisalabad 38000, Pakistan

Ataf Ali Altaf
Department of Chemistry, Faculty of Science, University of Gujrat, Hafiz Hayat Campus, Gujrat 50700, Pakistan

Syed Ghulam Musharraf
International Center for Chemical and Biological Sciences, Hussain Ebrahim Jamal Research Institute of Chemistry, University of Karachi, Karachi 75270, Pakistan

Muhammad Zubair
Department of Chemistry, Faculty of Science and Technology, Government College University Faisalabad, Faisalabad 38000, Pakistan

Faiz-Ul-Hassan Nasim
Department of Chemistry, Faculty of Science, Islamia University of Bahawalpur, Bahawalpur 63000, Pakistan

Asma Yaqoob
Department of Chemistry, Faculty of Science, Islamia University of Bahawalpur, Bahawalpur 63000, Pakistan

Vincenzo DeFeo
Department of Pharmaceutical and Biomedical Sciences, University of Salerno, Via Ponte don Melillo, Fisciano (Salerno) I-84084, Italy

Muhammad Zia-Ul-Haq
Offices of Research, Innovation and Commercialization, Lahore College for Women University, Lahore 54600, Pakistan

Bhuwan K. Chhetri
Department of Chemistry, University of Alabama in Huntsville, Huntsville, AL 35899, USA

Nasser A. Awadh Ali
Pharmacognosy Department, Faculty of Pharmacy, Sana'a University, Yemen
Pharmacognosy Department, Faculty of Clinical Pharmacy, Albaha University, Saudi Arabia

William N. Setzer
Department of Chemistry, University of Alabama in Huntsville, Huntsville, AL 35899, USA

Sidney A. Katz
Department of Chemistry, Rutgers University, Camden, NJ 08102-1411, USA

Yoshiko Miura
Department of Chemical Engineering, Graduate School of Engineering, Kyushu University, 744 Motooka, Nishi-ku, Fukuoka 819-0395, Japan
School of Materials Science, Japan Advanced Institute of Science and Technology, 1-1 Asahidai, Nomi, Ishikawa 923-1292, Japan

Shunsuke Onogi
School of Materials Science, Japan Advanced Institute of Science and Technology, 1-1 Asahidai, Nomi, Ishikawa 923-1292, Japan

Tomohiro Fukuda
School of Materials Science, Japan Advanced Institute of Science and Technology, 1-1 Asahidai, Nomi, Ishikawa 923-1292, Japan
Department of Applied Chemistry and Chemical Engineering, Toyama National College of Technology, Hongo-campus, 13 Hongo-Machi, Toyama, Toyama 939-8630, Japan

Ross B Mikkelsen
Department of Radiation Oncology, Virginia Commonwealth University, 401 College Street, Richmond, VA 23298, USA

Peter Wardman
Gray Cancer Institute, Mount Vernon Hospital, Northwood, Middlesex HA6 2JR, UK

Preface

Chemistry is the science that studies the composition and changes in composition of the substances around us. Biological chemistry is the study of chemical processes within and relating to living organisms. It is closely related to molecular biology, the study of the molecular mechanisms by which genetic information encoded in DNA is able to result in the processes of life. The text *Elements of General and Biological Chemistry* explains recent advances in biological chemistry. First chapter reviews the chemistry and biological activities of secondary metabolites from marine isolates. Second chapter aims to review the scientific findings and research reported to date on MS that prove many of the remarkable various biological actions, effects and some uses of species as a source of bioactive natural compound. In third chapter, recent studies on the novel biological activities of allosamidins have been reviewed. Fourth chapter focuses on biological activities of hydrazone derivatives. The purpose of fifth chapter is to synthesize novel N-(6-arylbenzo[d]thiazol-2-yl)acetamides employing the Pd(0)-catalyzed Suzuki cross coupling methodology. In sixth chapter, we present a survey of several aromatic plants used in traditional medicine in Yemen, their traditional uses, their volatile chemical compositions, and their biological activities. Seventh chapter reviews on the chemical and toxicological properties of depleted and natural uranium and some of the possible consequences from long term, low dose exposure to depleted uranium in the environment. In eighth chapter, a series of novel glycol-clusters containing sulfonated N-acetyl-D-glucosamine (GlcNAc) have been synthesized using click chemistry. Last chapter summarizes recent studies on the chemistry of radiation-induced ROS/RNS generation and emphasizes interactions between ROS and RNS and the relative roles of cellular ROS/RNS generators as amplifiers of the initial ionization events. Cellular mechanisms for regulating ROS/RNS levels are discussed.

Chapter 1

DIVERSITY OF SECONDARY METABOLITES FROM MARINE BACILLUS SPECIES: CHEMISTRY AND BIOLOGICAL ACTIVITY

Muhammad Abdul Mojid Mondol[1], Hee Jae Shin[2] and Mohammad Tofazzal Islam[3]

[1]School of Science and Technology, Bangladesh Open University, Board Bazar, Gazipur 1705, Bangladesh

[2]Marine Natural Products Chemistry Laboratory, Korea Institute of Ocean Science & Technology, Ansan, Seoul 425-600, Korea

[3]Department of Biotechnology, Bangabandhu Sheikh Mujibur Rahman Agricultural University, Gazipur 1706, Bangladesh

ABSTRACT

Marine *Bacillus* species produce versatile secondary metabolites including lipopeptides, polypeptides, macrolactones, fatty acids, polyketides, and isocoumarins. These structurally diverse compounds exhibit a wide range of biological activities, such as antimicrobial, anticancer, and antialgal activities. Some marine *Bacillus* strains can detoxify heavy metals through reduction processes and have the ability to produce carotenoids. The present article reviews the chemistry and biological activities of secondary metabolites from marine isolates. Side by side, the potential for application of these novel natural products from marine *Bacillus* strains as drugs, pesticides, carotenoids, and tools for the bioremediation of heavy metal toxicity are also discussed.

INTRODUCTION

Microorganisms especially bacteria and fungi are promising sources of structurally diverse and potent bioactive compounds [1,2,3]. Many of these bioactive compounds are used as chemotherapeutic agents for the treatment of human and animal diseases [4,5]. There is a continuing demand for novel bioactive compounds to treat drug-resistant human and animal pathogens

[6,7,8], and management of devastating pathogens of crops, which are insensitive or less sensitive to existing chemical pesticides [9,10,11,12].

Marine environments including the subsurface are believed to contain a total of approximately 3.67×10^{30} microorganisms [13]. About 70% of the earth's surface is covered by the ocean representing 80% of life on earth indicating an enormous pool of microbial biodiversity and potential discovery of new natural products [14]. Among diverse microbial species, isolates of marine *Bacillus* belong to phylogenetically and phenogenetically heterogeneous groups of bacteria. They are ubiquitous in the marine environment and can tolerate adverse conditions such as high temperature, pressure, salinity, and pH [15]. Generally, *Bacillus* strains need more nutrition and space to be the fastest growing bacteria for which they compete with other organisms. Due to the diluting effect of the ocean drives, marine organisms produce potent bioactive compounds to fight off their competitors or to escape from micropredation [16,17]. Metabolically marine strains are different from their terrestrial counterparts, and thereby, they may produce unique bioactive compounds, which are not found in their terrestrial counterparts [18,19]. The ability to produce diverse classes of antibiotics by *Bacillus* spp. has been evidence by several genomic studies. For example, the genome sequence of the widely distributed *Bacillus* strains revealed that about 8% of genome is devoted to synthesizing antibiotics [20,21]. Similarly, genome analysis of a marine *B. subtilis* subsp. *spiziseniistrain* gtP20b, isolated from the Indian Ocean, indicated the presence of huge number of genes for biosynthesis of secondary metabolites [22].

Marine *Bacillus* isolates produce structurally diverse classes of secondary metabolites, such as lipopeptides, polypeptides, macrolactones, fatty acids, polyketides, lipoamides, and isocoumarins [23,24] (Figure 1). These structurally versatile compounds exhibit a wide range of biological activities, such as antimicrobial, anticancer, antialgal, and antiperonosporomycetal [23,24]. As *Bacillus* strains rapidly grow in liquid media even under stressful conditions and readily forms resistant spores, it might be useful as an effective biocontrol agent against various phytopathogens [25]. Structures, syntheses, and specific functions of diverse antibiotics produced by *B. subtilis* have elaborately been reviewed [26].

Extensive use of pesticides during crop production and exposure of industrial toxic waste to the environment results in the deposition of heavy metals (Pd, Hg, Cu, Cd, Cr, Co) in soil and water bodies, and this practice can pose serious threats to crop production as well as to the health of all organisms in the aquatic environment. Although trace amount of many heavy metal ions are required for various biochemical activities of all living organisms, higher concentrations of these ions are generally toxic to their cells. Surprisingly, some bacterial species exhibit tolerance and resistance towards high concentrations of heavy metals. Marine strains, in general, thrive under harsh conditions when compared to the most of terrestrial environments, providing them with enormous tolerance, and thus, they are often considered as potential candidates for the detoxification of heavy metals [27]. Most of the toxic heavy metals are reduced rather than oxidized by native microbes, with a few exceptions, since the reduced forms are, generally, less toxic. Bacterial tolerance to heavy metals generally follows the pattern of efflux, accumulation, complexation, and reduction [28].

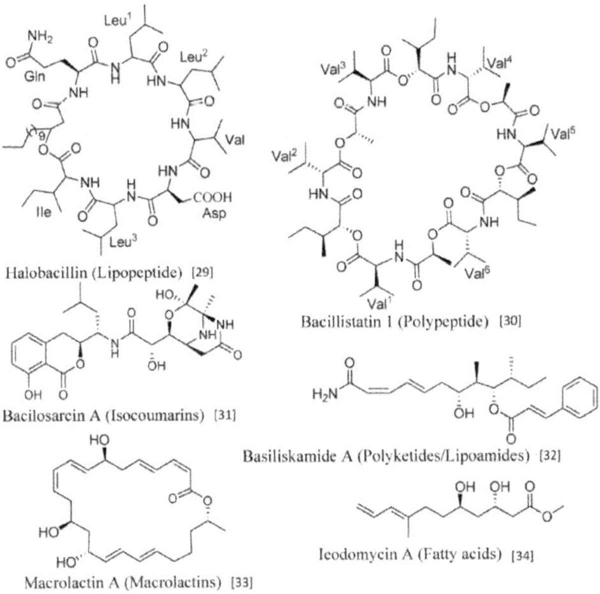

Figure 1: Major classes of secondary metabolites produced by *Bacillus* strains.

Nowadays, food growers heavily rely on chemical pesticides to prevent or control diseases in their crop plants. Deposition of these pesticide residues in food, soil, and water bodies is imposing enormous threats to human health, environment, and ecosystem [35]. Consequently, there is an increasing demand from consumers and environment conservationists to replace chemical pesticides with natural, environment-friendly antagonistic microorganisms with novel mode of action for sustainable production of crops [11,12].

As marine *Bacillus* isolates produce diverse bioactive secondary metabolites with novel modes of action, they may have great potential for the development of effective management strategies to combat human, animal and phytopathogens in biorational manners. Several general reviews on bioactive secondary metabolites from *Bacillus* isolates have been published [23,26], however, no report has so far been published on bioactive compounds from marine *Bacillus* species. This article comprehensively reviews the chemistry and biological activities of diverse secondary metabolites from marine *Bacillus*species, and discusses the potentials of these natural products for the development of effective drugs, agrochemicals, carotenoids, as well as tools for the bioremediation of environmental pollution due to heavy metal contamination.

BIOACTIVE COMPOUNDS

Lipopeptides

Cyclic lipopeptides (cLPs) are versatile metabolites produced by a variety of bacterial genera. They are composed of a short cyclic oligopeptide linked to a fatty acid tail, and exhibit potent surfactant properties [36]. cLPs are produced nonribosomally on large NRPS (nonribosomal peptide synthetase)—PKS (polyketide synthetase) hybrid synthetases. Endospore-forming rhizobacterium (terrestrial) *B. subtilis* produces varieties of antimicrobial peptides that are either ribosomally synthesized and post-translationally modified (e.g., lantibiotics and lantibiotic-like peptides), or nonribosomally generated. Ribosomally and nonribosomally synthesized peptides by terrestrial *Bacillus* spp. are elaborately reviewed by Stein [26]. cLPs have received considerable attention for their antibiotic activities against a range of human- and plant-pathogenic organisms, including enveloped viruses, mycoplasmas, trypanosomes, bacteria, fungi, and peronosporomycetes [37].

A large proportion of the secondary metabolites produced by the *Bacillus* isolates are cyclic lipopeptides belonging to three families: iturins, fengycins, and surfactins [38]. The chemical structure of c-LPs have a peptide backbone composed of seven (iturins and surfactins) or 10 (fengycins) amino acids connected to a β-hydroxy (fengycins and surfactins) or β-amino (iturins) fatty acid, which may vary from C-10 to C-16 for surfactins, C-14 to C-17 for iturins, and C-14 to C-18 for fengycins. Each lipopeptide family can be subdivided again based on position of specific amino acid in the peptide ring. For example, the fengycin family is subdivided into fengycin A and fengycin B, where D-alanine and D-valine is present in the sixth position, respectively. Due to difference in length, branching, and saturation of acyl chain, homologues are formed within each subdivision of cLPs (Figure 2). Based on the comparison of retention times and molecular masses with those of known compounds, all three families of lipopeptide antibiotics may be assigned using LC-MS with specific elution programs [39].

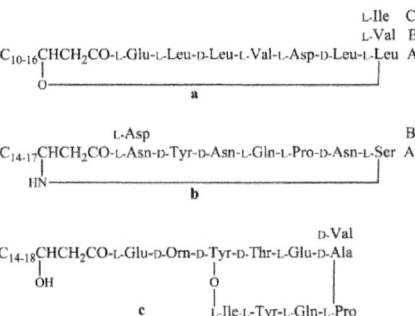

Figure 2: General structures of surfactin (a); iturin (b) and fengycin (c).

A cLPs producer, *B. amyloliquefaciens* strain GA1 was initially isolated from strawberry culture. To identify gene fragments responsible for synthesis of cLPs, partial DNA sequencing of strain GA1 has been done. Analysis of sequencing data revealed that fourteen gene fragments with homology toward gene clusters were involved in the synthesis of cLPs [38]. Out of fourteen, in eleven gene fragments, operons *srf* and *fen* had 80%–96% and 41%–92% amino acid identity directing the synthesis of surfactin and fengycin, respectively [38]. Further analysis of sequencing data revealed that three gene fragments have been involved in directing the synthesis of an iturin lipopeptide in the strain FZB42 and have 48%–82% amino acid similarity with the itu*DABC* operon encoding the iturin A synthetase in *B. subtilis* RB14 [40].

Five surfactin analogs (1–5) (Figure 3) have been isolated from the culture broth of a *Bacillus pumilus* (SP21) through bioassay-guided fractionation [41]. The producing strain was isolated from a sediment sample collected from the Bahamas. Compounds 1–4 showed selectively good inhibitory activity against *S. aureus*, *P. vulgaris*, and *E. faecalis* (6.5 to 25 μg/mL), but not against *P. aeruginosa* [41].

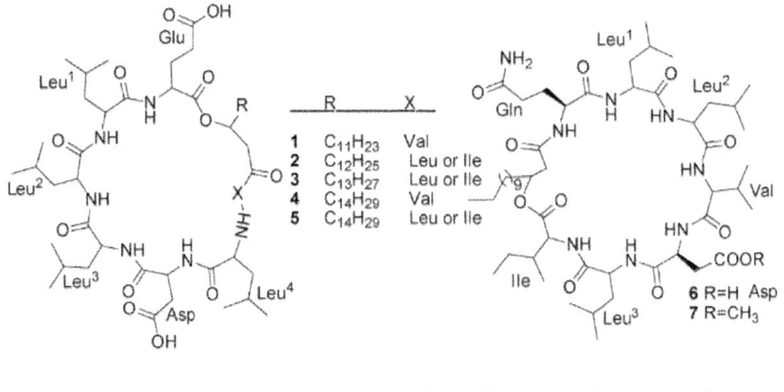

Figure 3: Structures of surfactin analogs (1–5) and halobacillins (6 and 7).

Halobacillin (6) and methyl halobacillin (7) (Figure 3), two cyclic acylpeptides have been isolated from the culture broth of a *Bacillus* sp. CND-914. The strain CND-914 was obtained from a deep-sea sediment core taken at near Guaymas Basins in Mexico [29]. Halobacillin, the first novel acylpeptide of the iturin class similar to surfactin, is one of the most effective known biosurfactants [42]. The major difference between halobacillin and surfactin is the replacement of glutamic acid of surfactin with a glutamine in halobacillin. Halobacillin inhibits the growth of human colon tumor cells (HCT-116) with an IC_{50} of 0.98 µg/mL but it does not exhibit antimicrobial activity like surfactin (Table 1).

Three lipopeptides (8–10) (Figure 4), iso-C16 fengycin B (8), anteiso-C17 fengycin B (9) and a new iturinic lipopeptide, mojavensin A (10), were obtained by bioactivity-guided fractionation from the fermentation broth of *B. mojavensis* B0621A, which was isolated from the mantle of a pearl oyster *P. martensii* collected from Weizhou Island in the South China Sea [43]. These lipopeptides displayed dose-dependent antifungal activity against a broad spectra of phytopathogens as well as being weakly antagonistic to *S. aureus*. Moreover, they all displayed cytotoxic activities against the human leukemia (HL-60) cell line with IC_{50} values of 100, 100, and 1.6 µM, respectively [43].

Table 1: List of some important bioactive compounds isolated from marine strains

Compounds	Producing strains	Test organisms/cell lines	Inhibitory concentrations	Nature of bioactivities	Ref.
Halobacillin (6)	*Bacillus* sp. CND-914	Human HCT-116 cancer cells	0.98 µg/mL (IC_{50})	Anticancer	[29]
Mixirin (11)	*Bacillus* sp.	Human HCT-116 cancer cells	0.68 µg/mL (IC_{50})	Anticancer	[44]
Bogorol A (15)	*Bacillus* sp.	MRSA	2 µg/mL (MIC)	Antibacterial	[45]
Loloatin B (18)	*Bacillus* sp.	MRSA, VRE	1–2 µg/mL (MIC)	Antibacterial	[46]
Bacillistatins 1 (19) and 2 (20)	*B. silvestris*	Human cancer cell line	10^{-4}–10^{-5} µg/mL (GI_{50})	Anticancer	[30]
Bacillamide (27)	*Bacillus* sp.	*C. polykrikoides*	LC_{50} after 6 h: 3.2 µg/mL	Antialgal	[47]
Bacilosarcin A (35)	*B. subtilis*	Barnyard millet sprouts	82% inhibition at 50 µM	Plant growth regulator	[31]
Macrolactin S (69)	*B. amyloliquefaciens*	*E. coli* and *S. aureus*	0.3 and 0.1 µg/mL (MIC)	Antibacterial	[48]
Macrolactin V (86)	*B. amyloliquefaciens*	*E. coli*, *B. subtilis* and *S. aureus*	0.1 µg/mL (MIC)	Antibacterial	[48]
Basiliskamides A (21) and B (22)	*B. laterosporus*	*C. albicans* and *A. fumigatus*	1.0 and 3.1 µg/mL 2.5 and 5.0 µg/mL	Antifungal	[32]

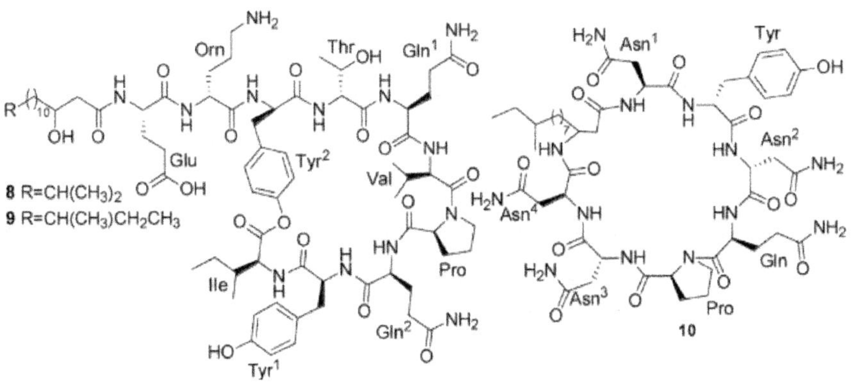

Figure 4: Structures of fengycins (8 and 9) and mojavensin (10).

Three new cyclic acylpeptides named mixirins A–C (11–13) (Figure 5) belonging to the iturin class have been isolated from marine bacterium *Bacillus* sp. [44]. This isolate was obtained from sea mud near the Arctic pole. Mixirins A, B, and C inhibited the growth of human colon tumor cells (HCT-116) with IC_{50} of 0.68, 1.6, 1.3 µg/mL, respectively (Table 1) [44].

A new cyclic lipopeptide, marihysin A (14) (Figure 5) was isolated from the fermentation broth of the marine *B. marinus* B-9987 isolated from the tissues of rhizophere of *Suaeda salsa* in the intertidal zone of the Bohai Bay, China, which exhibited a broad-spectrum activity against plant pathogens with minimum inhibitory concentrations (MICs) of 100–200 µg/mL [49].

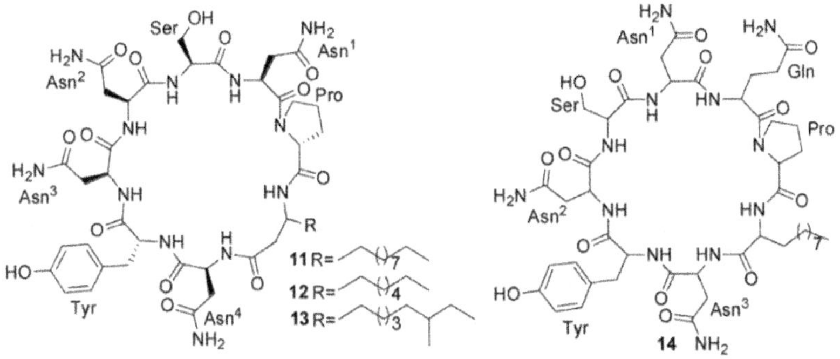

Figure 5: Structures of mixirins A–C (11–13) and marihysin A (14).

Polypeptides

Nonribosomal peptides (NRPs) are synthesized by large multimodular nonribosomal peptide-synthetase (NRPS) by elongation of activated monomers of amino acid building blocks. NRPSs are organized in modules responsible for the incorporation of a specific amino acid in polypeptide chain by three successive steps: adenylation, thiolation, and condensation [50]. Two types of polypeptides are generally produced by *Bacillus* strains are linear and desipeptides.

Bogorol A (15) (Figure 6), a novel peptide antibiotic has been obtained from the culture broth of a marine *Bacillus* sp. isolated from a tropical reef habitat in Papua New Guinea [51]. Bogorol A illustrates a new structural template for "cationic peptide antibiotics", and it showed good activity against methicillin-resistant *S. aureus* (MRSA) (MIC 2 µg/mL), vancomycin-resistant enterococcus (VRE) (MIC 10 µg/mL) and moderate activity against *E. coli* (MIC 35 µg/mL), and no activity against *S. maltophilia* (>200 µg/mL).

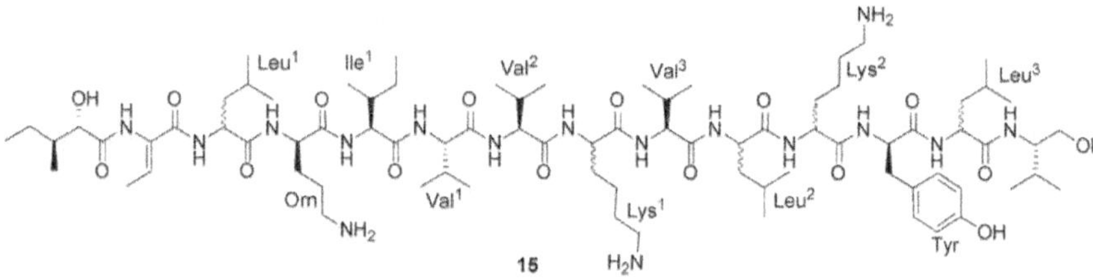

Figure 6: Structure of bogorol A (15).

Two new cyclic depsipeptides, turnagainolides A (16) and B (17) (Figure 7), have been isolated from laboratory cultures of a marine strain RJA2194 [51]. This strain was isolated from a sediment sample collected near Turnagain Island and identified as a *Bacillus* species. The structures of 16 and 17 are representative of epimers at the site of macrolactonization. Turnagainolide B (17) showed activity in a SHIP1 activation assay [51].

Loloatin B (18) (Figure 7) is a novel cyclic decapeptide antibiotic with potent antimicrobial activity against Gram-positive bacteria produced by a *Bacillus* sp. which was isolated from the tissues of a marine worm [46]. In screening tests for antimicrobial activity, loloatin B inhibited the growth of MRSA, VRE, and penicillin resistant *S. pneumoniae* with MICs of 1–2 µg/mL.

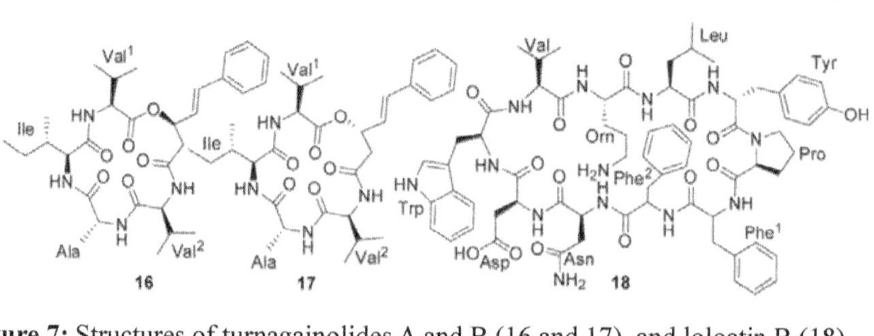

Figure 7: Structures of turnagainolides A and B (16 and 17), and loloatin B (18).

Two new cyclodepsipeptides designated as bacillistatins 1 (19) and 2 (20) (Figure 8) have been isolated from the culture broth of *Bacillus silvestris* that was obtained from a Pacific Ocean (southern Chile) crab [30]. Each 12-unit cyclodepsipeptide (19 and 20) strongly inhibited the growth of a human cancer cell line panel, with GI_{50}'s of 10^{-4}–10^{-5} µg/mL, and each compound was determined to be active against antibiotic-resistant *S. pneumoniae*.

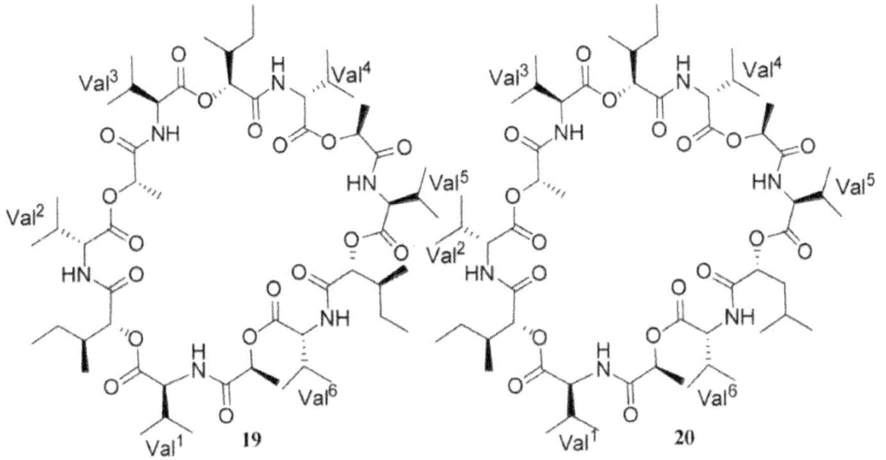

Figure 8: Structures of bacillistatins 1 (19) and 2 (20).

Polyketides/Lipoamides

Polyketides are extremely large classes of secondary metabolites that are assembled from simple acyl-coenzyme A and form the basis of numerous pharmaceuticals, agrochemicals, and veterinary agents. Polyketides are biosynthesized by polyfunctional megasynthases organized into repeated functional units known as modules. The modular megaproteins responsible for polyketide biosynthesis are known as polyketide synthases (PKSs) [50,52]. Due to their mechanism of versatile assemblage, the polyketides exhibit remarkable diversity both in terms of structure and biological activities.

A *B. laterosporus* isolate, obtained from coastal waters of Papua New Guinea, has been shown to produce some novel metabolites such as basiliskamide A (21), basiliskamide B (22), tupuseleiamide A (23), and tupusleiamide B (24) [32] (Figure 9). Compound 22 was simply a regioisomer of 21, and 24 was a regioisomer of 23. The observed MIC values of 21 and 22 against *C. albicans* and *A. fumigatus* were 1.0 and 3.1 µg/mL, and 2.5 and 5.0 µg/mL, respectively.

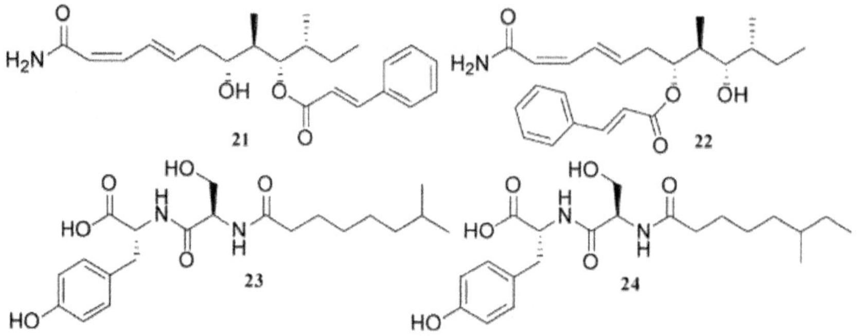

Figure 9: Structures of basiliskamides (21 and 22) and tupuseleiamides (23 and 24).

Two unique polyketides, ieodoglucomides A (25) and B (26) (Figure 10) were isolated from a marine-derived bacterium *B. licheniformis* [53]. This bacterium was isolated from a sediment sample collected from Ieodo in Republic of Korea's southern reef. Compounds 25 and 26 displayed moderate *in vitro* antimicrobial activity against both Gram-positive and Gram-negative pathogenic bacteria (MIC 8–32 µg/mL). Furthermore, ieodoglucomide B (26) exhibited cytotoxic activity against lung cancer and stomach cancer cell lines with GI_{50} values of 25.18 and 17.78 µg/mL, respectively [52].

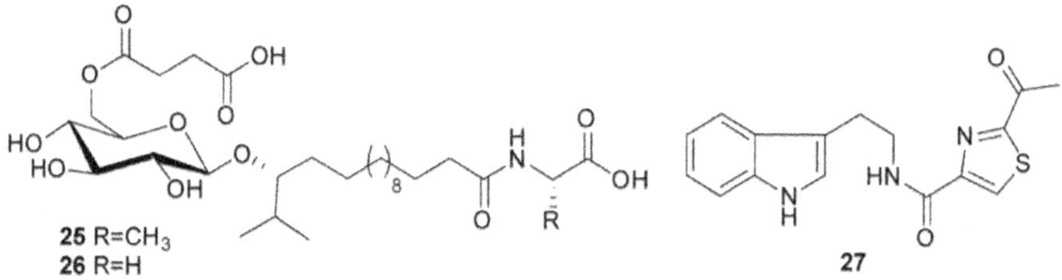

Figure 10: Structures of ieodoglucomides A and B (25 and 26), and bacillamide (27).

Bacillamide (27) (Figure 10) was isolated from *Bacillus* sp. SY-1, which was obtained from seawater during the blooming period of *C. polykrikoides* in Masan Bay [47]. Bacillamide (27) showed significant algicidal activity against *C. polykrikoides* (LC_{50} after 6 h: 3.2 µg/mL) and selective activity against dinoflagellates and raphidophytes. However, 27 showed neither algicidal activity against microalgae of other phyla such as diatom, green algae, and cyanobacteria, nor growth inhibitory effects against bacteria, fungi, or yeast. Therefore, 27 might be considered as a useful algicidal agent for regulating the blooms of harmful dinoflagellate species such as *C. polykrikoides*.

A novel thiazole alkaloid, neobacillamide A (28) together with three known related bacillamides A–C (29–31), were isolated from the bacterium *B.*

atrophaeus, which was associated with the South China Sea sponge *Dysidea avara* [54] (Figure 11). It is interesting to note that all bacillamides A–C (29–31) contain a common tryptamine moiety in their molecules while in 28 the amine portion is replaced by a phenethylamine. Neobacillamide A was the first member of *Bacillus* thiazole alkaloids, which has been determined to contain a phenethylamine moiety. Compounds 28 and 31 were evaluated for their inhibitory activity against HL60 human leukemia cells and A549 human lung cancer cells but both compounds were found inactive.

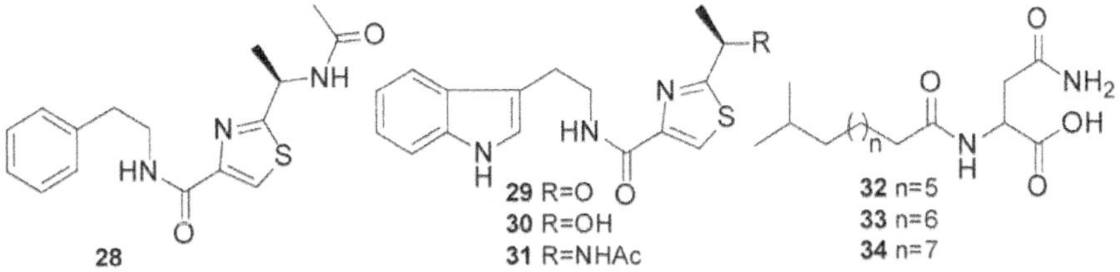

Figure 11: Structures of thiazole alkaloids (28–31) and lipoamides A–C (32–34).

Three new compounds named lipoamides A–C (32–34) (Figure 11) have been isolated from the culture broth of a sediment*B. pumilus* (SP21) [41]. Lipoamide A (32) showed weak antibacterial activity against *S. aureus* and *P. aeruginosa* (MIC > 100 µg/mL).

Isocoumarins

Isocoumarin-type metabolites from microorganisms are characterized by the amino-containing substituent, which is presumably derived from leucine, at the 3-position in the dihydrocoumarin core. There have been several such isocoumarin compounds as baciphelacin [55], amicoumacins [56,57,58,59], and xenocoumacins [60]. These compounds possess the common chromophore, 3,4-dihydro-8-hydroxyisocoumarin in their structures and many of them are produced by the genus*Bacillus*.

Two novel isocoumarins, bacilosarcins A (35) and B (36), and three known isocoumarins, amicoumacins A (37), B (38), and C (39), were isolated from the culture broth of a marine-derived bacterium *B. subtilis* TP-B0611 [31] (Figure 12). The strain B0611 was isolated from the intestine of a sardine (*S. melanosticta*) collected in Toyama Bay in Japan. Compound 35possesses an unprecedented 3-oxa-6,9-diazabicyclo[3.3.1]nonane ring system, where 36 has a 2-hydroxymorpholine moiety which is rare in nature. Compound 35 showed 82% inhibition at 50 µM against growth of barnyard millet sprouts and on the contrary 36 showed very weak activity at the same concentration. Of particular interest, amicoumacin A (37) showed more potent activity than 35. On the basis of biogenetic consideration, it is reasonable to assume that compound 35 was derived from amicoumacin A (37), forming a biacetyl equivalent C_4 unit and two water molecules. This may imply that 35 behaves as a prodrug of 37 in plant cells. The activity levels shown by 35 and 37 are higher than that of herbimycin A, a potent herbicidal compound from *Streptomyces* [61], suggesting that they may be lead molecules for plant growth regulators.

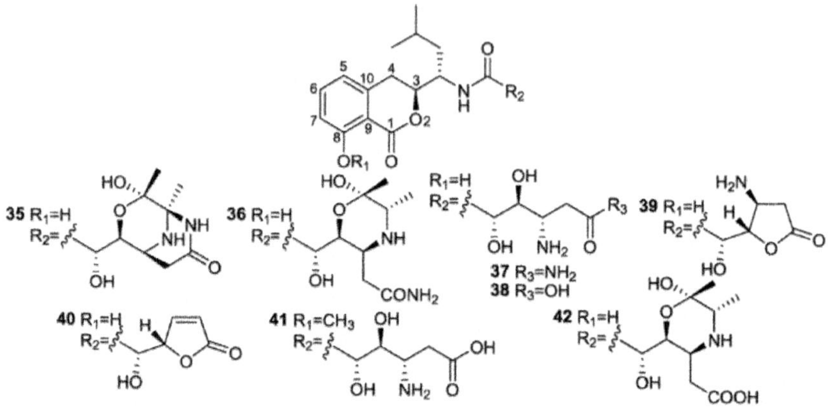

Figure 12: Structures of isocoumarins (35–42).

Six known analogs (36–41), one new bacisarcin C (42) (Figure 12) and four novel amicoumacins, lipoamicoumacins A–D (43–46) (Figure 13) were isolated from the culture broth of a marine-derived bacterium, *B. subtilis* B1779 [62]. The strain B1779 was isolated from a marine sediment sample, which was collected from the Red Sea in April 2010. All isolated compounds were evaluated for cytotoxicity and antibacterial activities. Only compounds 36 and 37, which have an amide functional group (-NH_2) exhibited cytotoxicity against HeLa cells with IC50 values of 33.60 and 4.32 μM, respectively, indicating that the amide group of amicoumacin plays a critical role in cytotoxicity. This was further supported by a comparison of cytotoxicity between compounds 36 and 42 and between 37 and 38. A similar antibacterial effect of amide functional group has been shown among amicoumacins [62].

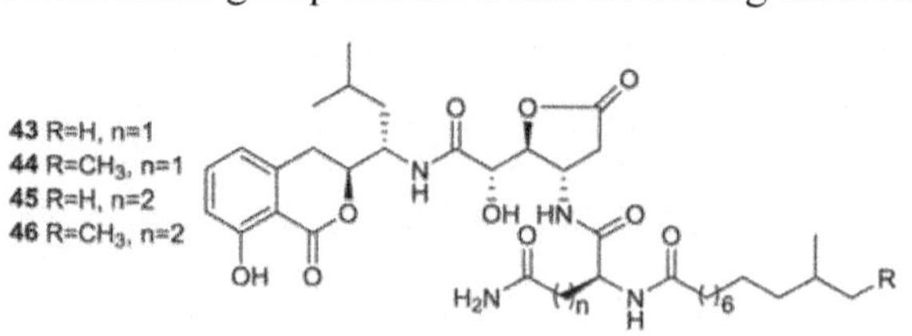

Figure 13: Structures of lipoamicoumacins A–D (43–46).

Fatty Acids

Bacteria are known to synthesize fatty acids via the classic fatty acid synthase (FAS) pathway with chain length ranging from C_{12} to C_{19} [63,64]. Fatty acid from acetyl-CoA and malonyl-CoA precursors through the action of enzymes called fatty acid synthases (FAS).

Methyl-branched fatty acids with methyl groups at even carbon atoms (methyl substituents on carbon 2, 4, 6) occur in several organisms, and originate from the selective incorporation of methylmalonyl-CoA by fatty acid synthases [65]. The *iso-anteiso* branching in the novel methyl-branched fatty acids is most probably derived from leucine and isoleucine followed by a series of elongations with malonyl-CoA. At either the last or penultimate elongation step methylmalonyl-CoA seems to be selectively incorporated by one of the fatty acid synthesizing enzymes from the bacterium resulting in the methyl-branched fatty acids. Whether this is a known or unknown, methyl-branched fatty acid synthase in bacteria is, as of yet, a matter of speculation.

A series of novel *iso-anteiso* fatty acids (47–52) (Figure 14) with chain lengths between C_{11} and C_{19} and an interesting series of linear alkylbenzene fatty acids with chain lengths between C_{10} and C_{14} were produced by a halophilic *Bacillus* sp. [66]. The biological activity of these novel fatty acids has not yet been reported.

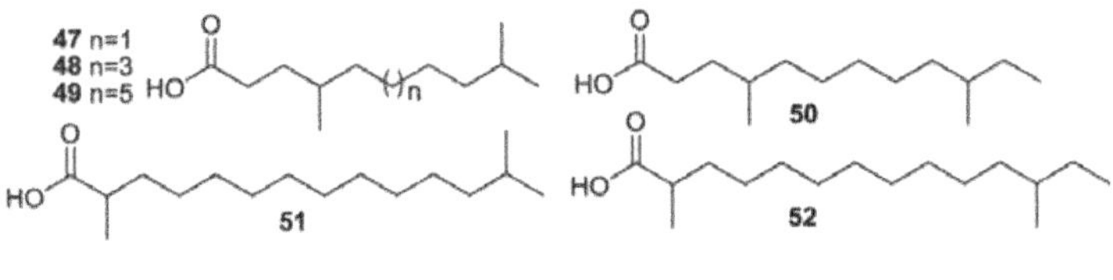

Figure 14: Structures of *iso-anteiso* fatty acids (47–52).

Bioassay-guided isolation of the EtOAc extract of a marine *Bacillus* sp. 09ID194 cultured in modified Bennett's broth medium, yielded six new unsaturated hydroxy fatty acids, named ieodomycins A–D (53–56) [34] and linieodolides A and B (57 and 58) [67] (Figure 15). The producing strain was isolated from a sediment sample collected from Ieodo, Republic of Korea's southern reef. Compounds 53–58 exhibited antimicrobial activity against *B. subtilis* and *E. coli* with MICs of 32–64 µg/mL, but showed weak growth inhibition against the yeast, *S. cerevisiae*, with an MIC of 128–256 µg/mL.

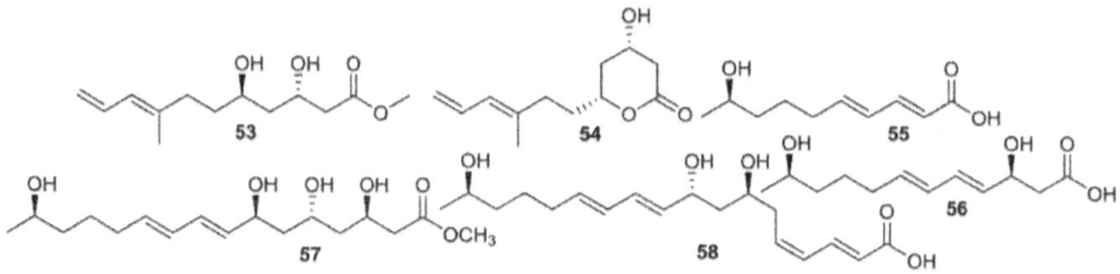

Figure 15: Structures of unsaturated fatty acids (53–58).

Macrolactins

The macrolactins are polyene cyclic macrolactones consisting of 24-membered ring lactones with modifications such as the attachment of glucose β-pyranoside, and they may also occur as linear analogs [33]. The macrolactin carbon skeleton contains three separate diene structure elements in a 24-membered lactone ring. The macrolactin class of macrolactones is mainly produced by both terrestrial and marine strains [23].

In the genome of *B. amyloliquefaciens* FZB42, a plant-root-colonizing environmental strain with the ability to stimulate plant growth and to suppress soil-borne plant pathogens in the rhizosphere [68,69], three PKS operons altogether about 196,340 pb were located at sites approximately 1.4 Mbp (*pks*2), 1.7 Mbp (*pks*1), and 2.3 Mbp (*pks*3) clockwise from the origin of replication in the genome which is 3916 kb in size [70,71]. These three-gene clusters show a modular organization which is typical for type I PKS systems, indicating that the strain FZB42 has the biosynthetic machinery for the production of at least three different kinds of polyketides. *pks*1 and *pks*3 have been attributed to the production of bacillaene and difficidin/oxydifficidin, respectively, where the *pks*2 is involved in macrolactins biosynthesis. The macrolactone rings of macrolactins are formed via cyclization of polyketide chains assembled by PKS type-I enzymes that perform repetitive decarboxylative condensations of carboxylic acids with an activated carboxylic acid starter unit [72].

Difficidin (59) and oxidifficidin (60) (Figure 16), detected in the culture broth of *B. amyloliquefaciens* (FZB42) are highly unsaturated 22-membered macrocyclic polyene lactone phosphate esters with broad-spectrum of antibacterial activity [73]. Difficidin (59) has recently shown promising suppressive activity against the enterobacterium *Erwinia amylovara*, a devastating plant pathogen which causes necrotrophic fire blight disease affecting apple, pear, and other rosaceous plants [74].

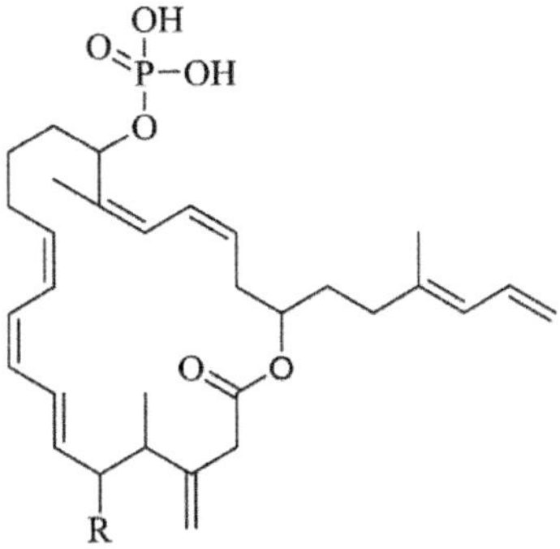

Figure 16: Structures of difficidin (59) and oxidifficidin (60).

The primary mode of action of difficidin (59) is the inhibition of protein synthesis. It has proven to be highly bactericidal to both growing and stationary phase cultures and inhibited protein synthesis more rapidly than RNA, DNA, or cell-wall synthesis in growing cells [75]. At least 32 macrolactins have been characterized so far including macrolactins A–Z, 7-O-succinyl macrolactin A, 7-O-succinyl macrolactin F, 7-O-malonyl macrolactin A, and three ether-containing macrolactin A. Most of them were produced by marine sediment isolates and few of them were by soil isolates [76,77]. The first macrolactin family, including macrolactins A–F (61, 64–66, 70, 74) (Figure 17 and Figure 18), as well as the open-chain macrolactinic and isomacrolactinic acid, were isolated from an unclassified deep-sea sediment bacterium [33]. Macrolactin A was of particular interest because of its biological activities. It shows selective antibacterial activity against *S. aureus* and *B. subtilis* at a concentration of 5 and 20 µg/disc, respectively, as well as the ability to inhibit B16-F10 murine melanoma cancer cells in *in vitro* assays, as well as mammalian *Herpes simplex* viruses. In addition, it protects lymphoblast cells against HIV by inhibiting virus replication [33].

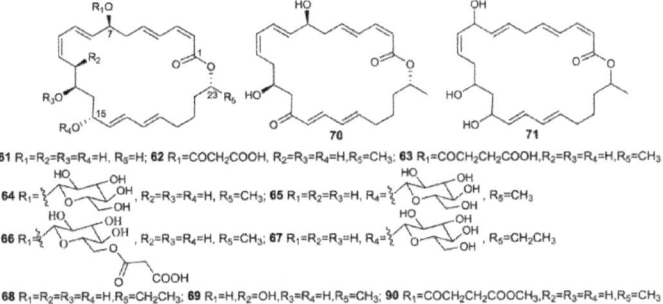

Figure 17: Structures of 24-membered macrolactin A (61) and its derivatives (62–71 and 90).

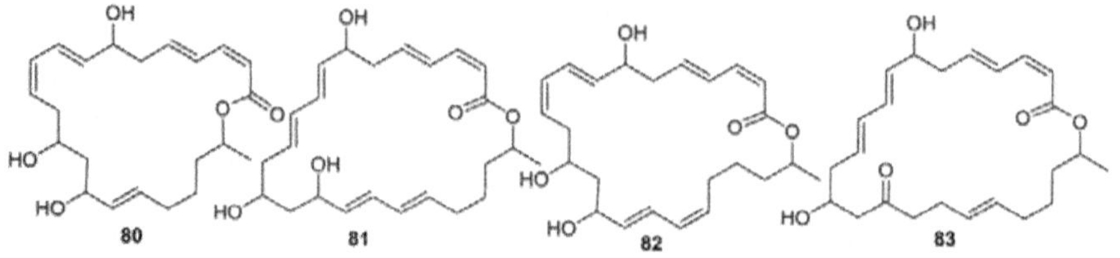

Figure 18: Structures of 22- (80) and 24-membered (81 and 82) including keto (83) macrolactins.

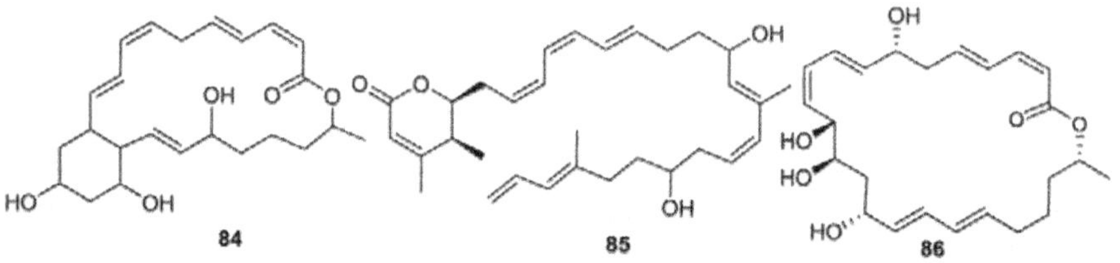

Figure 19: Structures of bicyclic (84) and polyene (85) macrolactins, and 12-hydroxy macrolactin A (86).

Macrolactins G–M (71, 80–84, 68) (Figure 17, Figure 18, Figure 19) were produced by a *Bacillus* sp. PP19-H3, which was isolated from a marine macroalga, *Schizymenia dubyi*, collected at the Omaezaki coast of Shizuoka prefecture in Japan [78]. These macrolactins include a 22-membered ring (80) or bicyclic lactone (84) in addition to their geometric isomers of macrolactins A and F. These macrolactins were more active against *S. aureus* (MIC 5–10 µg/mL) than *B. subtilis* (MIC 60 µg/mL), whereas macrolactin K (83), containing keto group (C=O) at C-15 showed very weak antibacterial activity against both tested pathogens (MIC > 100 µg/mL), indicating that the -OH group at C-15 may play an important role in the antibacterial activity of macrolactins [78].

With a view to find out peptide deformylase inhibitors, four glycosylated macrolactins O–R (73, 67, 77, and 79) (Figure 17, Figure 20 and Figure 21) were isolated from the liquid cultures of *Bacillus* sp. AH159-1 [72]. Macrolactins O–R inhibited *S. aureus* peptide deformylase (PDF) in dose-dependent manners with IC_{50} (µM) values of 53.5, 57.7, 12.1, and 61.5, respectively. All these compounds also inhibited bacterial growth against *E. coli* with an MIC of 100 µg/mL [77]. Macrolactin N (72) (Figure 18) also inhibited *S. aureus*'s peptide deformylase (PDF) in a similar fashion to macrolactins O–R [79].

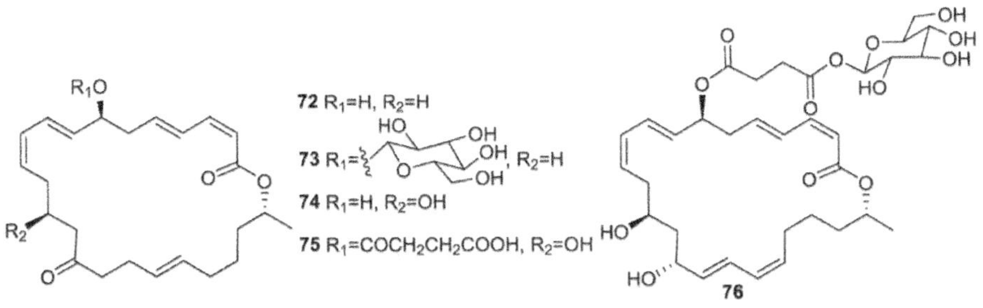

Figure 20: Structures of keto (72–75) and ester (76) macrolactins.

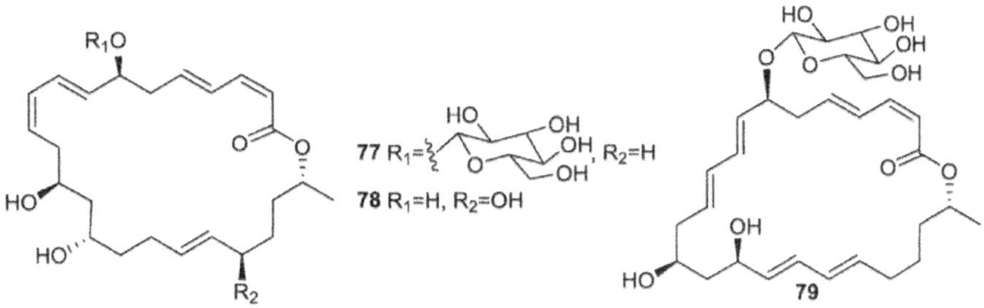

Figure 21: Structures of glycosylated (77 and 79) and a 24-membered (78) macrolactins.

A new macrolactin T (78) and a new polyene δ-lactone macrolactin U (85) (Figure 19 and Figure 21), along with known macrolactins A, B, D, O and S (69) were isolated from the culture broth of the bacterium *B. marinus*, which was isolated from *Suaeda salsa* collected on the coastline of the Bohai Sea of China [80]. Macrolactins T and B exhibited MIC activity toward *A. solani*, *P. oryzae*, and *S. aureus*, at concentrations of 0.8, 2.8, 5.5 and 7.5, 20.1, 4.5, respectively [80].

Macrolactins V (86) and S (69) (Figure 17 and Figure 21) were isolated from the culture broth of a marine bacterium, *B. amyloliquefaciens* SCSIO 00856, which was isolated from a South China Sea gorgonian, *Junceella juncea* [48]. Macrolactin V (86) exhibited potent antibacterial activity against *E. coli*, *B. subtilis* and *S. aureus* with an MIC value of 0.1 μg/mL, and no activity against *B. thuringiensis*. Macrolactin S (69) showed strong antibacterial activity against both *E. coli* and *S. aureus* with MIC values of 0.3 and 0.1 μg/mL, respectively, but weak activity against *B. subtilis* (MIC 100 μg/mL), which indicated that the configuration of 7-OH may affect the antibacterial activity of the epimers 86 and 69 [48]. 7-O-succinyl macrolactin A (63) and 7-O-succinyl macrolactin F (75) (Figure 17 and Figure 18) were isolated from a marine sediment *Bacillus* sp. Sc026 [81], and 7-O-malonyl macrolactin A (62) (Figure 17) from a *B. subtilis* [82]. The bacterium (Sc026) was isolated from marine sediment collected from nearby Sichang Island (at a 15-m depth), Chonburi, Thailand. 63 and 75 were tested in an agar diffusion assay at concentration of 50 and 100 μg/disk, respectively. The inhibition zones of 63 and 75 against *B. subtilis* and *S. aureus* were 10 and 9, and 24 and 8 mm, respectively. The minimum restrictive concentrations (MRCs) of 62 were between 1 and 64 μg/mL for *S. aureus* and MRSA strains, and between 0.06 and 4 μg/mL for *E. faecalis* and clinical isolates, VRAS (vancomycin-resistant/ ampicillin-sensitive) E305 and VRAR (vancomycin-resistant/ampicillin-resistant) E315 [82].

Macrolactins W (76), Y (91), Z (92), X (90), macrolactinic acid (93) and three ether-containing unique macrolactins (87–89) (Figure 17, Figure 18, Figure 22, and Figure 23) were isolated from a sediment *Bacillus* sp. 09ID194 [67,83,84]. Interestingly, these compounds were produced by this strain only in low salinity culture medium (12 g/L) but not in high salinity culture medium (32 g/L). These compounds showed MIC against pathogenic bacteria at a concentration range of 8–64 μg/mL. The position of ether group was important for the antimicrobial activity of these ether-containing macrolactins (87–89) [84].

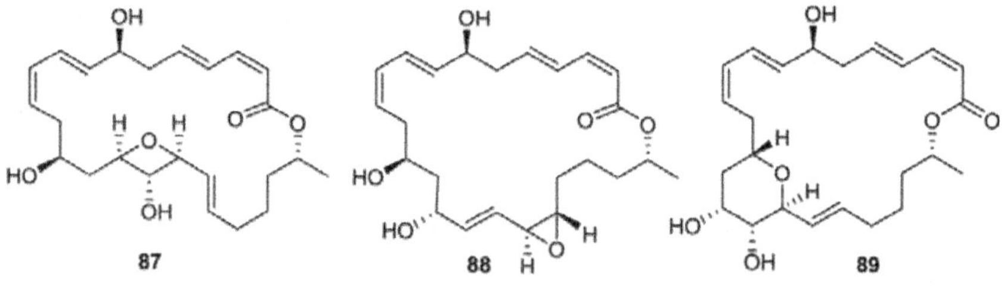

Figure 22: Structures of ether-containing macrolactins (87–89).

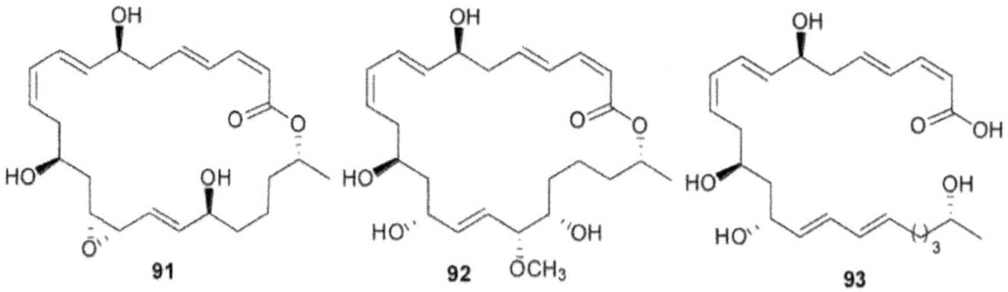

Figure 23: Structures of ether (91) and methoxy (92) containing macrolactins and macrolactinic acid (93).

DETOXIFICATION OF HEAVY METALS

Chromium (Cr) is one of the major causes of environmental contamination by heavy metals. The toxicity of chromium (Cr) lies in its oxidative states [85]. The higher oxidation states impose more toxicity (10–100 times) than the lower oxidation states [86]. Nutritionally, small amount of Cr(III) is essential for balanced human and animal diet for preventing adverse effects caused by glucose and lipid metabolisms [87,88]. Swallowing large amounts of Cr(III) may pose threats to human health [89,90]. Hexavalent Cr (Cr^{6+}) is extremely carcinogen and may be the cause of death after ingestion of a large dose [91]. The tolerable intake level of Cr(VI) and Cr(III) are 1 and 8 µg/L for freshwater life, 1 and 50 µg /L for marine life, and 8 and 5 µg/L for irrigation water, respectively [92,93].

A marine isolate, *B. licheniformis*, can reduce 10–500 mg/L of Cr(VI) to Cr(III) within 24–72 h in liquid medium [94]. When this strain was cultured in the liquid medium, it produces chromium reductase indicating that probably Cr(VI) may be reduced to Cr(III) by enzymatic activity. This strain also secretes an extracellular surface-active agent (biosurfactant) in the medium, which provides tolerance to the cells towards hexavalent chromium and protects the cells from oxidative stress.

The oxidation of soluble manganese(II) to insoluble Mn(III, IV) oxide plays an important role in removing environmental heavy metal hazards. These Mn oxides oxidize many organic and inorganic compounds, and to scavenge a variety of other metals on their highly charged surfaces. In addition to catalyzing important process, microorganisms capable of Mn(II) oxidation are potential candidates for the removal, detoxification, and recovery of metals from the environment. Mature spores of a marine *Bacillus* sp. strain SG1 oxidize Mn(II) to MnO_2 [95]. Vegetative cells of the same strain reduce MnO_2 under low-oxygen conditions. The rate of MnO_2 reduction was a function of cell density.

Although *Bacillus* strains play a central role in biogeochemical recycling of metals in marine environment the molecular and biochemical mechanisms for most of these recycling processes are still poorly understood. It is assumed that most of the heavy metals in the environment are detoxified by reduction process. However, it is important to recognize that biochemical recycling of heavy metals is a new field and is largely limited to studies of microbial species. To remediate a range of heavy metals, it is important to select microbial strains, which carry inherently genetic machinery to reduce multiple metals. The concentration of heavy metals (Cd, Co, Cr, Hg, Pb) around hydrothermal vents is high [96]. Therefore, *Bacillus* strains living around hydrothermal vents may capable of recycling a wide range of heavy metals. The *Bacillus* strains isolated from "hydrothermal vents" should therefore be good candidates for the detoxification of environmental pollution caused by toxic heavy metals.

MARINE *BACILLUS* STRAINS AS POTENTIAL BIOCONTROL AGENTS

Most of the microorganism-based biopesticides have been developed from terrestrial bacteria. Only a few fungi have been used as efficient biocontrol agents [97]. Presently, about half of the commercially available bacterial biocontrol agents are prepared from *Bacillus* strains [12] and among them, *B. thuringiensis* accounts for more than 70% of total sales. This bacterium, omnipresent in marine environment [98], produces two proteins named Cry and Cyt during its sporulation phase, each of which is highly toxic to insects, but not to mammals or to the environment. These toxic proteins kill insects by forming pores in the gut walls. Major insect families, which can be controlled by Cry/Cyt toxins include Coleoptera, Lepidoptera, and Diptera. A wide range of plant diseases (root rot, leaf spot, anthracnose, gray mold, early blight, late blight, powdery mildew, downy mildew, and bacterial spot) can also be controlled successfully by the antagonistic *Bacillus* isolates [99].

The genus *Bacillus* is present in every niche of terrestrial and marine environments, even in the hot springs [100]. Many *Bacillus* species can also be found in both terrestrial and marine environments. This bacterial genus could be considered as one of the major sources of potential microbial biopesticides because of its broad genetic biodiversity [22,36], its large body of literary evidence, as well as generally regarded as safe by the US Food and Drug Administration (USFDA) and capable of withstanding unfavorable conditions through the formation of resistant spores. The main specific mechanisms involved in biocontrol of plant diseases by this bacterial genus include: competition for ecological niche/substrate in the rhizosphere, production of inhibitory chemicals, and induction of so-called systemic resistance in host plants. Marine *Bacilli* appear to be more effective biocontrol agents, compared to their terrestrial counterparts [101].

And so, the question remains as to whether marine or terrestrial *Bacillus* strains will be better candidates as biocontrol agents. Marine environments (with a wide variation in temperature, pressure, salt concentration and pH) are different from the terrestrial ones. The active strain, with high resistance to salt, heat, pH and stress, can be used to prepare a biological pesticide. The marine strains carry these properties naturally [15]. Marine *Bacilli* forms resistant spores quickly in unfavorable conditions, and is easily converted to powder formulations, having longer shelf-life when compared to other products containing living organisms, and can be prepared commercially at a relatively low cost (relatively unspecialized culture procedures). However, these products have some limitations due to partial protection against pathogen and pests, inconsistent effects, and a lack of ecological knowledge, which warrants precaution in field applications. There are several approaches towards the improvement of biopesticide efficacy, including selection of suitable strains, combination of synergistic strains, combinations with other forms of biopesticides, chemical pesticides, plant fertilization and agricultural practices, as well as suitable formulation and application methods [11,12].

MARINE *BACILLUS* ISOLATES AS A POTENTIAL SOURCE OF NATURAL CAROTENOIDS

Carotenoids are yellow, orange, and red pigments, which are widely distributed in plants and microorganisms (photosynthetic organisms, bacteria, and fungi). Chemically carotenoids usually contain a polyene C_{40} carbon skeleton, and can be acyclic, or cyclic groups, may be present at one or both ends of the backbone and having many derivatives [102]. Plants, algae, and fungi produce carotenoids containing C_{40} carbon backbone, whereas bacteria can produce a diverse range of carotenoids with either C_{40} or C_{30} carbon backbone. Each double bond in the polyene chain of a carotenoid may exist in two forms: *trans* or *cis*. Carotenoids obtained from natural sources are predominantly or entirely in *trans* form. Carotenoids are widely used as antioxidants, pro-vitamin A, and food and feed additives. Presently, synthetic carotenoids are meeting the bulk of market demands, but due to the "green wave", coined to represent changes in consumers, industries are now looking for natural sources of carotenoids. Currently, there are no feasible sources of natural carotenoids including widely screened terrestrial microgranisns [102]. The recent screening of marine bacteria indicated that they are a promising source of natural carotenoids [103]. So far, more than 600 different carotenoids have been identified from natural sources, of which only 24 are available in human foodstuffs. The most widely used carotenoids as food are β-carotene, β-cryptoxanthin, lycopene, lutein, and violaxanthin. Carotenoids contain isoprene skeleton and are biosynthesized by tail-to-tail linkage of two C_{20} geranylgeranyl diphosphate molecules.

Irrespective of their sources, the effectiveness of carotenoinds largely depends on their bioavailablity. Due to unusual chemical structures of bacterial carotenoids, questions remained to be address bioavailability of these molecules in humans. Glycosylated carotenoids (Figure 24) isolated from marine spore-forming strains, *B. indicus* HU36 and *B. firmus* GB1, showed better bioavailability (about 4.5 times as high) than that of pure β-carotene *in vitro* digestion experiments [104]. All the marine isolates are not able to produce carotenoids. Marine yellow and orange spore-forming *Bacillus* strains have been shown to produce carotenoids [105]. Photosynthetic marine organisms (algae and plants), either independently or in symbiosis with microorganisms, produce carotenoids to prevent oxidation by sunlight [106]. Marine yellow, red, and orange spore-forming *Bacillus* strains as well as strains living in photosynthetic organisms may be a good source of carotenoids useful for human.

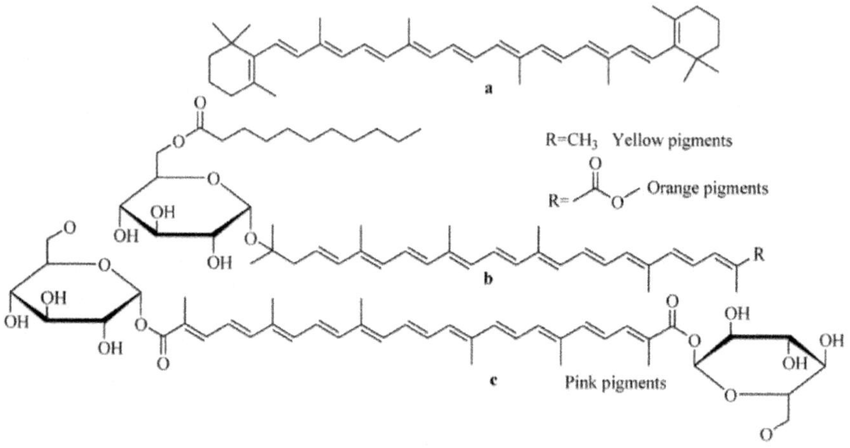

Figure 24: Structures of β-carotene (a) and main carotenoids synthesised by (b) HU36 and (c) GB1 spore-forming strains.

CONCLUSIONS AND FUTURE PERSPECTIVES

Marine *Bacillus* species represent a rich source of structurally diverse classes of secondary metabolites including lipopeptides, polypeptides, macrolactones, fatty acids, polyketides, lipoamides, isocoumarins, and carotenoids. These structurally versatile natural products of marine isolates are derived from complex biosynthetic pathways. Some of these bioactive compounds have high potentials for the development of effective pharmaceutical and agrochemical products. Due to having genetic capability to adapt extreme conditions, *Bacillus* strains isolated from unique niches of environments (e.g., hydrothermal vent, deep sea, pH > 9.0 and salt lakes) may produce useful bioactive compounds [8]. The silent cryptic biosynthetic gene clusters of marine isolates may be activated with a view to discover new natural products by culturing them under varying stressful conditions (e.g., nutrient, pH, salinity or temperature stresses). Another important feature of the genus *Bacillus* is that they can detoxify heavy metals through reduction processes, which might be considered as candidates for the bioremediation of heavy metal toxicity. Bioremediation is an eco-friendly and cost-effective strategy for eliminating xenobiotic compounds from polluted environments. Next-generation sequencing is providing crucial insights in the molecular and biological mechanisms involved in bioremediation of environmental pollutants like heavy metal contaminations. These insights will improve bacterial bioremediation strategies, monitoring their progress, and determining their success [107]. *Bacillus*-based biopesticides can improve plant health through unique modes of action, and thus, have a high potential for commercial applications. The frequent occurrence of *B. subtilis* in the natural environment and the production capability of a vast array of antibiotics must be considered for application in pest management. Marine *Bacillus*-based biopesticides have a great potential in sustainable agricultural practices in the future. To meet the consumers demand for fully natural practices and reduced environmental hazard, marine yellow and orange spore-forming *Bacillus* strains, as well as those which are symbiotic with photosynthetic marine organisms may be a good source of natural carotenoids.

ACKNOWLEDGMENTS

This research was supported in part by Korea Institute of Ocean Science and Technology (Grants PE99121 to H. J. S.). The author (MTI) is thankful to the World Bank for funding a sub-project CP # 2071 in the Department of Biotechnology of BSMR Agricultural University, Bangladesh under Higher Education Quality Enhancement Project (HEQEP) of University Grants Commission of Bangladesh.

REFERENCES

1. Lebar, M.D.; Heimbegner, J.L.; Baker, B.J. Cold-water marine natural products. *Nat. Prod. Rep.* 2007, *24*, 774–797.
2. Fenical, W. Chemical studies of marine bacteria: Developing a new resource. *Chem. Rev.* 1993, *93*, 1673–1683.
3. Laatsch, H. Marine Bacterial Metabolites. In *Frontiers in Marine Biotechnology*; Proksch, P., Muller, W.E.G., Eds.; Horizon Bioscience: Norfolk, UK, 2006; pp. 225–288. [Google Scholar]
4. Schwartsmann, G.; da Rocha, A.B.; Berlinck, R.G.S.; Jimeno, J. Marine organisms as a source of new anticancer agents.*Lancet Oncol.* 2001, *2*, 221–225.
5. Jha, R.K.; Zi-rong, X. Biomedical compounds from marine organisms. *Mar. Drugs* 2004, *2*, 123–146.
6. Nathan, C. Antibiotics at the crossroads. *Nature* 2004, *431*, 899–902.

7. Von Nussbaum, F.; Brands, M.; Hinzen, B.; Weigand, S.; Häbich, D. Antibacterial natural products in medicinal chemistry—Exodus or revival. *Angew. Chem. Int. Ed.* 2006, *45*, 5072–5129.

8. Li, J.W.-H.; Vederas, J.C. Drug discovery and natural products: End of an era or an endless frontier? *Science* 2009, *325*, 161–165.

9. Islam, M.T.; von Tiedemann, A.; Laatsch, H. Protein kinase C is likely to be involved in zoosporogenesis and maintenance of flagellar motility in the peronosporomycete zoospores. *Mol. Plant Microbe Interact.* 2011, *24*, 938–947.

10. Islam, M.T.; Hashidoko, Y.; Deora, A.; Ito, T.; Tahara, S. Suppression of damping-off disease in host plants by the rhizoplane bacterium *Lysobacter* sp. Strain SB-K88 is linked to plant colonization and antibiosis against soilborne peronosporomycetes. *Appl. Environ. Microbiol.* 2005, *71*, 3786–3796.

11. Islam, M.T. Potentials for Biological Control of Plant Disease by *Lysobacter* spp., with Special Reference to Strain SB-K88. In *Bacteria in Agrobiology: Plant Growth Responses*; Maheshwari, D.K., Ed.; Springer-Verlag: Berlin/Heidelberg, Germany, 2011; pp. 335–364.

12. Islam, M.T.; Hossain, M.M. Biological Control of Peronosporomycete Phytopathogens by Antagonistic Bacteria. In*Bacteria in Agrobiology: Plant Disease Management*; Maheshwari, D.K., Ed.; Springer-Verlag: Berlin/Heidelberg, Germany, 2013; pp. 167–218.

13. Whitman, W.B.; Coleman, D.C.; Wiebe, W.J. Prokaryotes: The unseen majority. *Proc. Natl. Acad. Sci.USA* 1998, *95*, 6578–6583.

14. Kennedy, J.; Marchest, J.R.; Dobson, A.D.W. Marine metagenomics: Strategies for the discovery of novel enzymes with biotechnological applications from marine environments. *Microb. Cell Fact.* 2008, *7*, 27–28.

15. Rampelotto, P.H. Resistance of microorganisms to extreme environmental conditions and its contribution to astrobiology. *Sustainability* 2010, *2*, 1602–1623.

16. Sayem, S.M.A.; Manzo, E.; Ciavatta, L.; Tramice, A.; Cordone, A.; Zanfardino, A.; de Felice, M.; Varcamonti, M. Anti-biofilm activity of an exopolysaccharide from a sponge-associated strain of *Bacillus licheniformis*. *Microb. Cell Fact.*2011, *10*, 74.

17. Paul, V.J.; Arthur, K.E.; Williams, R.R.; Ross, C.; Sharp, K. Chemical defenses: From compounds to communities. *Biol. Bull.* 2007, *213*, 226–251.

18. Jensen, P.R.; Fenical, W. Strategies for the discovery of secondary metabolites from marine bacteria: Ecological perspectives. *Annu. Rev. Microbiol.* 1994, *48*, 559–584.
19. Feling, R.H.; Buchanan, G.O.; Mincer, T.J.; Kauffman, C.A.; Jensen, P.R.; Fenical, W. Salinosporamide A: A highly cytotoxic proteasome inhibitor from a novel microbial source, a marine bacterium of the new genus *salinospora.Angew. Chem. Int. Ed. Engl.* 2003, *20*, 355–357.
20. Chen, X.H.; Koumoutsi, A.; Scholz, R.; Eisenreich, A.; Schneider, K.; Heinemeyer, I.; Morgenstern, B.; Voss, B.; Hess, W.R.; Reva, O.; et al. Comparative analysis of the complete genome sequence of the plant growth—Promoting bacterium *Bacillus amyloliquefaciens* FZB42. *Nat. Biotechnol.* 2007, *25*, 1007–1014.
21. Kunst, F.; Ogasawara, N.; Moszer, I.; Albertini, A.M.; Alloni, G.; Azevedo, V.; Bertero, M.G.; Bessières, P.; Bolotin, A.; Borchert, S.; et al. The complete genome sequence of the Gram-positive bacterium *Bacillus subtilis*. *Nature* 1997, *390*, 249–256.
22. Fan, L.; Bo, S.; Chen, H.; Ye, W.; Kleinschmidt, K.; Baumann, H.I.; Imhoff, J.F.; Kleine, M.; Cai, D. Genome Sequence of*Bacillus subtilis* subsp. *spizizenii* gtP20b, isolated from the Indian Ocean. *J. Bacteriol.* 2011, *193*, 1276 1277.
23. Hamdache, A.; Lamarti, A.; Aleu, J.; Collado, I.G. Non-peptide metabolites from the genus *Bacillus. J. Nat. Prod.* 2011,*74*, 893–899.
24. Baruzzi, F.; Quintieri, L.; Morea, M.; Caputo, L. Antimicrobial Compounds Produced by *Bacillus* spp. and Applications in Food. In *Science against Microbial Pathogens: Communicating Current Research and Technological Advances*; Vilas, A.M., Ed.; Formatex: Badajoz, Spain, 2011; pp. 1102–1111.
25. Shoda, M. Bacterial control of plant diseases. *Biosci. Bioeng.* 2000, *89*, 515–521.
26. Stein, T. *Bacillus subtilis* antibiotics: Structures, syntheses and specific functions. *Mol. Microbiol.* 2005, *56*, 845–857.
27. Dash, H.R.; Neelam, M.; Chakraborty, J.; Kumari, S.; Das, S. Marine bacteria: Potential candidates for enhanced bioremediation. *Appl. Microbiol. Biotechnol.* 2013, *97*, 561–571.
28. Nies, D.H. Microbial heavy metal resistance. *Appl. Microbiol. Biotechnol.* 1999, *51*, 730–750.
29. Trischman, J.A.; Jensen, P.R.; Fenical, W. Halobacillin: A cytotoxic cyclic acylpeptide of the iturin class produced by a marine *Bacillus. Tetrahedron*

Lett. 1994, *35*, 5571–5574.

30. Pettit, G.R.; Knight, J.C.; Herald, D.L.; Pettit, R.K.; Hogan, F.; Mukku, V.J.R.V.; Hamblin, J.S.; Dodson, M.J., II; Chapuis, J.C. Antineoplastic agents. 570. Isolation and structure elucidation of bacillistatins 1 and 2 from a marine *Bacillus silvestris. J. Nat. Prod.* 2009, *72*, 366–371.

31. Azumi, M.; Ogawa, K.; Fujita, T.; Takeshita, M.; Furumai, T.; Igarashi, Y.; Yoshida, R. Bacilosarcins A and B, novel bioactive isocoumarins with unusual heterocyclic cores from the marine-derived bacterium *Bacillus subtilis.Tetrahedron* 2008, *64*, 6420–6425.

32. Barsby, T.; Kelly, M.T.; Andersen, R.J. Tupuseleiamides and basiliskamides, new acyldipeptides and antifungal polyketides produced in culture by a *Bacillus laterosporus* isolate obtained from a tropical marine habitat. *J. Nat. Prod.*2002, *65*, 1447–1451.

33. Gustafson, K.; Roman, M.; Fenical, W. The macrolactins, a novel class of antiviral and cytotoxic macrolides from a deep-sea marine bacterium. *J. Am. Chem. Soc.* 1989, *111*, 7519–7524.

34. Mondol, M.A.M.; Kim, J.H.; Lee, M.A.; Tareq, F.S.; Lee, H.S.; Lee, Y.J.; Shin, H.J. Ieodomycins A–D, antimicrobial fatty acids from a marine *Bacillus* sp. *J. Nat. Prod.* 2011, *74*, 1606–1612.

35. Damalas, C.A.; Eleftherohorinos, I.G. Pesticide exposure, safety issues, and risk assessment indicators. *Int. J. Environ. Res. Public Health.* 2011, *8*, 1402–1419.

36. Ongena, M.; Jacques, P. *Bacillus* lipopeptides: Versatile weapons for plant disease biocontrol. *Trends Microbiol.* 2008,*16*, 115–125.

37. Raaijmakers, J.M.; de Bruijin, I.; de Kock, M.J. Cyclic lipopeptide production by plant-associated *Pseudomonas* spp.: Diversity, activity, biosynthesis, and regulation. *Mol. Plant Microbe Interact.* 2006, *19*, 699–710.

38. Arguelles-Arias, A.; Ongena, M.; Halimi, B.; Lara, Y.; Brans, A.; Joris, B.; Fickers, P. *Bacillus amyloliquefaciens* GA1 as a source of potent antibiotics and other secondary metabolites for biocontrol of plant pathogens. *Microb. Cell Fact.* 2009,*8*.

39. Malfanova, N.; Franzil, L.; Lugtenberg, B.; Chebotar, V.; Ongena, M. Cyclic lipopeptide profile of the plant-beneficial endophytic bacterium *Bacillus subtilis* HC8. *Arch. Microbiol.* 2012, *194*, 893–899.

40. Tsuge, K.; Akiyama, T.; Shoda, M. Cloning, sequencing, and characterization of the iturin A operon. *J. Bacterial.* 2001,*183*, 6265–6273.

41. Berrue, F.; Ibrahim, A.; Boland, P.; Kerr, R.G. Newly isolated marine *Bacillus pumilus* (SP21): A source of novel lipoamides and other antimicrobial agents. *Pure Appl. Chem.* 2009, *81*, 1027–1031.
42. Cooper, D.G.; MacDonald, C.R.; Duff, S.J.B.; Kosaric, N. Enhanced production of surfactin from *Bacillus subtilis* by continuous product removal and metal cation additions. *Appl. Environ. Microbiol.* 1981, *42*, 408–412.
43. Ma, Z.; Wang, N.; Hu, J.; Wang, S. Isolation and characterization of a new iturinic lipopeptide, mojavensin A produced by a marine-derived bacterium *Bacillus mojavensis*. *J. Antibiot.* 2012, *65*, 317–322.
44. Zhang, H.L.; Hua, H.M.; Pei, Y.H.; Yao, S. Three new cytotoxic cyclic acylpeptides from marine *Bacillus* sp. *Chem. Pharm. Bull.* 2004, *52*, 1029–1030.
45. Barsby, T.; Kelly, M.T.; Gagne, S.M.; Andersen, R.J. Bogorol A produced in culture by a marine *Bacillus* sp. reveals a novel template for cationic peptide antibiotics. *Org. Lett.* 2001, *3*, 437–440.
46. Gerard, J.; Haden, P.; Kelly, M.T.; Andersen, R.J. Loloatin B, cyclic decapeptide antibiotic, produced in culture by a tropical marine bacterium. *Tetrahedron Lett.* 1996, *37*, 7201–7294.
47. Jeong, S.Y.; Ishida, K.; Ito, Y.; Okada, S.; Murakami, M. Bacillamide, a novel algicide from the marine bacterium,*Bacillus* sp. SY-1, against the harmful dinoflagellate, *Cochlodinium polykrikoides*. *Tetrahedron Lett.* 2003, *44*, 8005–8007.
48. Gao, C.-H.; Tian, X.-P.; Qi, S.-H.; Luo, X.-M.; Wang, P.; Zhang, S. Antibacterial and antilarval compounds from marine gorgonian-associated bacterium *Bacillus amyloliquefaciens* SCSIO 00856. *J. Antibiot.* 2010, *63*, 191–193.
49. Liu, R.F.; Zhang, D.J.; Li, Y.G.; Tao, L.M.; Tian, L. A new antifungal cyclic lipopeptide from *Bacillus marinus* B-9987.*Helv. Chim. Acta* 2010, *93*, 2419–2425.
50. Fickers, P. Antibiotic compounds from *Bacillus*: Why are they so amazing? *Am. J. Biochem. Biotechnol.* 2012, *8*, 40–46.
51. Li, D.; Carr, G.; Zhang, Y.; Williams, D.E.; Amlani, A.; Bottriell, H.; Mui, A.L.F.; Andersen, R. Turnagainolides A and B, cyclic depsipeptides produced in culture by a *Bacillus* sp.: Isolation, structure elucidation, and synthesis. *J. Nat. Prod.* 2011, *74*, 1093–1099.
52. Cane, D.E.; Walsh, C.T. The parallel and convergent universes of polyketide synthases and nonribosomal peptide synthetases. *Chem.*

Biol. 1999, *6*, 319–325.

53. Tareq, F.S.; Kim, J.-H.; Lee, M.A.; Lee, H.-S.; Lee, Y.-J.; Lee, J.-S.; Shin, H.J. Ieodoglucomides A and B from a marine-derived bacterium *Bacillus licheniformis*. *Org. Lett.* 2012, *14*, 1464–1467.
54. Yu, L.-L.; Li, Z.-Y.; Peng, C.-S.; Li, Z.-Y.; Guo, Y.-W. Neobacillamide A, a novel thiazole-containing alkaloid from the marine bacterium *Bacillus vallismortis* C89, associated with South China Sea Sponge *Dysidea avar*. *Helv. Chim. Acta* 2009, *92*, 607–612.
55. Okazaki, H.; Kishi, T.; Beppu, T.; Arima, K. A new antibiotic baciphelacin. *J. Antibiot.* 1975, *28*, 717–719.
56. Itoh, J.; Omoto, S.; Shomura, T.; Nishizawa, N.; Miyado, S.; Yuda, Y.; Shibata, U.; Inouye, S. Amicoumacin-A, a new antibiotic with strong antiinflammatory and antiulcer activity. *J. Antibiot.* 1981, *34*, 611–613.
57. Itoh, J.; Omoto, S.; Shomura, T.; Nishizawa, N.; Miyado, S.; Yuda, Y.; Shibata, U.; Inouye, S. Chemical structures of amicoumacins produced by *Bacillus pumilus*. *Agric. Biol. Chem.* 1982, *46*, 1255–1259.
58. Shimojima, Y.; Hayashi, H.; Ooka, T.; Shibukawa, M. (Studies on AI-77s, microbial products with pharmacological activity) structures and the chemical nature of AI-77s. *Tetrahedron Lett.* 1982, *23*, 5435–5438.
59. Shimojima, Y.; Hayashi, H.; Ooka, T.; Shibukawa, M.; Iitaka, Y. Studies on AI-77s, microbial products with gastroprotective activity. Structures and the chemical nature of AI-77s. *Tetrahedron* 1984, *40*, 2519–2527.
60. McInerney, B.V.; Taylor, W.C.; Lacey, M.J.; Akhurst, R.J.; Gregson, R.P. Biologically active metabolites from *Xenorhabdus* spp., Part 2. Benzopyran-1-one derivatives with gastroprotective activity. *J. Nat. Prod.* 1991, *54*, 785–795.
61. Omura, S.; Iwai, Y.; Takahashi, Y.; Sadakane, N.; Nakagawa, A.; Oiwa, H.; Hasegawa, Y.; Ikai, T. Herbimycin, a new antibiotic produced by a strain of *Streptomyces*. *J. Antibiot.* 1979, *32*, 255–261.
62. Li, Y.; Xu, Y.; Liu, L.; Han, Z.; Lai, P.Y.; Guo, X.; Zhang, X.; Lin, W.; Qian, P.Y. Five new amicoumacins isolated from a marine-derived bacterium *Bacillus subtilis*. *Mar. Drugs* 2012, *10*, 319–328.
63. Fulco, A.J. Fatty acid metabolism in bacteria. *Prog. Lipid Res.* 1983, *22*, 133–160.
64. Fang, J.; Kato, C. FAS or PKS, lipid biosynthesis and stable carbon isotope fractionation in deep-sea piezophilic bacteria. *Commun. Curr. Res. Educ. Top. Trends Appl. Microbiol.* 2007, *1*, 190–200.
65. Kolattukudy, P.E.; Bohnet, S.; Sasaki, G.; Rogers, L. Developmental

changes in the expression of *S*-acyl fatty acid synthase thioesterase gene and lipid composition in the uropygial gland of mallard ducks (*Anas platyrhynchous*). *Arch. Biochem. Biophys.* 1991, *284*, 201–206.

66. Carballeira, N.M.; Ilieva, M.; Miranda, C.; Tzvetkova, I.; Lozano, C.M.; Nechev, J.T.; Ivanova, A.; Stefanov, K. Characterization of novel methyl-branched chain fatty acids from a halophilic *Bacillus* species. *J. Nat. Prod.* 2001, *64*, 256–259.

67. Mondol, M.A.M.; Tareq, F.S.; Kim, J.-H.; Lee, M.A.; Lee, H.-S.; Lee, Y.-J.; Lee, J.-S.; Shin, H.J. New antimicrobial compounds from a marine-derived *Bacillus* sp. *J. Antibiot.* 2013, *66*, 89–95.

68. Idriss, E.E.S.; Bochow, H.; Ross, H.; Borriss, R.Z. Use of *Bacillus subtilis* as biocontrol agent. VI. Phytohormone-like action of culture filtrates prepared from plant growth-promoting *Bacillus amyloliquefaciens* FZB24, FZB42, FZB45 and*Bacillus subtilis* FZB37. *J. Plant Dis. Prot.* 2004, *111*, 583–597.

69. Krebs, B.; Höding, B.; Kübart, S.M.; Workie, A.; Junge, H.; Schmiedeknecht, G.; Grosch, P.; Bochow, H.; Heves, M.Z. Use of *Bacillus subtilis* as biocontrol agent: Activities and characterization of *Bacillus subtilis* strains. *J. Plant Dis. Prot.*1998, *105*, 181–197.

70. Koumoutsi, A.; Chen, X.-H.; Henne, A.; Liesegang, H.; Hitzeroth, G.; Franke, P.; Vater, J.; Borriss, R. Structural and functional characterization of gene clusters directing nonribosomal synthesis of bioactive cyclic lipopeptides in*Bacillus amyloliquefaciens* strain FZB42. *J. Bacteriol.* 2004, *184*, 1084–1096.

71. Chen, X.H.; Vater, J.; Piel, J.; Franke, P.; Scholz, R.; Schneider, K.; Koumoutsi, A.; Hitzeroth, G.; Grammel, N.; Strittmatter, A.W.; *et al.* Structural and functional characterization of three polyketide synthase gene clusters in*Bacillus amyloliquefaciens* FZB42. *J. Bacteriol.* 2006, *188*, 4024–4036.

72. Schneider, K.; Chen, X.-H.; Vater, J.; Franke, P.; Nicholson, G.; Borriss, R.; Süssmuth, R.D. Macrolactin is the polyketide biosynthesis product of the pks2 cluster of *Bacillus amyloliquefaciens* FZB42. *J. Nat. Prod.* 2007, *70*, 1417–1423.

73. Zimmerman, S.B.; Schwartz, C.D.; Monaghan, R.L.; Pelak, B.A.; Weissberger, B.; Gilfillan, E.C.; Mochales, S.; Hernandez, S.; Currie, S.A.; Tejera, E.; *et al.* Difficidin and oxydifficidin: Novel broad spectrum antibacterial antibiotics produced by *Bacillus subtilis*. I. Production, taxonomy and antibacterial activity. *J. Antibiot.* 1987, *40*, 1677–1681.

74. Chen, X.H.; Scholz, R.; Borriss, M.; Junge, H.; Moegel, G.; Kunz, S.;

Borriss, R. Difficidin and bacilysin produced by plant-associated *Bacillus amyloliquefaciens* are efficient in controlling fire blight disease. *J. Biotechnol.* 2009, *140*, 38–44.

75. Zweerink, M.M.; Edison, A. Difficidin and oxydifficidin: Novel broad spectrum antibacterial antibiotics produced by*Bacillus subtilis*. III. Mode of action of difficidin. *J. Antibiot.* 1987, *40*, 1692–1697.

76. Lu, X.L.; Xu, Q.Z.; Liu, X.Y.; Cao, X.; Ni, K.Y.; Jiao, B.H. Marine Drugs—Macrolactins. *Chem. Biodivers.* 2008, *5*, 1669–1674.

77. Zheng, C.-J.; Lee, S.; Lee, C.-H.; Kim, W.-G.J. Macrolactins O–R, glycosylated 24-membered lactones from *Bacillus* sp. AH159-1. *J. Nat. Prod.* 2007, *70*, 1632–1635.

78. Nagao, T.; Adachi, K.; Sakai, M.; Nishijima, M.; Sano, H. Novel macrolactins as antibiotic lactones from a marine bacterium. *J. Antibiot.* 2001, *54*, 333–339.

79. Yoo, J.; Zheng, C.; Lee, S.; Kwak, J.; Kim, W. Macrolactin N, a new peptide deformylase inhibitor produced by *Bacillus subtilis*. *Bioorg. Med. Chem. Lett.* 2006, *16*, 4889–4892.

80. Xue, C.M.; Tian, L.; Xu, M.J.; Deng, Z.W.; Lin, W.H. A new 24-membered lactones and a new polyene δ-lactone from the marine bacterium *Bacillus marinus*. *J. Antibiot.* 2008, *61*, 668–674.

81. Jaruchoktaweechai, C.; Suwanborirux, K.; Tanasupawatt, S.; Kittakoop, P.; Menasveta, P. New macrolactins from a marine *Bacillus* sp. Sc026. *J. Nat. Prod.* 2000, *63*, 984–986.

82. Romero-Tabarez, M.; Jansen, R.; Sylla, M.; Lunsdorf, H.; Haubler, S.; Santosa, D.A.; Timmis, K.N.; Molinari, G. 7-*O*-malonyl macrolactin A, a new macrolactin antibiotic from *Bacillus subtilis* active against methicillin-resistant*Staphylococcus aureus*, vancomycin-resistant enterococci, and a small-colony variant of *Burkholderia cepacia*. *Antimicrob. Agents Chemother.* 2006, *50*, 1701–1709.

83. Mondol, M.A.M.; Kim, J.-H.; Lee, M.A.; Lee, H.-S.; Lee, Y.-J.; Shin, H.J. Macrolactin W, a new antibacterial macrolide from a marine *Bacillus* sp. *Bioorg. Med. Chem. Lett.* 2011, *21*, 3832–3835.

84. Mondol, M.A.M.; Tareq, F.S.; Kim, J.-H.; Lee, M.A.; Lee, H.-S.; Lee, Y.-J.; Lee, J.-S.; Shin, H.J. Cyclic ether-containing macrolactins, antimicrobial 24-membered isomeric macrolactones from a marine *Bacillus* sp. *J. Nat. Prod.* 2011, *74*, 2582–2587.

85. Losi, E.; Amrhein, C.; Frankenberger, W.T.J. Environmental biochemistry of chromium. *Rev. Environ. Contam. Toxicol.*1994, *136*, 91–121.

86. Katz, S.A.; Salem, H. *The Biological and Environmental Chemistry of Chromium*; VCH Publishers, Inc.: New York, NY, USA, 1994.
87. Anderson, R.A. Essentiality of Cr in humans. *Sci. Total Environ.* 1989, *86*, 75–81.
88. Anderson, R.A. Chromium as an essential nutrient for humans. *Regul. Toxicol. Pharmacol.* 1997, *26*, 35–41.
89. Costa, M. Toxicity and carcinogenicity of Cr(VI) in animal models and humans. *Crit. Rev. Toxicol.* 1997, *27*, 431–442.
90. Zhitkovich, A.; Voitkun, V.; Costa, M. Formation of the aminoacid-DNA complexes by hexavalent and trivalent chromium *in vitro*: Importance of trivalent chromium and the phosphate group. *Biochemistry* 1996, *35*, 7275–7282.
91. Syracuse Research Corporation, *Toxicological Profile for Chromium*; Prepared for U.S. Department Health and Human Services, Public Health Service, Agency for Toxic Substances and Disease Registry, under Contract No. 205-88-0608; ATSDR: Atlanta, GA, USA, 1993.
92. Krishnamurthy, S.; Wilkens, M.M. Environmental chemistry of Cr. *Northeast. Geol.* 1994, *16*, 14–17.
93. Pawlisz, A.V. Canadian water quality guidelines for Cr. *Environ. Toxicol. Water Qual.* 1997, *12*, 123–161.
94. Kavitha, V.; Radhakrishnan, N.; Gnanamani, A.; Mandal, A.B. Management of chromium induced oxidative stress by marine *Bacillus licheniformis*. *Biol. Med.* 2011, *3*, 16–26.
95. De Vrind, J.P.M.; Boogerd, F.C.; de Vrind-de Jong, E.W. Manganese reduction by a marine *Bacillus* species. *J. Bacteriol.* 1986, *167*, 30–34.
96. Jorge, R.-I.; Soto, L.A.; Federico, P.-O. Heavy-metal accumulation in the hydrothermal vent clam *Vesicomya gigas* from Guaymas basin, Gulf of California. *Deep Sea Res.* 2003, *50*, 675–761.
97. Shoresh, M.; Harman, G.; Mastouri, F. Induced systemic resistance and plant responses to fungal biocontrol agents.*Annu. Rev. Phytopathol.* 2010, *48*, 21–43.
98. Baig, D.N.; Mehnaz, S. Determination and distribution of cry-type genes in halophilc *Bacillus thuringiensis* isolates of Arabian Sea sedimentary rocks. *Microbiol. Res.* 2010, *165*, 376–383.
99. Cawoy, H.; Bettiol, W.; Fickers, P.; Ongena, M. *Bacillus-Based Biological Control of Plant Diseases*; InTech: New York, NY, USA, 2011 October 21; ISBN 978-953-307-459-7.
100. Hoch, J.; Sonenshein, A.; Losick, A. *Bacillus subtilis and Other Gram-*

Positive Bacteria: Biochemistry, Physiology and Molecular Genetics; American Society for Microbiology: Washington, DC, USA, 1993.

101. Devaraja, T.; Banerjee, S.; Yusoff, F. A holistic approach for selection of *Bacillus* spp. as a bioremediator for shrimp postlarvae culture. *Turk. J. Biol.* 2013, *37*, 92–100.

102. Nells, H.J.; de Leenheen, A.P. Microbial source of carotenoid pigments used in foods and feeds. *J. Appl. Bacteriol.* 1991, *70*, 181–191.

103. Stafsnes, M.H.; Josefsen, K.D.; Kildahl-Andersen, G.; Valla, S.; Ellingsen, T.E.; Bruheim, P. Isolation and characterization of marine pigmented bacteria from Norwegian coastal waters and screening for carotenoids with UVA-blue light absorbing properties. *J. Microbiol.* 2010, *48*, 16–23.

104. Sy, C.; Gleize, B.; Chamot, S.; Dangles, O.; Carlin, F.; Veyrat, C.C.; Borel, P. Glycosyl carotenoids from marine spore-forming *Bacillus* sp. strains are readily bioaccessible and bioavailable. *Food Res. Int.* 2013, *51*, 914–923.

105. Duc, L.H.; Fraser, P.D.; Tam, N.K.M.; Cutting, S.M. Carotenoids present in halotolerant *Bacillus* spore formers. *FEMS Microbiol. Lett.* 2006, *255*, 215–224.

106. Nugraheni, S.A.; Khoeri, M.M.; Kusmita, L.; Widyastut, Y.; Radjasa, O.K. Characterization of carotenoid pigments from bacterial symbionts of seagrass *Thalassia hemprichii. J. Coast. Dev.* 2010, *14*, 51–60.

107. Desal, C.; Pathak, H.; Madamwar, D. Advances in molecular and "omics" technologies to gauge microbial communities and bioremediation at xenobiotic/anthropogen contaminated sites. *Bioresour. Technol.* 2010, *101*, 1558–1569.

Chapter 2

MENTHA SUAVEOLENS EHRH. (LAMIACEAE) ESSENTIAL OIL AND ITS MAIN CONSTITUENT PIPERITENONE OXIDE: BIOLOGICAL ACTIVITIES AND CHEMISTRY

Mijat Božović, Adele Pirolli and Rino Ragno

Rome Center for Molecular Design, Department of Drug Chemistry and Technology, Sapienza University, P.le Aldo Moro 5, 00185 Rome, Italy

ABSTRACT

Since herbal medicines play an important role in the treatment of a wide range of diseases, there is a growing need for their quality control and standardization. *Mentha suaveolens* Ehrh. (MS) is an aromatic herb with fruit and a spearmint flavor, used in the Mediterranean areas as a traditional medicine. It has an extensive range of biological activities, including cytotoxic, antimicrobial, antioxidant, anti-inflammatory, hypotensive and insecticidal properties, among others. This study aims to review the scientific findings and research reported to date on MS that prove many of the remarkable various biological actions, effects and some uses of this species as a source of bioactive natural compounds. On the other hand, piperitenone oxide (PO), the major chemical constituent of the carvone pathway MS essential oil, has been reported to exhibit numerous bioactivities in cells and animals. Thus, this integrated overview also surveys and interprets the present knowledge of chemistry and analysis of this oxygenated monoterpene, as well as its beneficial bioactivities. Areas for future research are suggested.

INTRODUCTION

Essential oils are volatile, natural, complex compound mixtures characterized by a strong odor. They arise from the secondary metabolism of the plant, normally formed in special cells or groups of cells or in glandular hairs found on many leaves and stems. Essential oils are variable mixtures composed principally of terpenoids, including monoterpenes and sesquiterpenes (diterpenes may also be present), and their oxygenated derivatives. A variety of other molecules may also occur, such as aliphatic hydrocarbons, acids, alcohols, aldehydes, acyclic esters or lactones, and exceptionally nitrogen- and sulphur-containing compounds, coumarins and phenylpropanoid homologues. Known for their antiseptic (*i.e.*, bactericidal, virucidal and fungicidal), medicinal properties and their fragrance, they are used in embalmment, preservation of foods and as antimicrobial, analgesic, sedative, anti-inflammatory, spasmolytic and local anesthetic remedies [1].

The spread of drug-resistant pathogens is nowadays one of the most serious threats to the successful treatment of microbial diseases. It has been well-known since ancient times that certain plants and spices have antimicrobial activity [2,3]. They produce an enormous array of secondary metabolites, and it is commonly accepted that a significant part of this chemical diversity serves to protect plants against microbial pathogens. The World Health Organization (WHO) has noted that a majority of the World's population depends on traditional medicine for its primary healthcare [4].

Many essential oils and their ingredients have been shown to exhibit a range of biological activities, including antibacterial and antifungal activity. As a result, essential oils and/or their components are becoming increasingly popular as natural antimicrobial agents used for a wide variety of purposes. Their preparations find applications as naturally occurring antimicrobial agents in pharmacology, pharmaceutical botany, phytopathology, medical and clinical microbiology and food preservation. This review focuses on the essential oil of *Mentha suaveolens* Ehrh. (EOMS) and one of its main chemical constituents—piperitenone oxide (PO) (Figure 1).

A

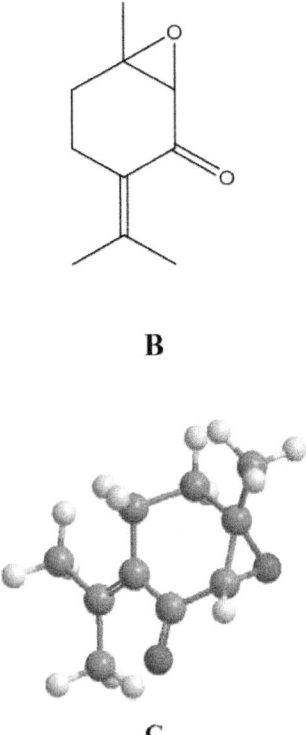

Figure 1: Wild MS (A) and PO 2D (B) and 3D (C) structure depictions.

MENTHA: SPECIES, TAXONOMY, OCCURRENCE AND USES

Based on phylogenetic analysis of morphology, chromosome numbers and major essential oil constituents the genus *Mentha* (mint), an important member of the *Lamiaceae* family, is highly diverse [5,6]. It is represented by about 19 species and 13 natural hybrids, mainly perennial herbs, growing wildly in damp or wet places throughout the temperate regions of Europe, Asia, Africa, Australia and North America. Mints are fast growing, invasive and generally tolerate a wide range of agro-climatic conditions [7].

Mentha identification is difficult, since in addition to much phenotypic plasticity and genetic variability, most of the species are capable of hybridization with each other. Hybrids are frequent in Nature but can usually be recognized by their intermediate appearance and sterility, although fertile hybrid swarms occasionally occur [8]. The present literature suggests the classification of the genus *Mentha* into the three basic lines named: *capitatae*, *spicatae* and *verticillatae*, based on the characteristic inflorescence. The *capitatae* line includes all species with compact, head-like inflorescences; the type species is *M. aquatica*. The *spicatae* species have a spike, as shown by *M. spicata*, *M. longifolia* and MS. The third line, represented by *M. arvensis*, has an inflorescence vertically partitioned into whorls [9]. Mints are also classified based on the dominant monoterpene compound prevailing in the essential oil resulting from three different metabolic pathways. Thus, the production of linalool and linalyl acetate is typical for the linalool pathway; menthol, menthone and menthofuran are constituents of the menthol pathway, and carvone, dihydrocarvone and carveol characterize the carvone pathway (Scheme 1) [5,9,10].

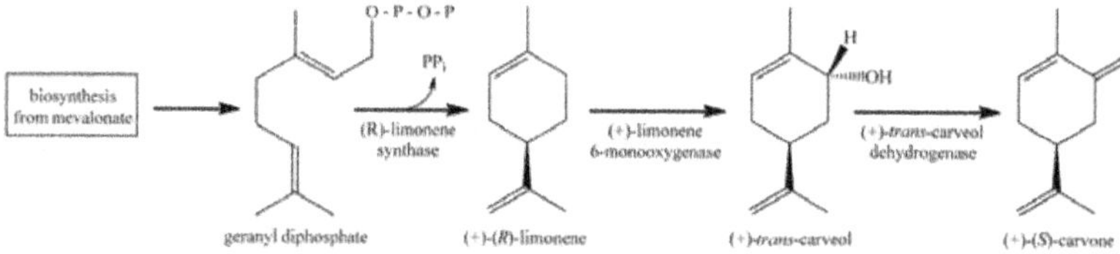

Scheme 1: Carvone biosynthetic pathway.

Eight species are reported in the Italian flora: *M. requienii*, *M. pulegium*, *M. arvensis* (and hybrids), *M. aquatica* (and hybrids) and the *M. spicata* group which includes *M. spicata*, *M. longifolia*, *M. microphylla* and MS [11]. *Mentha* species have been known for their medicinal and aromatherapeutic properties since ancient times. The ancient Egyptians, Romans and Greeks used peppermint as a flavoring agent for food and as a medicine, while mint essential oils have been used as perfumes, food flavors, deodorants and pharmaceuticals [7]. During the Middle Ages, powdered mint leaves were used to whiten the teeth [12]. Leaves, flowers and stems of *Mentha* spp. are frequently used in herbal teas or as additives in commercial spice mixtures for many foods to offer aroma and flavor. In addition, mints have been used as a folk remedy for treatment of nausea, bronchitis, flatulence, anorexia, ulcerative colitis and liver complaints, due to their anti-inflammatory, carminative, antiemetic, diaphoretic, antispasmodic, analgesic, stimulant, emmenagogue and anticatarrhal activities [13,14,15,16]. Different mint species are also used for rheumatism, dysentery, dyspepsia, skin allergies, chills, jaundice, throat infections, constipation, spasms, bladder stones, gall stone, diarrhea, toothache, stomach aches, dyspnea, gastrodynia, and as stimulant, diaphoretic, diuretic, reconstituent, stomach tonic, anti-infective, sedative, insect repellent, antimycobacterial, antifungal, antiallergic, virucidal, radioprotective, cyclooxygenase inhibitor, anti-inflammatory and hemostatic agents [17,18].

MENTHA ESSENTIAL OILS

Recently, essential oils and various extracts of plants have provoked interest as sources of natural products. They have been screened for their potential uses as alternative remedies for the treatment of many infectious diseases and the preservation food from the toxic effects of oxidants. Research on plants from different regions has led to innovative ways to use the essential oils [19]. Particularly, the antimicrobial activities of plant oils and extracts have formed the basis of many applications, including raw and processed food preservation, pharmaceuticals, alternative medicine and natural therapies [15].

Members of the genus *Mentha* produce some of the most widely used essential oils. Different species vary in their essential oil content and composition. The biosynthesis and metabolism of essential oils are strongly influenced by environmental factors, such as temperature, photoperiod, nutrition and salinity [20]. Plant chemotypes, cultivation practices and method of extraction also lead to variations in oil content and composition. Other factors affecting essential oil composition relate to agronomic and genotype conditions, such as harvesting time, plant age and crop density [7].

The essential oil isolated from mint leaves has economic importance and is widely used in the food industry, cosmetics, confectionary and pharmaceutical industries [5,21] Mint is widely cultivated for this oil, produced in many countries, such as America, India, China and Canada [5]. Commercially, the most important mint species are peppermint (*M. x piperita*), spearmint (*M. spicata*) and corn (American wild) mint (*M. canadensis*). Many species have been studied experimentally and the efficiency of some traditional applications were confirmed in many reports.

MENTHA SUAVEOLENS EHRH.: TAXONOMIC CHARACTERIZATION, DISTRIBUTION AND USES

MS, apple mint, woolly mint or round-leafed mint (synonyms: *M. macrostachya* Ten., *M. insularis* Req.) is an herbaceous, perennial herb with a sickly sweet scent that grows up to 100 cm in height (Figure 1). The stem is erect, quadrangular, and sparsely hairy to densely white-tomentose. It is monopodially branched, with short internodes. The foliage is light green with opposite, wrinkled, sessile or very short petiolate leaves which are ovate-oblong to suborbicular, 3 to 4.5 cm long and 2 to 4 cm broad. Obtuse, cuspidate or rarely acute, the leaves are widest near the base, serrate, with 10–20 teeth, hairy above, usually grey or white-tomentose to lanate. A prostrate branch (creeping sucker) growing from the axil of the leaves at the base of the flowering stem propagates below the level of the ground then gives root and turns upwards to give a new shoot. Many verticillasters, usually congested, form a terminal spike 4 to 9 cm long, consisting of a number of white or pinkish flowers [8,11,22].

This species is native to Southern and Western Europe, extending northwards to The Netherlands, cultivated as a pot-herb and naturalized in Northern and Central parts of Europe. It is generally found along streams, bogs and humid places [23]. MS has been used in the traditional medicine of Mediterranean areas and has a wide range of effects: hypotensive, stimulating, stomachic, carminative, choleretic, antispasmodic, sedative, tonic, anti-convulsive, insecticidal, *etc*.

It is also useful in cases of cough, nausea, anorexia and bronchitis [22] and finds application in digestion problems, influenza, respiratory ailments, rheumatism, skin diseases and irritation [24]. It shows depressor, analgesic, anti-inflammatory, cytotoxic, hepatoprotective and antifungal activities [10,14,25]. On reviewing the current literature on the phytochemistry of MS, flavonoids were the major constituents isolated from this species [10,26]. Concerning the biological activities, it was found that MS has antihypertensive [27], antioxidant and acetylcholinesterase inhibitory activities [28] and monoamine oxidase inhibitory activity [29]. The essential oil of MS was also found to have candidacidal activity [30,31] and a significant virucidal activity [32].

ESSENTIAL OIL COMPOSITION OF MS

There is an ongoing effort to screen plants used therapeutically in different regions of the World. However, it is well-known that the same taxon growing in different areas may have widely differing chemical components and hence differing biological properties [23]. Ingredients of EOMS have been subjected to a number of studies which have shown a difference in its constituents depending on the region of origin [33,34,35,36,37].

In general, investigations on the chemical composition of the essential oil from different populations collected in various regions showed high percentages of oxides. These include piperitone oxide and piperitenone oxide (PO) as major components [5,10,23,37]. Other chemotypes of this species showed high percentage of alcohols such as menthol [10] or ketones such as pulegone, piperitenone and dihydrocarvone (Table 1) [23,37]. Accordingly, three profiles of EOMS have been defined previously: the first profile is rich in pulegone, the second in PO and the third one contains similar quantities of PO and piperitone oxide [23].

Table 1: Chemical structures, names, MWs and CAS numbers of some of the most common constituents in EOMS

Chemical Structure	Chemical Name	MW	CAS Number
	piperitenone	150.22	491-09-8
	piperitenone oxide	166.22	35178-55-3
	piperitone oxide	168.23	5286-38-4
	pulegone	152.23	15923-80-6
	carvone	150.22	6485-40-1

Table 2: Example of EOMS chemical composition from two Moroccan localities [38]

N	Identified Compound	Kováts Index KI	Area%	
			Azrou	M'rirt
1	α-Pinene	939	0.36	0.36
2	Camphene	954	-	0.03
3	β-Pinene	979	0.65	0.65
4	meta-Mentha-1(7),8-diene	1000	0.18	0.02
5	α-Terpinene	1017	0.07	-
6	p-Cymene	1024	0.13	-
7	Limonene	1029	1.85	0.56
8	γ-Terpinene	1059	0.13	-
9	cis-Sabinene hydrate	1070	0.53	0.06
10	trans-Sabinene hydrate	1098	0.06	-
11	1-Octen-3-yl-acetate	1112	0.13	0.16
12	Dehydrosabinaketone	1120	0.05	-
13	4-Acetyl-1-methylcyclohexene	1137	0.08	-
14	Nopinone	1140	0.05	0.05
15	Borneol	1169	0.27	0.29
16	Terpinen-4-ol	1177	0.71	0.52
17	p-Cymen-8-ol	1182	0.12	-
18	α-Terpineol	1188	0.25	0.34
19	Coahuilensol methylether	1221	0.14	-
20	Pulegone	1237	2.34	0.47
21	cis-Carvone oxide	1263	0.44	0.19
22	Geranial	1267	-	0.2
23	Perillaaldehyde	1271	0.17	0.04
24	Isobornylacetate	1285	-	0.1
25	3'-Methoxyacetophenonone	1298	-	0.14
26	Piperitenone	1343	1.17	10.14
27	Piperitenone oxide	1368	74.69	81.67
28	Daucene	1381	0.11	-
29	β-Elemene	1390	0.16	-
30	4a-α,7-β,7a-α-Nepetalactone	1392	1.81	0.42
31	Longifolene	1407	0.27	-
32	β-Caryophyllene	1419	1.68	0.91
33	cis-Muurola-3,5-diene	1450	0.09	-
34	Spirolepechinene	1451	0.16	0.09
35	Khusimene	1455	0.68	-
36	cis-cadina-1(6),4-diene	1463	0.81	0.29
37	γ-Muurolene	1479	5.53	0.5
38	γ-Amorphene	1495	0.30	0.04

39	Aciphyllene	1501	0.10	-
40	γ-Cadinene	1513	0.11	0.1
41	trans-Calamenene	1522	0.77	-
42	α-Cadinene	1538	0.09	0.05
43	Spathulenol	1578	0.60	0.25
44	Caryophyllene oxide	1582	0.26	0.21
45	Globulol	1590	0.23	0.06
46	Ledol	1602	-	0.23
47	1,10-di-epi-Cubenol	1619	0.43	0.25
48	10-epi-α-Cadinol	1640	0.28	0.04
49	Torreyol	1646	0.05	-
50	α-Cadinol	1654	0.35	0.09
51	Germacra-4(15),5,10(14)trien-1-α-ol	1686	0.07	-
52	Shyobunol	1689	0.10	-

The populations of MS collected in various regions of Morocco showed a high percentage of oxides (PO and piperitone oxide), terpenic alcohols (fenchol, p-cymen-8-ol, geraniol, terpineol and borneol) and terpenic ketones (pulegone and piperitenone) [23]. Another investigation on Moroccan plant material from Azrou, Tetouan and Meknès confirmed the prevalence of PO (74.69%, 41.84% and 34%, respectively), while that from M'rirt is rich both in PO (81.67%) and piperitenone (10.14%) (Table 2) [38,39].

On the other hand, the composition of the essential oils from Béni-Mellal and Boulmane is totally different, with pulegone (85.5%) and menthol (40.50%) as the major compounds, as well as oil from the Oulmès region (Rabat) where piperitenone and pulegone are the main compounds (33.03% and 17.61%, respectively) [38,40]. Dominance of menthol (48.32%) and pulegone (20.27%) is also the characteristic of the material from Córdoba, Argentina [41]. The authors explained that this oil composition was unusual and closely related to that of *Mentha arvensis* var. *piperascens* which raises the possibility that the studied sample was not really MS but rather a hybrid (*M. arvensis* var. *piperiscens* x *M. suaveolens*) [41]. The oil obtained from aerial parts of the wild MS ssp. *timija*, an endemic species of Morocco, was characterised by a very high content of oxygen- containing monoterpenes with menthone (39.4%–10.8%), pulegone (62.3%–34.3%) and isomenthone (9.3%–7.8%) as the main constituents [42].

Hydrodistilled oils of the fresh aerial parts of MS cultivated in Egypt were characterized by carvone and limonene as the major constituents [10,43], while the cultivated material from Beheira (Egypt) showed dominance of linalool (35.32%), p-menth-1-en-8-ol (11.08%) and geranyl acetate (10.86%) [20,38]. Wild material analysis from the Alexandria-Cairo desert road (Egypt) showed PO (35.14%), germacrene-D (22.65%), o-menth-8-ene (8.98%), *trans*-β-farnesene (6.92%), veridiflorol (7.67%) and l-limonene (5.89%) as the main constituents [6].

MS is also widespread in Corsica, where it is represented by two subspecies: *suaveolens* and *insularis*. The latter species is endemic to the occidental Mediterranean Sea islands of Corsica, Sardinia, and the Balearic

Islands. These two subspecies are botanically close, but various morphological characteristics allow their differentiation. Analysis of 59 oil samples isolated from MS wild plants growing in Corsica, followed by statistical analysis of the data, allowed a clear differentiation of both subspecies with respect to the composition of their essential oils: the subspecies *suaveolens* yielded oils dominated either by piperitenone (73.5%) or PO (72%), while all the samples of the *insularis* subspecies contained pulegone and *cis-cis-p*-menthenolide as main components [37]. Similarly, plant material of *insularis* subspecies from Sardinia is also characterized by prevalence of pulegone [44]. Samples originating from Uruguay and Greece have shown a preponderance of PO that reached 62.4% and 80.8% [33]. A predominance of PO (55%) was also seen in materials from the Czech Republic [45]. However, the same species in northern Algeria contained three different chemotypes: a first one characterized by the predominance of PO (29.36%) and piperitone oxide (19.72%); the second one, conversely, with piperitone oxide (31.4%) as the most abundant component, followed by PO (27.79%), and the third contains piperitenone as major constituent (54.91%) [38]. A prevalence of piperitone oxide (40.5%) followed by hydroxy-*p*-menth-3-one (23.9%) was seen in a sample of cultivated MS in Padua (Italy) [46].

Analysis of the essential oil obtained from wild-type plants grown in the Tarquinia forests (Viterbo, Italy) showed a predominance of PO (>90%), with limonene and 1,8-cineole among minor constituents [25,30,32]. Another study on the same material was performed with the aim to analyze in details the essential oil extraction procedure in term of optimal period and extraction time. The material was submitted to hydrodistillation and the oil was collected at different times (1, 2, 3, 6, 12 and 24 h) on three different days of different months (July, August, September). The amount of the oil varied in function by both day of extraction and separation intervals. In August and September the oil yields were more than 2.5 times that of July. The maximum quantities of the oil are obtained in the first three and during the last twelve hours. In July, in the first 3 h almost 70% of the oil was extracted, while in August and September only 54% and 51%, respectively. In general, the most abundant constituent was PO, which percentage was maximum during the first three hours, disappearing in the last three extractions (after 6, 12 and 24 h). From the point of view of the extraction period, the extracted amounts were higher in the July and August samples than in the September sample. Other constituents are characteristic of the period and become important only after the fourth daily fraction. For example, in the month of July, demelverine and eucalyptol had a percentage of about 0.5%–2.0% in the first 3 h of extraction and increased to reach a maximum of about 29% after the 12-hour extraction. In August, β-caryophyllene oxide was always present and increased from about 0.4% after the 1-hour extraction to a maximum of 15.4% after the 12-hour extraction; cinerolone, that was absent in the first 3 h of extraction, showed high percentage in the next 3 h to reach a maximum of about 38.3% in the 24-hour extraction. In the September period, although PO was still the most abundant compound, the other important components seemed randomly distributed: in the first hour, contrarily from the previous period, β-caryophyllene oxide (13.5%) and cinerolone (18%) were at medium-high percentages, but then β-caryophyllene oxide decreased its percentage while cinerolone was absent but appeared again after 5 h till the last extraction (after 24 h), with a percentage varying from 2.4 to 17.5 [47].

ANTIMICROBIAL ACTIVITY OF MS

Essential oils and extracts have been used for thousands of years in food preservation, pharmaceuticals, alternative medicine and natural therapies [48]. They are potential sources of novel antimicrobial compounds, especially against bacterial pathogens and, in recent years, a large of number of investigations has been performed on their antimicrobial activities. Antimicrobial evaluations of essential oils are generally difficult because of their volatility, insolubility in water and complex chemistry [13,49].

Because of the mode of extraction, mostly by distillation from aromatic plants, essential oils contain a variety of volatile molecules such as terpenes and terpenoids, phenol-derived aromatic compounds and aliphatic components. The antimicrobial activity of essential oils has been extensively studied and demonstrated against a number of microorganisms, usually using direct-contact antimicrobial assays, such as different types of diffusion or dilution methods, as reviewed by many authors [50]. In these tests, essential oils are brought into direct contact with the selected microorganisms. However, due to high hydrophobicity and volatility of the essential oils, the direct-contact assays face many problems. In diffusion assays, the essential oils components are partitioned through the agar according to their affinity with water, and in dilution methods low water solubility has to be overcome by addition of emulsifiers or solvents (such as DMSO or ethanol) which may alter the activity [51]. Antimicrobial action is often determined by more than one component; each of them contributes to the beneficial or adverse effects. The major component may not be the only one responsible for the antimicrobial activity but a synergistic effect may take place with other oil components [50].

According to a literature survey, different mint species have been investigated in search of antimicrobial activities [13,15,48,51,52,53,54], inclusing several analyses performed on MS with an aim to investigate its antibacterial, antifungal or antiviral effects. Essential oils of MS grown in several regions in Morocco were tested for their activity against 19 bacteria, including Gram-positive and Gram-negative ones, and against three fungi, using solid phase and microtitration assays [23]. The antibacterial and antifungal activities of three types of EOMS were analyzed. Essential oil rich in pulegone strongly inhibited all bacteria, while the activity of that rich in PO was weaker. The activity of the third one (rich in both PO and piperitone oxide) had a tendency to be less important. Those results indicated that the efficacy of essential oils depends on their particular chemical composition, which was confirmed with another analysis where the major aromatic components of those oils were tested against the same organisms–pulegone was the most active against all bacteria, followed by PO; the activity of piperitone oxide seems to be two-fold lower than that of PO against a number of microorganisms, in particular, yeasts. These results indicate that pulegone may be the most active aromatic component in EOMS [23].

To study the importance of the chemical structure of the major constituents of EOMS, the antimicrobial activity of a series of aromatic components was analyzed and the activities of pulegone, PO and piperitone oxide were compared with carvone, limonene and menthone [55]. These latter aromatic components are also synthesized by *Mentha* species and have less antimicrobial activity. Menthone, which results from the reduction of pulegone, was less active than its precursor. Limonene and carvone also had moderate antimicrobial activity compared with pulegone and PO. Similar results were obtained by other authors [55] who found that pulegone showed a more potent biocidal activity than limonene, carvone and menthone. Compared with pulegone, limonene does not possess the extra cyclic double bond between C4 and C7, which results in the loss of the antimicrobial activity of this compound. Similarly, piperitone oxide does not possess this double bond and had in general a lower activity compared with PO and pulegone. Thus, this double bond seems to be important for the antimicrobial activity of the monoterpenes, possibly by favoring an active configuration of the molecule. The presence of an epoxide between C1 and C2 in PO decreases its antimicrobial activity compared with that of pulegone [23].

Investigation of the aerial parts of MS growing in Egypt showed moderate inhibitory activity against the tested human pathogenic bacteria. In that study, the antimicrobial screening of the ethanolic extract and its subfractions were performed [10]. The oil of the fresh aerial parts showed a potent antifungal activity against *Candida albicans, Saccharomyces cerevisiae* and*Aspergillus niger* [43]. Other studies on the Egyptian plant material showed a strong antibacterial activity of the essential oil, especially against *Staphylococcus aureus* [6]. The oils obtained from aerial parts of the wild and cultivated MS ssp. *timija*, an endemic species of Morocco, have been screened for antimicrobial activity. Both oils displayed good to excellent activity against all microorganisms tested, with the oil of the cultivated form being more active [42]. Both essential oils exhibited marked antifungal activity on all the *Candida*species tested, especially against *C. glabrata*. The antimicrobial activities of *timija* mint essential oil can be attributed to the presence of high concentrations of pulegone and menthone, two oxygenated monoterpenes with well-documented antibacterial and antifungal potential [22,37]. However, the comparison of the antimicrobial activity between these two major compounds showed that pulegone has more potent biocidal activity than menthone, which can explain the more potent antimicrobial activity of oil obtained from the cultivated *timija* mint [23,42].

Antimycotic analysis on the oil from the plant material from Tarquinia (Italy) displayed high activity against all strains of*Cryptococcus neoformans* and different dermatophyte strains, such as *Trycophyton mentagrophite, T. rubrum, T. violacee,Microsporum canis* and *M. gypseum*, where in all cases a good antimycotic activity was observed [22]. Another study on that material assessed the *in vitro* and *in vivo* antifungal activity of essential oil in an experimental vaginal candidiasis infection model. That study showed that EOMS was both candidastatic and candidacidal *in vitro*, as demonstrated in an *in vivo*monitoring imaging system [30]. The results obtained from another study demonstrated both the effects of the essential oil on *C. albicans* yeast cells and biofilms, and the synergism of the oil when used in combination with conventional antifungal drugs like fluconazole (FLC) and micafungin (MCFG) [31]. The antifungal activity of this oil was investigated for differences resulting from extraction time and period of material collection. That analysis showed that the oil extracted in the first 3 h showed good antifungal activities, but decreasing activity was evident with the oils from the 6 to 24-hour extractions. The study also confirmed PO as the principal active chemical constituent responsible for the essential oil's biological activity [47].

Determination of the antibacterial activity of three essential oils was performed *in vitro* against strains of *Pseudomonas syringe* pv. *actinidiae* (PSA), the causal agent of bacterial kiwifruit canker. That study included oil of MS from Tarquinia, which showed significantly higher activity than other two essential oils (*Rosmarinus officinalis* and *Melaleuca alternifolia*). Also, synergistic effects of those three essential oils were analyzed. Treating PSA with the mixture caused a significant decrease in the MIC, compared to their individual values, indicating a significant synergistic effect of the three essential oils combined. Results in that study clearly demonstrated that the mixture was able to kill PSA at the concentration about 16 times lower than the MIC values of the individual oils after 1 h of exposure [56]. This oil was also investigated against *Chlamydia trachomatis*, the most common sexually-transmitted bacterial infection worldwide. Those results showed effectiveness towards *C. trachomatis*, whereby it did not only inactivate infectious elementary bodies but it also inactivated chlamydial replication. The study also revealed the efficacy of the oil in combination with erythromycin: the combination inhibited *C. trachomatis* replication with a considerable four-fold reduction in the minimum effective dose of antibiotic [57].

The effects of EOMS derived from Tarquinia and its active principle PO were also tested in an *in vitro* experimental model of infection with herpes simplex virus type 1 (HSV-1), an important human pathogen. Moreover, a synergistic action was observed in combination with acyclovir. This study demonstrated that the oil, as well as its main compound PO, exerted stronger effects when added post-infection. When the oil, or the oxide, was preincubated with the virus before infection, both showed a significant virucidal activity, thus interfering directly with the viral envelope [32].

The antimicrobial activity of the essential oil from MS growing in the north of Morocco was evaluated on *Salmonella enterica*, *Listeria monocytogenes* and *Escherichia coli*, as well as for antiviral activity on the cytopathogenic murine norovirus (MNV-1). The results showed weak activity against *E. coli* and *L. monocytogenes*, but moderate activity was observed against *S. enterica*. The oil was mainly composed of PO, which was considered to have a low antimicrobial activity due to the presence of an epoxide between C1 and C2. This study showed that the oil tested had low antiviral activity against MNV-1 [39].

The *in vitro* antimicrobial activity of the essential oil of the aerial parts of MS ssp. *insularis* grown in Sardinia was assayed against six *Lactobacillus* species (including four probiotic strains), *Lactococcus lactis* ssp. *lactis* and *Staphylococcus xylosus*. Agar diffusion test results indicated that the essential oil exhibited a low antibacterial activity potential against all tested bacteria, except to *L. lactis* ssp. *lactis* and *S. xylosus* upon which it exerted a slight antibacterial activity. On the other hand, all yeasts strains tested (*Saccharomyces cerevisiae*, *Kloeckera apiculata*, *Candida zemplinina*, *Metschnikowia pulcherrima* and *Tetrapisispora phaffii*) were inhibited, and the oil exhibited slight to strong antifungal activity [44].

The mechanisms by which essential oils inhibit microorganisms involve different modes of action, but may be due in part to their hydrophobicity. As a result, they cause lipid partitioning of bacterial cell membranes and mitochondria, disturbing the cell structures and rendering them more permeable [58,59]. Extensive leakage from bacterial cells or the exit of critical molecules and ions, will lead to death [4]. Impairment of bacterial enzyme systems may also be a potential mechanism of antimicrobial action [60]. The antimicrobial activity of the essential oils can also be explained by the lipophilic character of the monoterpenes contained. The monoterpenes act by disrupting the microbial cytoplasmic membrane, which thus loses its high impermeability for protons and bigger ions. If the membrane integrity is disrupted, then its functions are compromised not only as a barrier but also as a matrix for enzymes and as an energy transducer. However, specific mechanisms involved in the antimicrobial action of monoterpenes remain poorly characterized [31]. According to a number of authors, Gram-negative bacteria are generally less susceptible than Gram-positive bacteria to the actions of essential oils, due to their outer membrane surrounding the cell wall which restricts diffusion of hydrophobic compounds through its polysaccharide covering [61,62,63,64]. According to some authors, this effect seems to be dependent on lipid composition and net surface charge of microbial membranes [62]. This statement is not always true; indeed different authors found no differences or greater sensibility of Gram-negative bacteria than Gram-positive to essential oils [44,52].

ANTIOXIDANT PROPERTIES OF EOMS

Free radicals are considered to initiate oxidation reactions that lead to aging and cause diseases in human beings. Moreover, activated oxygen incorporates reactive oxygen species (ROS) which consists of free (hydroxyl radicals, superoxide anion radicals) or non-free radicals (peroxide) [65]. These ROS are liberated by virtue of stress, and thus, an imbalance is developed in the body that damages cells in it and causes health problems [18]. On the other hand, restrictions have been imposed on the use of synthetic antioxidants because of their carcinogenicity and other toxic properties, which has considerably increased interest in natural antioxidants [18,19,66].

The active ingredients of a medicinal plant are mainly its secondary metabolites which are naturally produced during the metabolic processes of the plant's growth [18]. Natural antioxidants can be phenolic compounds (tocopherols, flavonoids and phenolic acids) and carotenoids (lutein, lycopene and carotene) [19,67]. Growing evidence has shown an inverse correlation between the intake of dietary antioxidants and the risk of chronic diseases such as coronary heart disease, cancer and several other aging-related health concerns [68,69]. There are several methods to determine the antioxidant capacity of plant extracts. However, the chemical complexity of extracts could lead to scattered results obtained from different techniques, depending on the test employed. Therefore, an approach with multiple assays in the screening work is highly advisable [70]. Investigation of the antioxidant activity of MS extracts from Morocco showed that the phenol extract presented a high antioxidant activity equivalent to that of butylated hydroxytoluene (BHT), an inhibitor well-known for its antioxidant activity [70,71]. Effectively, phenolic compounds are considered a major group of compounds that contribute to the antioxidant activities of botanical materials because of their scavenging ability on free radicals due to their hydroxyl groups [72]. The antioxidant activity of phenolic compounds is described as being largely influenced by the number of hydroxyl groups on the aromatic ring [73]. This activity is also due to their ability to scavenge free radicals, donate hydrogen atoms or electrons, or chelate metal cations [74]. The highest ferrous ion chelating activity was found in phenol and methanol extracts [71]. There is a linear correlation between the content of total phenolic compounds and their antioxidant capacity [29,72,75,76]. The results of this study indicate that phenolic compounds present in MS could be the major contributors of antioxidant capacities of this species [70].

Nine mint species from Pakistan were investigated as new potential sources of natural antioxidants. The methanolic extract assays revealed that significantly higher activity (82%) was observed in MS [18], which showed appreciable antioxidant activity only in the polar fractions while its decoction was also very effective in the inhibition of AChE and as a scavenger of radicals [28]. *Mentha* species prevent cell damage through their strong antioxidant activity, by scavenging free radicals and neutralizing bacterial invaders. They also promote the release of superoxide dismutase, a powerful antioxidant especially potent in destroying free radicals caused by imbalanced oxidation. Radical scavenging activity was observed when discoloration occurred, and MS were observed to produce high discoloration, followed by the other investigated species [18].

Analysis of some Egyptian mint species gave different results, since the lowest antioxidant activity was found in EOMS—only 6% [6]. Other analysis of the Egyptian EOMS showed potent *in vivo* (96% relative to vitamin E) and moderate *in vitro* antioxidant activities [43]. Ethanolic extracts of aerial parts of MS cultivated in Egypt and its subfractions (*n*-hexane, chloroform, ethyl acetate and *n*-butanol) were also evaluated. The ethyl acetate fraction showed the highest antioxidant activity: *in vivo* (as it restored the glutathione level in diabetic rats by 98%) and *in vitro* as it had the highest free radical scavenging activity (IC_{50} = 31 µg·mL^{-1}) [77]. Extracts of MS gathered from the interior of Portugal only showed appreciable antioxidant activity in the polar fractions [28].

INSECTICIDAL ACTIVITY OF EOMS

Plant insecticides have been used to fight pests for centuries. For instance, the use of plant extracts and powdered plant parts as insecticides was widespread during the Roman Empire. However, after the Second World War the few plants and plant extracts that had shown promising effects and were of widespread use were replaced by synthetic chemical insecticides. Later on, the adverse effect of chemical insecticides was realized with the appearance of problems like environmental contamination, residues in food and feed and pest resistance. Since the majority of plant insecticides are biodegradable, this has led to a revival of growing interest in the use of either plant extracts or essential oils. More than 1500 plant species have been reported to have insecticidal value [7]. Many plant secondary metabolites, such as alkaloids, monoterpenoids or phenylpropanoids are toxic to insects; in addition, essential oils extracted from plants have been widely investigated for pest control properties, with some providing to be toxic [78].

Mentha has historical significance as a medicinal and insecticidal plant in the traditional knowledge system. In the last few decades, many studies have been reported on the insecticidal activity of several *Mentha* species, largely in terms of adulticidal activity. EOMS from Azrou (Morocco) was investigated for its insecticidal activity against adults of *Sitophilus oryzae*. Considerable differences in insect mortality due to essential oil fumigation were observed using different concentrations and exposure times. Results showed that the essential oil was very toxic against *S. oryzae*, but the degree of this toxicity was influenced by the concentration applied and the exposure time [38]. Another study of the oil from the Moroccan material was carried out on two species of devastating insects of stored foodstuffs: *Sitophilus oryzae* and *Rizopertha dominica*. Mortality was 100% for amounts of 50 µL and 12 µL of the oil, while for the amount of 3 µL an acute toxicity was observed causing the mortality of 85% on the first and 100% on the second day [40]. To assess the biological activity of EOMS (also from Morocco), four concentrations were tested as fumigants against *Callosbruchus maculatus* reared on chickpea seeds. Great effectiveness of the oil was noticed in that study [79]. Oil from the Czech Republic had larvicidal activity against *Culex quinquefasciatus* [45].

The insecticidal activity of essential oils depends closely on their chemical composition. Monoterpenes have been well-documented as active fumigants and insecticides [80] and EOMS contains up to 86.2% of monoterpenes such as PO, pulegone, limonene, piperitenone, β-pinene, α-pinene and *p*-cymene. Their toxicity was proved toward stored product pests [34,40,79,80,81,82,83,84]. The fumigant toxicity of tested oil can be correlated with the abundance of PO. This oxygenated monoterpene possessed high toxicity against pests [45,80,82,84]. Differences in the chemical structures of monoterpenes are another factor influencing biological potency. Oxygenated terpenoids are more toxic than the non-oxygenated ones, and further, even among oxygenated ones, biological activity is differentiated by their other chemical groups and saturation [45]. Other components are present at low levels but may exert a synergistic effect [38].

The mode of action of essential oils and their constituents as insecticides is not known. The lipophilic nature of plant essential oils allows them to interfere with basic metabolic, biochemical, physiological and behavioral functions of insects [7]. Recent studies reported that essential oils and their constituents affect biochemical processes, which specifically disrupt the endocrine balance of insects. They may be neurotoxic or may act as insect growth regulators, disrupting the normal process of morphogenesis [85]. Further, monoterpenes have been investigated for their neurotoxicity [7]; they are typically volatile and rather lipophilic compounds that can penetrate into insects rapidly and interfere with their physiologic functions [86] by inhibiting acetylcholinesterase activity [80,87] and acting on insects' octopaminergic sites [88].

ADDITIONAL BIOACTIVITIES OF EOMS

EOMS from Egypt was screened for certain other biological activities. It exhibited analgesic and acute anti-inflammatory activities (75% and 82% relative to indomethacin) [43]. In addition, it exerted moderate cytotoxic and hepatoprotective activities [43]. Further investigation included different fractions of the ethanolic extract of the aerial parts of MS growing in Egypt. Analgesic and acute anti-inflammatory activities of the oral administration of the ethanolic extract and its subfractions (*n*-hexane, chloroform, ethyl acetate and *n*-butanol) were evaluated, using indomethacin as a standard drug. As a result, the ethanolic extract showed the most potent analgesic activity (78.5% potency compared to indomethacin), followed by the ethyl acetate and *n*-butanol fractions whose potency percentages were 66.3% and 54.7%, respectively. On the other hand, the ethyl acetate fraction was the most potent anti-inflammatory (88% potency) as compared to indomethacin, followed by the ethanolic extract (82.9%) and *n*-butanol fraction (62.6%) [10]. It is obvious that both analgesic and anti-inflammatory activities were exerted by the ethanol extract, the ethyl acetate and *n*-butanol fractions. It could be concluded that these activities may be attributed to their phenolic contents. The potent anti-inflammatory activities of the ethanolic extract may be due to its content of sterols, triterpenes, phenolic acids and flavonoids which have been proved to exert anti-inflammatory activity [10]. The hepatoprotective activity of the ethanolic extract and its subfractions was also evaluated. It revealed that the ethyl acetate fraction had the highest activity, as it prevented the increase caused by CCl_4 in the levels of aspartate amino transferase (AST), alanine amino transferase (ALT) and alkaline phosphatase (ALP) enzymes by

51.6%, 57.0% and 56.7%, respectively. The ethanolic extract, as well as its subfractions, were also tested for their cytotoxic activity. The results showed a significant activity of the ethanolic extract on liver carcinoma and larynx cancer cell lines (IC_{50} = 7.28 and 7.35 µg·mL^{-1}, respectively). The ethyl acetate fraction showed the highest activity against human liver carcinoma cell line (IC_{50} = 5.1 µg·mL^{-1}), the chloroform fraction was the most potent on the colon carcinoma cell line (IC_{50} = 14.40 µg·mL^{-1}) while *n*-hexane was the most active regarding the breast carcinoma cell line (IC_{50} = 13.5 µg·mL^{-1}) [77].

Essential oils, ethanolic extracts and decoction of 10 plants from interior Portugal were analyzed for their activity towards acetylcholinesterase (AChE) enzyme. MS showed AChE inhibitory capacity higher than 50% in the essential oil fraction, while the ethanolic extract was less potent. A high value of AChE inhibitory activity was found in a decoction of MS [28].

The pharmacological activity of a methanol extract of the leaves and stems of MS was analyzed in *in vivo* and *in vitro* models. The extract exhibited a central nervous system depressant action but no anticonvulsive activity. In order to determine the type of analgesia induced, the activity was evaluated using three different pain stimuli, *i.e.*, heat, mechanical and chemical agents such as acetic acid. The extract evaluated lacked any analgesic effect arising from CNS action since it did not increase the response time in the hot plate test. However, the extract showed a significant effect on mechanical and chemical stimuli, thus suggesting the induction of a peripheral analgesic effect. The extract also showed significant anti-inflammatory action inhibiting the rat paw oedema induced by carrageenan. Moreover, the *in vitro* studies showed a significant diminution in the contractile effects induced by histamine, serotonin and acetylcholine [14].

Since MS preparations have been applied in the traditional medicine of the Mediterranean areas as a hypotensive, the methanol and dichloromethanol extracts of the leaves and stems of this plant were tested for their effects on resting arterial blood pressure, heart rate and noradrenaline induced hypertension. Both extracts reduced the mean arterial blood pressure and heart rate, while only the dichloromethanol extract prevented the noradrenaline induced hypertension [27]. Mutagenicity tests revealed that the oil was not endowed with any particular toxic effect [47].

PIPERITENONE OXIDE—THE MAIN EOMS CHEMICAL COMPONENTS

1,2-Epoxypulegone (PO) is an important chemical constituent of the essential oils of many *Mentha* species, where it is formed by epoxidation of piperitenone

(Scheme 2). It was firstly named rotundifolone, since it was isolated from the essential oil of *M. rotundifolia* cultivated in Japan [89,90]. The taxonomic status of this species is confusing. Namely, according to Flora Europea [8], this was a misapplied name for MS (*Mentha rotundifolia* auct., non (L.) Huds.) and some authors consider it as its synonym [91], as mentioned in some articles [33,35,92]. On the other hand, there is a hybrid of MS and *M. longifolia*. named *Mentha* × *rotundifolia*. This problem has been discussed and clarified by some authors [93].

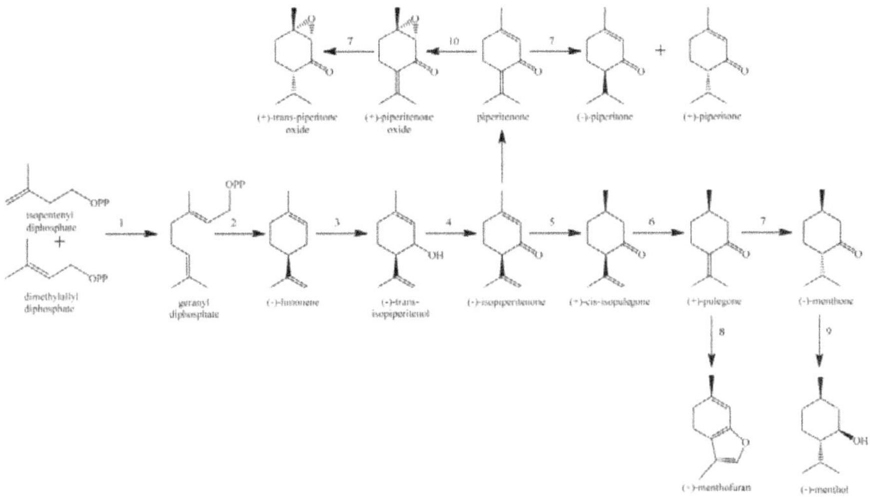

Scheme 2: The principal pathways for monoterpene biosynthesis in peppermint. The responsible enzymes are as follows: geranyl diphosphate synthase (1); (−)-limonene synthase (2); cytochrome P450 (−)-limonene-3-hydroxylase (3); (−)-trans-isopiperitenol dehydrogenase (4); (−)-isopiperitenone reductase (5); (+)-cis-isopulegone isomerase (6); (+)-PR (7); cytochrome P450 (+)-MFS (8); (−)-menthone reductase (9); and the terpenoid epoxidase (10). OPP denotes the diphosphate moiety [94].

This monoterpenoid ketone ($C_{10}H_{14}O_2$, molecular weight 166) is highly dextro-rotatory and has a low melting point (27.5 °C). The wavelengths of the ultraviolet absorption of PO and its semicarbazone are 260 and 273 nm, respectively, which indicates that this oxide has an α,β-unsaturated carbonyl system. PO does not give a ferric chloride reaction in methanol nor in aqueous suspension, but shows a positive Malaprade reaction. It reduces ammoniacal silver nitrate and Fehling solution. PO is soluble in ethanol, methanol, benzene and petroleum ether, but not in water and aqueous alkali [91]. PO is considered as one of the major common constituents of EOMS. This oxygenated monoterpene exhibits interesting activities, including cardiovascular, antimicrobial and insecticidal [38,79,82].

ALTERNATIVE SOURCES OF PO

Some other mint species are also rich in PO, which is often the major constituent of the essential oil of *Mentha x villosa*. The percentage varies from 55.4 to even 98% [95,96,97,98,99,100,101,102,103]. The oil of *M. microphylla* contains also PO in amounts of 32.9% [104] or even 65% [105]. Different investigations indicate high level of PO in the oils of *M. spicata* (up to 94.8%) [45,106,107,108,109,110,111] and *M. rotundifolia* (80.8%) [33,35,112], respectively. The last one was actually *Mentha ×rotundifolia* (the hybrid) as explained by authors themselves. The analysis of the *M. longifolia* oils also showed a significant amount of PO: ranging from 14.7% to 83.7% [12,19,45,52,113,114,115,116]. *M. aquatica* var. *crispa* oil from South Vietnam is also recognized as a rich source of PO (up to 91.2%) [117].

According to recent investigations, the plant species *Lippia pedunculosa* (Verbenaceae) may be considered as the new important source of PO since it is a predominant constituent ($\approx$72%) in its essential oil [118]. For sure, PO is not characteristic of the genus *Lippia*, but some of the other species also contains it as a minor or significant constituent [119]. Thus, *L. turbinate*essential oil contains 30% [120], the amount in *L. juneliana* oil is from 22.9% to 47.7% [121], while the oil of *L. alnifolia* was shown to have 44.6% of PO [122].

PO is also found to be the main constituent of some *Satureja* oils (Lamiaceae). Thus, *S. parvifolia* from Argentina contains 69.8% of PO [123,124], while the endemic *S. kallarica* from Iran has a 71.2% content [125]. Some *Calamintha* species (Lamiaceae) are also recognized to be quite rich in PO. The oils of *C. incana* from Turkey [126] and *C. nepeta* ssp. *glandulosa*from Belgium [127,128] contain up to 66.6% and 52% of PO, respectively.

Besides being available from natural sources, PO can be also produced by chemical synthesis. One of the ways is the epoxidation of piperitenone [119]. The monoterpenoid piperitenone (1) is of interest because it can be converted to a wide variety of important compounds. It has been obtained by several chemical routes [129,130]. The best yield was achieved with the process involving condensation of mesityl oxide (2) in tetrahydrofuran (THF) with methyl vinyl ketone (3) in THF and a solution of Triton-B (Scheme 3) [131].

Another synthetic procedure involves the same condensation, but uses sodium *tert*-butoxide as condensing agent in toluene [129]. However, this process mainly yields isoxylitone, the self-condensation product of 2 and less than 8% of 1. Subsequently, numerous investigations have attempted to attain a high yield of 1 by suppressing simultaneously the formation of isoxylitone. It had been found that when an alkali metal alkoxide was used as condensing agent, the self-condensation of 2 occurred in an aprotic solvent (e.g., *n*-hexane, benzene) but little or no reaction occurred if the solvent was replaced by a protic solvent (e.g., ethanol), tetrahydrofuran (THF) or a hydrous aprotic solvent. A heterogeneous system of potassium hydroxide and THF restricts the formation of isoxylitone considerably [130]. It has been reported that **1** is capable of forming a water-soluble sodium bisulfite addition compound. Treatment of the reaction mixture from the condensation reaction with bisulfite followed by ether extraction gave an aqueous solution from which **1** could be regenerated and isolated in 54% overall yield. Another condensation includes sodium hydride or a potassium 2-butoxide as catalysts [130].

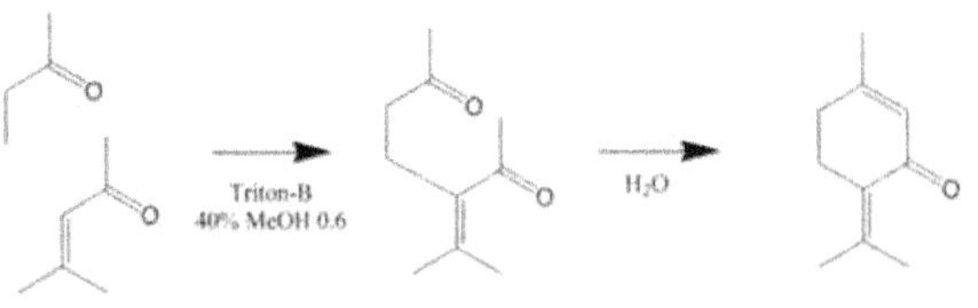

Scheme 3: Synthesis of piperitenone from mesityloxide and methyl vinyl ketone.

In the epoxidation step 30% aqueous hydrogen peroxide and 10% potassium hydroxide should be added to 1 in isopropyl alcohol. Two main fractions are obtained: the unreacted **1** mixed with PO, which could be separated by the column chromatography (Scheme 4) [119].

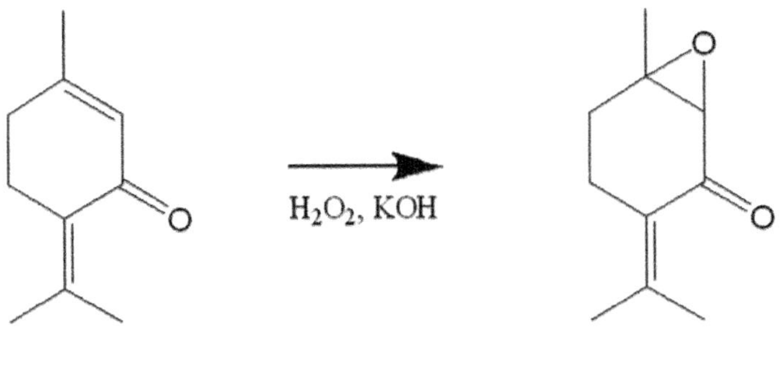

Scheme 4: Epoxidation of piperitenone.

BIOACTIVITIES OF PO

The monoterpenic ketone PO has been evaluated in relation to the following different biological activities: cardiovascular, hypotensive, bradycardic, insecticidal, trypanocidal, schistosomicidal, antimicrobial and antinociceptive properties. It has been shown that PO exhibited central analgesic activity in mice and rats [97]. PO was analyzed for potential activity on smooth muscle.

The experimental model was the guinea pig ileum, and the main conclusion of that study was that PO had a relaxant, depressant activity on intestinal smooth muscle [132].

PO exhibited strong toxic effect against the larvae of the mosquito species *Aedes aegypti* [103] but also against other mosquito species [82,84]. Although it has shown a good larvicidal effect against the *Ae. aegypti*, it was less potent than its essential oil of origin, which may be due to a synergistic interaction between PO and other compounds. Comparing PO and its analogues confirmed that the different functional groups and their positions in the *p*-menthane skeleton influence the larvicidal activity. In general, replacement of C=C double bonds by epoxide groups decreases the larvicidal potency. It can be concluded that with appropriate structural modification in the monoterpenes it may be possible to develop new larvicidal agents [103].

The essential oil from the leaves of *Lippia pedunculosa* and its main compounds, the monoterpenes PO and (*R*)-limonene, were evaluated for their trypanocidal activity against epimastigote and trypomastigote forms of *Trypanosoma cruzi*. PO was the most active compound. The effects of the oil and isolated compounds on the intracellular form of the parasite were also evaluated in cultures of macrophages infected with *T. cruzi*, but the treatment with (*R*)-limonene and PO caused a moderate reduction in the percentage of macrophages [118].

Biological effects of the *M. x villosa* essential oil and its main constituent PO were evaluated on adult worms of *Schistosoma mansoni* (Plathelminths). Worms were incubated with different concentrations of the oil and PO, which resulted in decreased worm motility continuing until 96 h of observation. At higher concentrations (100 and 70.96 µg·mL^{-1}, respectively), both the essential oil and PO caused mortality among adult *S. mansoni* worms [101].

Cardiovascular effects of intravenous treatment with the essential oil of *M. x villosa* were investigated. Additionally, that study examined whether the major constituent PO was the active principle mediating changes in mean aortic pressure and heart rate. The study showed that the treatment with oil in rats induced hypotensive and bradycardic effects, which appeared mostly attributed to the action of the main compound of the oil, PO [95]. The acute cardiovascular effects of PO were also investigated in rats by using a combined (*in vivo* and *in vitro*) approach. The acute administration of PO induced a short-lasting and dose-related decrease in arterial pressure, followed by a significant bradycardia, probably due to a non-specific muscarinic receptor stimulation. Furthermore, *in vitro* studies suggested that PO induced vasodilatation [97]. The effects of PO on vascular smooth muscle were analyzed, and the major finding was that PO-induced vasodilatation of the rat aorta was apparently mediated by an inhibitory effect on Ca^{2+} influx and inhibition of intracellular Ca^{2+} release from stores [96]. The authors concluded that the hypotensive effect was possibly due to a reduction in heart rate associated to a reduction of peripheral vascular resistance, both due to muscarinic activation [133]. PO induces muscle contraction in both depolarized and non-depolarized sartorius muscle of toad (*Bufo paracnemis*). The study demonstrated that PO induced muscle contraction by releasing calcium from the sarcoplasmic reticulum, probably by the activation of the ryanodine receptor [134].

Assessment of the antinociceptive activity of PO and the analogous compounds was performed using the acetic acid-induced writhing model in mice. All compounds showed to be more antinociceptive than PO against the pain response induced by acetic acid. It was found that the functional groups and their position on the ring of PO contributed to its antinociceptive activity [135]. All those hypotheses were subsequently strengthened by studies characterizing the molecular mechanism of action involved in relaxation produced by PO. The findings suggested that PO induced vasodilatation through two distinct but complementary mechanisms that clearly depended on the concentration used [136]. Another study was performed to evaluate the vasorelaxant effects of different monoterpenes and establish the structure-activity relationship of PO and its structural analogues. The results showed that both oxygenated and non-oxygenated monoterpenes exhibited relaxation activity. The absence of an oxygenated molecular structure was not a critical requirement for the molecule to be bioactive. It was also found that the position of ketone and epoxide groups in the monoterpene structures influenced the vasorelaxant potency and efficacy [137].

The oil of *M. x villosa* and its major component PO together with four similar analogues (limonene oxide, pulegone oxide, carvone epoxide and (+)-pulegone) were evaluated in relation to the antimicrobial activity against *Staphylococcus aureus*,*Escherichia coli*, *Pseudomonas aeruginosa*, *Candida albicans* and a strain of meticilin-resistant *Staphylococcus aureus* (MRSA). The essential oil, PO and the analogues showed antibacterial activity on *S. aureus* and antifungal activity on *C. albicans*. Limonene oxide and carvone epoxide were the substances with the lowest antimicrobial potential. None of the products showed antimicrobial activity against strains of the Gram negative bacteria *E. coli* and *P. aeruginosa* [138].

CONCLUSIONS AND FUTURE PERSPECTIVES

In general, plants have provided a source of inspiration for novel drug compounds. The increased interest in alternative natural substances is driving the research community to find new uses and applications for these substances and has led to a considerable increase in the use of medicinal plants. The results of the cited studies indicate that MS and PO, as the main compound of EOMS, show a wide range of biological activities. Keeping in mind that the biological properties can be the result of synergism, investigation of the main compounds alone seems questionable. However, PO is usually the predominant compound (sometimes more than 90%) in EOMS and was found to reflect quite well the biophysical and biological features of the whole oil. PO seems to be responsible for a lot of bioactivities although it is possible that its activity is slightly modulated by the other minor molecules.

Neither the essential oil nor PO have very strong antibacterial effects. Pulegone is the most important *Mentha* constituent responsible for antibacterial activity, but the oil of this species is usually not rich in it. The effect of the oil is significant only in cases where other compounds are present in reasonable amounts, such as pulegone, menthone or β-cymene. On the other hand, the oil and PO seem to be potent antifungal agents. In that sense, further investigations of PO can be proposed. There are some data indicating certain antiviral effects of the oil, as well as PO. Thus, that field is also interesting for the future examinations. It should be added that the further studies are needed to evaluate the *in vivo* potential in animal experimental models since there are little data about that aspect.

The essential oil has strong antioxidant potential. In some cases, it can be compared with butylated hydroxytoluene which is a well-known synthetic antioxidant additive. However, to the best of our knowledge, investigation of PO in this field is missing.

Insecticidal properties of EOMS and PO were evaluated. Since they usually exhibit strong toxic effects, this can definitely be a field for future study. The oil contains a huge amount of monoterpenes (often, even more than 85%) which have been well-documented as active fumigants and insecticides. On the other hand, oxygenated terpenoids are more toxic than the non-oxygenated ones. Thus, the fumigant toxicity of this oil can be justified by the high content of the monoterpene PO, which has already been well explored in that sense. What can be emphasized is that appropriate structural modification in the monoterpenes may lead to the development of new insecticidal agents.

Additional bioactivities of EOMS, as well as PO, have been investigated, and the results showed good potential in relation to analgesic, anti-inflammatory, antihypertensive and AChE inhibitory activities. There are also a significant number of studies on the different extracts, not the essential oil, that can be continued avenues of study in the future. Future studies should further explore the possible beneficial synergistic properties of combining PO (or the whole oil) with other natural or synthetic compounds.

ACKNOWLEDGMENTS

The authors are grateful to the ERASMUS program for supporting this study.

AUTHOR CONTRIBUTIONS

M.B, A.P. and R.R. together initiated and designed the work. R.R. contributed to literatures collection and drew the chemical compound structures. M.B. drafted the manuscript. M.B., A.P. and R.R. finalized and critically edited the manuscript before submission.

REFERENCES

1. Bakkali, F.; Averbeck, S.; Averbeck, D.; Idaomar, M. Biological effects of essential oils—A review. *Food Chem. Toxicol.* 2008, *46*, 446–475.
2. Nascimento, G.; Locatelli, J.; Freitas, P.; Silva, G. Antibacterial activity of plant extracts and phytochemicals on antibioticresistant bacteria. *Braz. J. Microbiol.* 2000, *31*, 247–256.
3. Valgas, C.; de Souza, S.; Smania, E.; Smania, A. Screening methods to determine antibacterial activity of natural products. *Braz. J. Microbiol.* 2007, *38*, 369–380.
4. Prabuseenivasan, S.; Jayakumar, M.; Ignacimuthu, S. *In vitro* antibacterial activity of some plant essential oils. *BMC Complement. Altern. Med.* 2006, *6*, 39.
5. *Mint—The Genus Mentha Medicinal and Aromatic Plants—Industrial Profiles*; Lawrence, B.M., Ed.; CRC Press: Boca Raton, FL, USA, 2007; pp. 1–547.
6. Elansary, H.O.; Ashmawy, N.A. Essential oils of mint between benefits and hazards. *J. Essent. Oil Bear. Plants* 2013, *16*, 429–438.
7. Kumar, P.; Mishra, S.; Malik, A.; Satya, S. Insecticidal properties of *Mentha* species: A review. *Ind. Crops. Prod.* 2011, *34*, 802–817.
8. Tutin, T.G.; Heywood, V.H.; Burges, N.A.; Moore, D.M.; Valentine, D.H.; Walters, S.M.; Webb, D.A. *Flora Europaea*; Cambridge University Press: Cambridge, UK, 1968; pp. 707–709.
9. Šarić-Kundalić, B.; Fialová, S.; Dobeš, C.; Ölzant, S.; Tekel'ová, D.; Grančai, D.; Reznicek, G.; Saukel, J. Multivariate numerical taxonomy of *Mentha* species, hybrids, varieties and cultivars. *Sci. Pharm.* 2009, *77*, 851–876.
10. El-Kashoury, E.A.; El-Askary, H.I.; Kandi, Z.A.; Ezzat, S.M.; Salem, M.A.; Sleem, A.A. Chemical and biological study of *Mentha*

suaveolens Ehrh. cultivated in Egypt. *J. Med. Plants Res.* 2014, *8*, 747–755.
11. Pignatti, S. *Flora d'Italia 2*; Edagricole: Bologna, Italy, 1982; pp. 494–499.
12. Hajlaoui, H.; Snoussi, M.; Jannet, H.B.; Mighri, Z.; Bakhro, A. Comparison of chemical composition and antimicrobial activities of *Mentha longifolia* L. ssp. *Longifolia* essential oil from two Tunisian localities (Gabes and Sidi Bouzid). *Ann. Microbiol.* 2008, *58*, 513–520.
13. Işcan, G.; Kirimer, N.; Kürkcüoglu, K.; Hüsnü, C.; Başer, K.; Demirci, F. Antimicrobial screening of *Mentha piperita*essential oil. *J. Agric. Food Chem.* 2002, *50*, 3943–3946.
14. Moreno, L.; Bello, R.; Primo-Yúfera, E.; Esplugues, J. Pharmacological properties of the methanol extract from *Mentha suaveolens* Ehrh. *Phytother. Res.* 2002, *16*, 10–13.
15. Hajlaoui, H.; Trabelsi, N.; Noumi, E.; Snoussi, M.; Fallah, H.; Ksouri, R.; Bakhrouf, A. Biological activities of the essential oils and methanol extract of tow cultivated mint species (*Mentha longifolia* and *Mentha pulegium*) used in the Tunisian folkloric medicine. *World J. Microbiol. Biotechnol.* 2009, *25*, 2227–2238.
16. Ceker, S.; Agar, G.; Alpsoy, L.; Nardemir, G.; Kizil, H.E.; Mete, E. Protective role of *Mentha longifolia* L. ssp. *longifolia*against Aflatoxin B_1. *J. Essent. Oil Bear. Plants* 2013, *16*, 600–607.
17. Naghibi, F.; Mosaddegh, M.; Motamed, S.M.; Ghorbani, A. *Labiatae* family in folk medicine in Iran: From ethnobotany to pharmacology. *Iran. J. Pharm. Res.* 2005, *2*, 63–79.
18. Ahmad, N.; Faza, H.; Ahmad, I.; Abbasi, B.H. Free radical scavenging (DPPH) potential in nine *Mentha* species.*Toxicol. Ind. Health* 2012, *28*, 83–89.
19. Iqbal, T.; Hussain, A.I.; Chatha, S.A.S.; Naqvi, S.A.R.; Bokhari, T.H. Antioxidant activity and volatile and phenolic profiles of essential oil and different extracts of wild mint (*Mentha longifolia*) from the Pakistani flora. *J. Anal. Methods Chem.* 2013.
20. Aziz, E.E.; Craker, L.E. Essential oil constituents of peppermint, pennyroyal and apple mint grown in a desert agrosystem. *J. Herbs Spices Med. Plants* 2010, *15*, 361–367.
21. Clark, G.S. An Aroma Chemical Profile: Menthol. *Perfumer Flavorist* 1998, *23*, 33–46.
22. El-Kashoury, E.; El-Askary, H.I.; Kandil, Z.A.; Salem, M.A. Botanical

and genetic characterization of *Mentha suaveolens*Ehrh cultivated in Egypt. *Pharacogn. J.* 2013, *5*, 228–237.

23. Oumzil, H.; Ghoulami, S.; Rhajaoui, M.; Ilidrissi, A.; Fkih-Tetouani, S.; Faid, M.; Benjouad, A. Antibacterial and antifungal activity of essential oils of *Mentha suaveolens*. *Phytother. Res.* 2002, *16*, 727–731.

24. Karousou, R.; Balta, M.; Hanlidou, E.; Kokkini, S. "Mints", smells and traditional uses in Thessaloniki (Greece) and other Mediterranean countries. *J. Ethnopharmacol.* 2007, *109*, 248–257.

25. Angiolella, L.; Vavala, E.; Sivric, S.; Diodata, D.A.F.; Ragno, R. *In vitro* activity of *Mentha suaveolens* essential oil against *Cryptococcus neoformans* and *dermatophytes*. *Int. J. Essent. Oil Ther.* 2010, *4*, 35–36.

26. Zaidi, F.; Voirin, B.; Jay, M.; Viricel, M.R. Free flavonoid aglycones from leaves of *Mentha pulegium* and *Mentha suaveoles* (Labiateae). *Phytochemistry* 1998, *48*, 991–994.

27. Bello, R.; Calatayud, S.; Beltrán, B.; Primo-Yúfera, E.; Esplugues, J. Cardiovascular effects of the methanol and dichloromethanol extracts from *Mentha suaveolens* Ehrh. *Phytother. Res.* 2001, *15*, 447–448.

28. Ferreira, A.; Proença, C.; Serralheiro, M.L.M.; Araújo, M.E.M. The *in vitro* screening for acetylcholinesterase inhibition and antioxidant activity of medicinal plants from Portugal. *J. Ethnopharmacol.* 2006, *108*, 31–37.

29. López, V.; Martín, S.; Gómez-Serranillos, M.P.; Carretero, M.E.; Jäger, A.K.; Calvo, M.I. Neuroprotective and neurochemical properties of mint extracts. *Phytother. Res.* 2010, *24*, 869–874. [PubMed]

30. Pietrella, D.; Angiolella, L.; Vavala, E.; Rachini, A.; Mondello, F.; Ragno, R.; Bistoni, F.; Vecchiarelli, A. Beneficial effect of *Mentha suaveolens* essential oil in the treatment of vaginal candidiasis assessed by real-time monitoring of infection. *BMC Complement. Altern. Med.* 2011, *11*, 18.

31. Stringaro, A.; Vavala, E.; Colone, M.; Pepi, F.; Mignogna, G.; Garzoli, S.; Cecchetti, S.; Ragno, R.; Angiolella, L. Effects of *Mentha suaveolens* essential oil alone or in combination with other drugs in *Candida albicans*. *Evid.-Based Complement. Altern. Med.* 2014, *2014*, 125904:1–125904:9.

32. Civitelli, L.; Panella, S.; Marcocci, M.E.; de Petris, A.; Garzoli, S.; Pepi, F.; Vavala, E.; Ragno, R.; Nencioni, L.; Palamara, A.T.; *et al*. *In vitro* inhibition of herpes simplex virus type 1 replication by *Mentha suaveolens* essential oil and its main component piperitenone oxide. *Phytomedicine* 2014, *21*, 857–865.

33. Lorenzo, D.; Paz, D.; Dellacassa, E.; Davies, P.; Vila, R.; Cañigueral, S. Essential oils of *Mentha pulegium* and *Mentha rotundifolia* from Uruguay. *Braz. Arch. Biol. Technol.* 2002, *45*, 519–524.
34. El Arch, M.; Satrani, B.; Farah, A.; Bennani, L.; Boriky, D.; Fechtal, M.; Blaghen, M.; Talbi, M. Composition chimique et activités antimicrobienne et insecticide de l'huile essentielle de *Mentha rotundifolia* du Maroc. *Acta Bot. Gallica* 2003, *150*, 267–274.
35. Brada, M.; Bezzina, M.; Marlier, M.; Carlier, A. Variabilité de la composition chimique des huiles essentielles de*Mentha rotundifolia* du nord de l'Algérie. *Biotechnol. Agron. Soc. Environ.* 2007, *11*, 3–7.
36. Abbaszadeh, B.; Valadabadi, S.A.; Farahani, H.A.; Darvishi, H.H. Studying of essential oil varations in leaves of*Mentha* species. *Afr. J. Plant Sci.* 2009, *3*, 217–221.
37. Sutour, S.; Bradesi, P.; Casanova, J.; Tomi, F. Composition and chemical variability of *Mentha suaveolens* ssp. *suaveolens*and *M. suaveolens* ssp. *insularis* from Corsica. *Chem. Biodivers.* 2010, *7*, 1002–1008.
38. Zekri, N.; Sabri, H.; Khannouchi, S.; El Belghiti, M.A.; Zair, T. Phytochemical study and fumigant toxicity of *Mentha suaveolens* Ehrh essential oil from Morocco against adults of *S. oryzae* (L.). *Aust. J. Basic Appl. Sci.* 2013, *7*, 599–606.
39. El Moussaoui, N.; Sanchez, G.; Khay, E.O.; Idaomar, M.; Ibn Mansour, A.; Abrini, J.; Aznar, R. Antibacterial and antiviral activities of essential oils of northern Moroccan plants. *Br. Biotechnol. J.* 2013, *3*, 318–331.
40. Benayad, N.; Ebrahim, W.; Hakiki, A.; Mosaddak, M. Chemical characterization and insecticidal evaluation of the essential oil of *Mentha suaveolens* Ehrh. and *Mentha pulegium* L. growing in Morocco. *Food Ind.* 2012, *13*, 27–32.
41. Velasco-Negueruela, A.; Perez-Alonso, M.J. Essential oils of *Calamintha nepeta* (L.) Savi and *Mentha aff suaveolens* Ehrh. grown in Córdoba, Argentina. *J. Essent. Oil Res.* 1996, *8*, 81–84.
42. Kasrati, A.; Jamalia, C.A.; Bekkouchea, K.; Lahcenb, H.; Markouka, M.; Wohlmuthc, H.; Leachc, D.; Abbada, A. Essential oil composition and antimicrobial activity of wild and cultivated mint timija (*Mentha suaveolens* subsp. *timija*(Briq.) Harley), an endemic and threatened medicinal species in Morocco. *Nat. Prod. Res.* 2013, *27*, 1119–1122.
43. El-Kashoury, E.A.; El-Askary, H.I.; Kandil, Z.A.; Salem, M.A.; Sleem, A.A. Chemical composition and biological activities of the essential oil of *Mentha suaveolens* Ehrh. *Z. Naturforsch.* 2012, *67c*, 571–579.

44. Petretto, G.L.; Fancello, F.; Zara, S.; Foddai, M.; Mangia, N.P.; Sanna, M.L.; Omer, E.A.; Menghini, L.; Chessa, M.; Pintore, G. Antimicrobial activity against beneficial microorganisms and chemical composition of essential oil of *Mentha suaveolens* ssp. *insularis* grown in Sardinia. *J. Food Sci.* 2014, *79*, 369–377.

45. Pavela, R.; Kaffková, K.; Kumšta, M. Chemical composition and larvicidal activity of essential oils from different*Mentha* L. and *Pulegium* species against *Culex quinquefasciatus* say (Diptera: Culicidae). *Plant Prot. Sci.* 2014, *50*, 1–36.

46. Avato, P.; Sgarra, G.; Casadoro, G. Chemical composition of the essential oils of *Mentha* species cultivated in Italy. *Sci. Pharm.* 1995, *63*, 223–230.

47. Garzoli, S.; Pirolli, A.; Vavala, E.; Di Sotto, A.; Sartorelli, G.; Božović, M.; Angiolella, L.; Mazzanti, G.; Pepi, F.; Ragno, R. Multidisciplinary approach to determine the optimal time and period to extract the essential oil from *Mentha suaveolens* Ehrh. *Molecules* 2015. submitted.

48. Hammer, K.A.; Carson, C.F.; Riley, T.V. Antimicrobial activity of essential oils and other plant extracts. *J. Appl. Microbiol.* 1999, *86*, 985–990.

49. Simić, D.; Vuković-Gačić, B.; Knežević-Vukčević, J.; Đarmati, Z.; Jankov, R.M. New assay system for detecting bioantimutagens in plant extracts. *Arch. Biol. Sci.* 1994, *46*, 81–85.

50. Burt, S. Essential oils: Their antibacterial properties and potential application in foods—A review. *Int. J. Food Microbiol.* 2004, *94*, 223–253.

51. Nedorostova, L.; Kloucek, P.; Kokoska, L.; Stolcova, M.; Pulkrabek, J. Antimicrobial properties of selected essential oils in vapour phase against foodborne bacteria. *Food Control* 2009, *20*, 157–160.

52. Gulluce, M.; Sahin, F.; Sokmen, M.; Ozer, H.; Daferera, D.; Sokmen, A. Antimicrobial and antioxidant properties of the essential oils and methanol extract from *Mentha longifolia* L. ssp. *Longifolia*. *Food Chem.* 2007, *103*, 1449–1456.

53. Mahboubi, M.; Haghi, G. Antimicrobial activity and chemical composition of *Mentha pulegium* L. essential oil. *J. Etnopharmacol.* 2008, *119*, 325–327.

54. Arman, M.; Yousefzadi, M.; Khademi, S.Z. Antimicrobial activity and composition of the essential oil from *Mentha mozaffarianii*. *J. Essent. Oil Bear. Plants* 2011, *14*, 131–135.

55. Flamini, G.; Cioni, P.L.; Puleio, R.; Morelli, I.; Panizzi, L. Antimicrobial

activity of the essential oil of *Calamintha nepeta* and its constituent pulegone against bacteria and fungi. *Phytother. Res.* 1999, *13*, 349–351.

56. Vavala, E.; Passariello, C.; Pepi, F.; Colone, M.; Garzoli, S.; Ragno, R.; Pirolli, P.; Stringaro, A.; Angiolella, L. Antimicrobial activity of essential oils against *Pseudomonas syringae* pathovar *actinidiae* (PSA) agent of canker kiwifruit. *Nat. Prod. Res.* 2015. accepted.

57. Sessa, R.; Di Pietro, M.; de Santis, F.; Ragno, R.; Angiolella, L. Effects of *Mentha suaveolens* essential oil on *Chlamydia trachomatis*. *BioMed Res. Int.* 2015, *2015*, 508071.

58. Helander, I.M.; Alakomi, H.; Latva-Kala, K.; Mattila-Sandholm, T.; Pol, I.; Smid, E.J.; Gorris, L.G.M.; von Wright, A. Characterization of the action of selected essential oil components on gram-negative bacteria. *J. Agric. Food Chem.* 1998, *46*, 3590–3595.

59. Gill, A.O.; Holley, R.A. Disruption of *Escherichia coli, Listeria monocytogenes* and *Lactobacillus sakei* cellular membranes by plant oil aromatics. *Int. J. Food Microbiol.* 2006, *15*, 1–9.

60. Lakušić, B.; Slavkovska, V.; Pavlović, M.; Milenković, M.; Antić-Stanković, J.; Couladis, M. Chemical composition and antimicrobial activity of the essential oil from *Chaerophyllum aureum* L. (Apiaceae). *Nat. Prod. Commun.* 2009, *4*, 115–118. [PubMed]

61. Cosentino, S.; Tuberoso, C.I.; Pisano, B.; Satta, M.; Mascia, V.; Arzedi, E. In vitro antimicrobial activity and chemical composition of Sardinian *Thymus* essential oils. *Lett. Appl. Microbiol.* 1999, *29*, 130–135.

62. Trombetta, D.; Castelli, F.; Sarpietro, M.G.; Venuti, V.; Cristani, M.; Daniele, C.; Saija, A.; Mazzanti, G.; Bisignano, G. Mechanisms of antibacterial action of three monoterpenes. *Antimicrob. Agents Chemother.* 2005, *49*, 2474–2478.

63. Cos, P.; Vlietinck, A.J.; Berghe, D.V.; Maes, L. Anti-infective potential of natural products: How to develop a stronger *in vitro* 'proof-of-concept'. *J. Ethnopharmacol.* 2006, *106*, 290–302.

64. Hyldgaard, M.; Mygind, T.; Meyer, R.L. Essential oils in food preservation: mode of action, synergies, and interactions with food matrix components. *Front. Microbiol.* 2012, *3*.

65. Ramarathnam, N.; Osawa, T.; Ochi, H.; Kawakishi, S. The contribution of plant food antioxidants to human health. *Trends Food Sci. Technol.* 1995, *6*, 75–77.

66. Sultana, B.; Anwar, F.; Przybylski, R. Antioxidant activity of phenolic

components present in barks of *Azadirachta indica*, *Terminalia arjuna*, *Acacia nilotica* and *Eugenia jambolana* Lam. trees. *Food Chem.* 2007, *104*, 1106–1114.

67. Huda-Faujan, N.; Noriham, A.; Norrakiah, A.S.; Babji, A.S. Antioxidant activity of plants methanolic extracts containing phenolic compounds. *Afr. J. Biotechnol.* 2009, *8*, 484–489.

68. Hussain, A.I.; Anwar, F.; Chatha, S.A.S.; Jabbar, A.; Mahboob, S.; Nigam, P.S. *Rosmarinus officinalis* essential oil: Antiproliferative, antioxidant and antibacterial activities. *Braz. J. Microbiol.* 2010, *41*, 1070–1078.

69. Hussain, A.I.; Anwar, F.; Sherazi, S.T.H.; Przybylski, R. Chemical composition, antioxidant and antimicrobial activities of basil (*Ocimum basilicum*) essential oils depends on seasonal variations. *Food Chem.* 2008, *108*, 986–995.

70. Bichra, M.; El-Modafar, C.; El-Abbassi, A.; Bouamama, H.; Benkhalti, F. Antioxidant activities and phenolic profiles of six Moroccan selected herbs. *J. Microbiol. Biotechnol. Food Sci.* 2013, *2*, 2320–2338.

71. Bichra, M.; El Modafar, C.; El-Boustani, E.; Benkhalti, F. Antioxidant and anti-browning activities of *Mentha suaveolens* extracts. *Afr. J. Biotechnol.* 2012, *11*, 8722–8729.

72. Djeridane, A.; Yousfi, M.; Nadjemi, B.; Boutassouna, D.; Stocker, P.; Vidal, N. Antioxidant activity of some Algerian medicinal plants extracts containing phenolic compounds. *Food Chem.* 2006, *97*, 654–660.

73. Wang, M.; Simon, J.E.; Aviles, I.F.; He, K.; Zheng, Q.Y.; Tadmor, Y. Analysis of antioxidative phenolic compounds in Artichoke (*Cynara scolymus* L.). *J. Agric. Food Chem.* 2003, *51*, 601–608.

74. Amarowicz, R.; Pegg, R.B.; Rahimi-Moghaddam, P.; Barl, B.; Weil, J.A. Free-radical scavenging capacity and antioxidant activity of selected plant species from the Canadian prairies. *Food Chem.* 2004, *84*, 551–562.

75. Katsube, T.; Tabata, H.; Ohta, Y.; Yamasaki, Y.; Anuurad, E.; Shiwaku, K. Screening for antioxidant activity in edible plant products: Comparison of low-density lipoprotein oxidation assay, DPPH radical scavenging assay and Folin–Ciocalteu assay. *J. Agric. Food Chem.* 2004, *52*, 2391–2396.

76. Katalinić, V.; Miloš, M.; Kulišić, T.; Jukić, M. Screening of 70 medicinal plant extracts for antioxidant capacity and total phenols. *Food Chem.* 2006, *94*, 550–557.

77. El-Askary, H.I.; El-Kashoury, E.A.; Kandil, Z.A.; Salem, M.A.; Ezzat, S.M. Biological activity and standardization of the ethanolic extract of the aerial parts of *Mentha suaveolens* Ehrh. *World J. Pharm. Pharm.*

Sci. 2014, 3, 223–241.

78. Aziz, E.E.; Abbass, M.H. Chemical composition and efficiency of five essential oils against the Pulse Beetle*Callosobruchus maculatus* (F.) on *Vigna radiata* Seeds. *Am.-Eurasian J. Agric. Environ. Sci.* 2010, *8*, 411–419.

79. Boughdad, A.; Elkasimi, R.; Kharchafi, M. Activité insecticide des huiles essentielles de *Mentha.* sur *Callosobrochus maculatus* (F) (Coleoptera, Bruchidae). In Proceedings of the AFPP—Neuvième Conférence Internationale sur les ravageurs en Agriculture, Montpellier, France, 26–27 October 2011.

80. Abdelgaleil, S.A.; Mohamed, M.I.; Badawy, M.E.; El-Arami, S.A.A. Fumigant and contact toxicities of monoterpenes to*Sitophilus oryzae* (L.) and *Tribolium castaneum* (Herbst) and their inhibitory effects on acetylcholinesterase activity. *J. Chem. Ecol.* 2009, *35*, 518–525.

81. Franzios, G.; Mirotsou, M.; Hatziapostolou, E.; Kral, J.; Scouras, G.Z.; Mavragani-Tsipidou, P. Insecticidal and genotoxic activities of mint essential oils. *J. Agric. Food Chem.* 1997, *45*, 2690–2694.

82. Tripathi, A.K.; Prajapati, V.; Ahmad, A.; Aggarwal, K.K.; Khanuja, S.P.S. Piperitenone oxide as toxic, repellent, and reproduction retardant toward malarial vector *Anopheles Stephensi* (Diptera: Anophelinae). *J. Med. Entomol.* 2004, *41*, 691–698.

83. Liska, A.; Rozman, V.; Kalinović, I.; Ivezić, M.; Balićević, R. Contact and fumigant activity of 1,8-cineole, eugenol and camphor against *Tribolium castaneum* (Herbst). *Julius.-Kühn.-Arch.* 2010, *425*, 716–720.

84. Koliopoulos, G.; Pitarokili, D.; Kioulos, E.; Michaelakis, A.; Tzakou, O. Chemical composition and larvicidal evaluation of *Mentha, Salvia* and *Melissa* essential oils against the West Nile virus mosquito *Culex pipiens*. *Parasitol. Res.* 2010, *107*, 327–335.

85. Pavlidou, V.; Karpouhtsis, I.; Franzios, G.; Zambetaki, A.; Scouras, Z.; Mavragani-Tsipidou, P. Insecticidal and genotoxic effects of essential oils of Greek sage, *Salvia fruticosa*, and mint, *Mentha pulegium*, on *Drosophila melanogaster*and *Bactrocera oleae* (Diptera: Tephritidae). *J. Agric. Urban. Entomol.* 2004, *21*, 39–47.

86. Lee, S.; Peterson, C.J.; Coats, J.R. Fumigation toxicity of monoterpenoids to several stored product insects. *J. Stored Prod. Res.* 2002, *39*, 349–355.

87. Praveena, A.; Sanjayak, K.P. Inhibition of acetylcholinesterase in three insects of economic importance by linalool, a monoterpene phytochemical. In *Insect Pest Management, A Current Scenario*, 2011 ed.; Ambrose, D.P., Ed.; Entomology Research Unit, St. Xavier's College:

Palayamkottai, India, 2011; pp. 340–345.
88. Price, D.N.; Berry, M.S. Comparison of effects octopamine and insecticidal essential oils on activity in the nerve cord, foregut, and dorsal unpaired median neurons of cockroaches. *J. Insect. Physiol.* 2006, *52*, 309–319.
89. Reitsema, R.H. A new ketone from Oil of *Mentha rotundifolia*. *J. Am. Chem. Soc.* 1956, *78*, 5022–5025.
90. Shimizu, S. Studies of the essential oil of *Mentha rotundifolia*, Part I. Isolation of rotundifolone, a new terpenic ketone.*Bull. Agric. Chem. Soc. Jpn.* 1956, *20*, 84–88.
91. Hendriks, H.; van Os, F.H.L. Essential oil of two chemotypes of *Mentha suaveolens* during ontogenesis. *Phytochemistry*1976, *15*, 1127–1130.
92. Sutour, S. Etude de la Composition Chimique d'huiles Essentielles et d'extraits de Menthe de Corse et de Kumquats. Ph.D. Thesis, University of Corsica, Corsica, France, 2010.
93. Denslow, M.W.; Poindexter, D.B. *Mentha suaveolens* and *M. x rotundifolia* in North Carolina: A clarification of distribution and taconomic identity. *J. Bot. Res. Inst. Texas* 2009, *31*, 383–389.
94. Mahmoud, S.S.; Croteau, R.B. Menthofuran regulates essential oil biosynthesis in peppermint by controlling a downstream monoterpene reductase. *Proc. Nat. Acad. Sci. USA* 2003, *100*, 14481–14486.
95. Lahlou, S.; Carneiro-Leão, R.F.L.; Leal-Cardoso, J.H.; Toscano, C.F. Cardiovascular Effects of the Essential Oil of*Mentha x villosa* and its Main Constituent, Piperitenone Oxide, in Normotensive Anaesthetised Rats: Role of the Autonomic Nervous System. *Planta Med.* 2001, *67*, 638–643.
96. Guedes, D.N.; Silva, D.F.; Barbosa-Filho, J.M.; Medeiros, I.A. Calcium antagonism and the vasorelaxation of the rat aorta induced by rotundifolone. *Braz. J. Med. Biol. Res.* 2004, *37*, 1881–1887.
97. Guedes, D.N.; Silva, D.F.; Barbosa-Filho, J.M.; Medeiros, I.A. Muscarinic Agonist Properties Involved in the Hypotensive and Vasorelaxant Responses of Rotundifolone in Rats. *Planta Med.* 2002, *68*, 700–704.
98. Almeida, R.N.; Hiruma, C.A.; Barbosa-Filho, J.M. Analgesic effect of rotundifolone in rodents. *Fitoterapia* 1996, *4*, 334–338.
99. De Abreu Matos, F.J.; Machado, M.I.L.; Craveiro, A.A.; Alencar, J.; Barbosa, J.M.; Da Cunha, E.V.L.; Hiruma, C.A. Essential oil of *Mentha x villosa* from northeastern Brazil. *J. Essent. Oil Res.* 1999, *11*, 41–44.
100. Sousa, P.J.C.; Linard, C.F.B.M.; Azevedo-Batista, D.; Oliveira, A.C.;

Coelho-de-Souza, A.N.; Leal-Cardoso, J.H. Antinociceptive effects of the essential oil of *Mentha x villosa* leaf and its major constituent piperitenone oxide in mice.*Braz. J. Med. Biol. Res.* 2009, *42*, 655–659. [PubMed]

101. Matos-Rocha, T.J.; Cavalcanti, M.G.; Barbosa-Filho, J.M.; Lucio, A.S.S.C.; Veras, D.L.; Feitosa, A.P.S.; Siqueira Junior, J.P.; Almeida, R.N.; Marques, M.O.M.; Alves, L.C.; et al. In vitro evaluation of schistosomicidal activity of essential oil of *Mentha x villosa* and some of its chemical constituents in adult worms of *Schistosoma mansoni*. *Planta Med.* 2013, *79*, 1307–1312.

102. Teles, S.; Pereira, J.A.; Santos, C.H.B.; Menezes, R.V.; Malheiro, R.; Lucchese, A.M.; Silva, F. Effects of geographical origin on the essential oil content and composition of fresh and dried *Mentha x villosa* Hudson leaves. *Ind. Crops Prod.*2013, *46*, 1–7.

103. Lima, T.C.; Da Silva, T.K.M.; Silva, F.L.; Barbosa-Filho, H.M.; Mayo Marques, M.O.; Santos, R.L.C.; Cavalcanti, S.C.D.H.; de Sousa, D.P. Larvacidal activity of *Mentha x villosa* Hudson essential oil, rotundifolone and derivates.*Chemosphere* 2014, *104*, 37–43.

104. Tomei, P.E.; Manganelli, R.E.U. Composition of the essential oil of *Mentha microphylla* from the gennargentu mountains (Sardinia, Italy). *J. Agric. Food Chem.* 2003, *51*, 3614–3617.

105. Halim, A.F.; Mashaly, M.M.; Sandra, P. Constituents of the essential oil of *Mentha microphylla* C. Koch. *Egypt J. Pharm. Sci.* 1990, *31*, 437–441.

106. Pino, J.A.; Rosado, A.; Sanchez, E. Essential oil of *Mentha spicata* from Cuba. *J. Essent. Oil Res.* 1998, *10*, 657–659.

107. Garg, S.N.; Bahl, J.R.; Bansal, R.P.; Mathur, A.K.; Kumar, S. Piperitenone oxide and/or 1,8-cineole rich essential oils produced by seed progeny clones of *Mentha spicata* accession grown in Indo-Gangetic plains. *J. Med. Aromat. Plant Sci.*2000, *22*, 755–759.

108. Sartoratto, A.; Machado, A.L.M.; Delarmelina, C.; Figueira, G.M.; Duarte, M.C.T.; Rehder, V.L.G. Composition and antimicrobial activity of essential oils from aromatic plants used in Brazil. *Braz. J. Microbiol.* 2004, *35*, 275–280.

109. Hua, C.X.; Wang, G.R.; Lei, Y. Evaluation of essential oil composition and DNA diversity of mint resources from China. *Afr. J. Biotechnol.* 2011, *10*, 16740–16745.

110. Orav, A.; Kapp, K.; Raal, A. Chemosystematic markers for the essential oils in leaves of *Mentha* species cultivated or growing naturally in Estonia. *Proc. Estonian Acad. Sci.* 2013, *62*, 175–186.

111. Padalia, R.C.; Verma, R.S.; Chauhan, A.; Sundaresan, V.; Chanotiya, C.S. Essential oil composition of sixteen elite cultivars of *Mentha* from western Himalayan region, India. *Maejo. Int. J. Sci. Technol.* 2013, *7*, 83–93.

112. Riahi, L.; Elferchichi, M.; Ghazghazi, H.; Jebali, J.; Ziadi, S.; Aouadhi, C.; Chograni, H.; Zaouali, Y.; Zoghlami, N.; Mliki, A. Phytochemistry, antioxidant and antimicrobial activities of the essential oils of *Mentha rotundifolia* L. in Tunisia. *Ind. Crops Prod.* 2013, *43*, 883–889.

113. Maffei, M. A chemotype of *Mentha longifolia* (L.) Hudson particularly rich in piperitenone oxide. *Flavour. Frag. J.* 1988,*3*, 23–26.

114. Abu-Al-Futuh, I.M.; Abdelmageed, O.H.; Jamil, R.M.; Avato, P. A piperitenone oxide chemotype of *Mentha longifolia*(L.) Huds growing wild in Jordan. *J. Essent. Oil Res.* 2000, *12*, 530–532.

115. Kakhky, A.M.; Rustaiyan, A.; Masoudi, S.; Tabatabaei-Anaraki, M.; Salehi, H.R. Composition of the essential oil of*Perovskia abrotanoides* Karel and *Mentha longifolia* L. from Iran. *J. Essent. Oils Bear. Plants* 2009, *12*, 205–212.

116. Hussain, A.I.; Anwar, F.; Nigam, P.S.; Ashraf, M.; Gilani, A.H. Seasonal variation in content, chemical composition and antimicrobial and cytotoxic activities of essential oils from four *Mentha* species. *J. Sci. Food Agric.* 2010, *90*, 1827–1836. [PubMed]

117. Thach, L.N.; Nhung, T.H.; My, V.T.N.; Tran, H. The new rich source of rotundifolone: *Mentha aquatica* Linn. var. *crispa*oil from microwave-assisted hydrodistillation. *J. Essent. Oil Res.* 2013, *25*, 39–43.

118. Menezes, L.R.A.; Santos, N.N.; Meira, C.S.; Ferreira Dos Santos, J.A.; Guimarães, E.T.; Soares, M.B.P.; Nepel, A.; Barison, A.; Costa, E.V. A New Source of (R)-Limonene and Rotundifolone from Leaves of *Lippia pedunculosa*(Verbenaceae) and their Trypanocidal Properties. *Nat. Prod. Commun.* 2014, *9*, 737–739. [PubMed]

119. Reitsema, R.H. The synthesis of racemic piperitenone oxide and diosphenolene. *J. Am. Chem. Soc.* 1957, *79*, 4465–4468.

120. Juliani, H.R.; Koroch, A.; Simon, J.E.; Biurrun, F.N.; Castellano, V.; Zygadlo, J.A. Essential Oils from Argentinean Aromatic Plants. *Acta Hortic.* 2004, *629*, 491–498.

121. Duschatzky, C.; Bailac, P.; Carrascull, A.; Firpo, N.; Ponzi, M. Essential Oil of Lippia aff. juneliana Grown in San Luis, Argentina. Effect of Harvesting Period on the Essential Oil Composition. *J. Essent. Oil Res.* 1999, *11*, 104–106.

122. Mates, F.J.A.; Machado, M.I.L.; Silva, M.G.V.; Craveiro, A.A.; Alencar, J.W. Essential Oils of *Lippia alnifolia* Schau. (Verbenaceae) and *Lippia aff Gracillis* H.B.K., two Aromatic Medicinal Shrubs from Northeast Brazil. *J. Essent. Oil Res.* 2000, *12*, 295–297.

123. Viturro, C.I.; Molina, A.; Guy, I.; Charles, B.; Guinaudeau, H.; Fournet, A. Essential oils of *Satureja Boliviana* and *S. parvifolia* growing in the region of Jujuy, Argentina. *Flavour Fragance J.* 2000, *15*, 377–382.

124. Cabana, R.; Silva, L.R.; Valenão, P.; Viturro, C.I.; Andrade, P.B. Effect of different extraction methodologies on the recovery of bioactive metabolites from *Satureja parvifolia* (Phil.) Epling (Lamiaceae). *Ind. Crops Prod.* 2013, *48*, 49–56.

125. Pirbalouti, A.G.; Vosoghi, N.; Craker, L.; Shirmardi, H. Chemical composition of the essential oil of *Satureja kallarica*Jamzad. *J. Essent. Oil Res.* 2014, *26*, 228–231.

126. Tümen, G.; Baser, K.H.C.; Kürkcüoglu, M.; Demircakmak, B. Composition of the essential oil of *Calamintha incana*(Sm.) Boiss. from Turkey. *J. Essent. Oil Res.* 1995, *7*, 679–680.

127. De Pooter, H.L.; de Buyck, L.F.; Schamp, N.M. The volatiles of *Calamintha nepeta* subsp. *glandusola*. *Phytochemistry* 1986, *25*, 691–694.

128. Krimer, N.; Baser, K.H.C.; Özek, T.; Kürkçüoglu, M. Composition of the essential oil of *Calamintha nepeta* subsp.*glandusola*. *J. Essent. Oil Res.* 1992, *4*, 189–190.

129. Bergmann, E.D.; Bracha, P. A simple synthesis of piperitenone. *J. Org. Chem.* 1959, *24*, 994–995.

130. Beereboom, J.J. The Synthesis of Piperitenone via Mesityl Oxide and Methyl Vinyl Ketone. *J. Org. Chem.* 1966, *31*, 2026–2027.

131. Nakanishim, O.; Fujitani, M.; Ichimoto, I.; Ueda, H. An Improved Process for the Synthesis of Piperitenone from Mesityl oxide and Methyl Vinyl Ketone. *Agric. Biol. Chem.* 1980, *44*, 1667–1668.

132. Sousa, P.J.C.; Magalhães, P.J.C.; Lima, C.C.; Oliveira, V.S.; Leal-Cardoso, J.H. Effects of piperitenone oxide on the intestinal smooth muscle of the guinea pig. *Braz. J. Med. Biol. Res.* 1997, *30*, 787–79.

133. Santos, M.R.V.; Moreira, F.V.; Fraga, B.P.; de Sousa, D.P.; Bonjardim, L.R.; Quintans-Junior, L.J. Cardiovascular effects of monoterpenes: A review. *Rev. Bras. Farmacogn. Braz. J. Pharmacogn.* 2011, *21*, 764–771.

134. Kassouf-Silva, I.; Leal-Cardoso, J.H.; Damiani, C.E.N.; Fogaça, R.T.H. Effect of Piperitenone Oxide on the Skeletal muscle of Toad. *J. Nat.*

Prod. 2011, *4*, 65–70.

135. De Sousa, D.P.; Júnior, E.V.M.; Oliveira, F.S.; de Almeida, R.N.; Nunes, X.P.; Barbosa-Filho, J.M. Antinociceptive Activity of Structural Analogues of Rotundifolone: Structure-Activity Relationship. *Z. Naturforsch.* 2007, *62c*, 39–42.

136. Silva, D.F.; Araújo, I.G.A.; Albuquerque, J.G.F.; Porto, D.L.; Dias, K.L.G.; Cavalcante, K.V.M.; Veras, R.C.; Nunes, X.P.; Barbosa-Filho, J.M.; Araújo, D.A.M.; *et al*. Rotundifolone-Induced Relaxation is Mediated by BKCa Channel Activation and Cav Channel Inactivation. *Basic Clin. Pharmacol. Toxicol.* 2011, *109*, 465–475.

137. Lima, T.C.; Mota, M.M.; Barbosa-Filho, J.M.; Dos Santos, M.R.V.; de Sousa, D.P. Structural relationships and vasorelaxant activity of monotrpenes. *DARU J. Pharm. Sci.* 2012, *20*, 23.

138. Arruda, T.A.; Antunes, R.M.P.; Catão, R.M.R.; Lima, E.O.; Sousa, D.P.; Nunes, X.P.; Pereira, M.S.V.; Barbosa-Filho, J.M.; da Cunha, E.V.L. Preliminary study of the antimicrobial activity of *Mentha x villosa* Huds. essential oil, rotundifolone and its analogues. *Rev. Bras. Farmacogn.* 2006, *6*, 307–311.

Chapter 3

NOVEL BIOLOGICAL ACTIVITIES OF ALLOSAMIDINS

Shohei Sakuda [1], Hiromasa Inoue [2] and Hiromichi Nagasawa [1]

[1]Department of Applied Biological Chemistry, the University of Tokyo, Bunkyo-ku, Tokyo 113-8657, Japan
[2]Department of Pulmonary Medicine, Kagoshima University, 8-35-1 Sakuragaoka, Kagoshima 890-8520, Japan

ABSTRACT

Allosamidins, which are secondary metabolites of the *Streptomyces* species, have chitin-mimic pseudotrisaccharide structures. They bind to catalytic centers of all family 18 chitinases and inhibit their enzymatic activity. Allosamidins have been used as chitinase inhibitors to investigate the physiological roles of chitinases in a variety of organisms. Two prominent biological activities of allosamidins were discovered, where one has anti-asthmatic activity in mammals, while the other has the chitinase-production- promoting activity in allosamidin-producing S*treptomyces*. In this article, recent studies on the novel biological activities of allosamidins are reviewed.

INTRODUCTION

Chitin, a polymer of β-1,4 linked *N*-acetyl-D-glucosamine, is an important biomass, second only to cellulose in abundance in Nature. Its occurrence in living organisms is specialized, such as in insect cuticles, fungal cell walls, and crab shells [1]. In these chitin-containing organisms, chitin is a polysaccharide constituent that is essential for their growth, therefore, chitin is an ideal target for developing insecticides or fungicides with high selectivity [2,3]. Inhibitors of enzymes responsible for chitin metabolism in the chitin-containing organisms are possible candidates for useful drugs. Chitin synthase and chitinase are key enzymes for chitin synthesis and degradation, respectively.

Chitin synthase is present only in chitin-containing organisms and is critical for their growth. Therefore, inhibitors of this enzyme such as polyoxins and nikkomycins are used practically as high selectivity insecticides or fungicides [4]. On the other hand, chitinases are ubiquitously present not only in chitin-containing organisms, but also in non-chitin-containing organisms such as bacteria, plants, or mammals. Chitinases have a variety of physiological roles, and therefore basic studies on the specific chitinase present in each organism are very important for understanding these properties and physiological roles. Specific chitinase inhibitors are useful not only as potential insecticide or fungicide candidates, but also as probes for basic research [5].

Allosamidin (Figure 1), a metabolite of a soil bacterium *Streptomyces* sp., was discovered in 1986 as the first chitinase inhibitor [6,7]. It has a unique pseudotrisaccharide structure consisting of two *N*-acetyl-D-allosamine moieties and one allosamizoline moiety [8,9]. Allosamizoline is a unique five-membered cyclitol derivative fused with a dimethylaminooxazoline ring. Biosynthesis of allosamidin was studied using incorporation experiments with a variety of labeled precursors (Figure 1) [10,11]. Both allosamine and the 2-aminocyclitol skeleton originate from the D-glucosamine molecule, and the $(CH_3)_2N-C$ moiety of allosamizoline comes from a methyl group of L-methionine and a guanidino group of L-arginine. The cyclopentane ring of allosamizoline was shown to be likely formed through a 6-aldehyde intermediate [12]. To date, seven natural allosamidins have been isolated (Figure 2) [13]. The chemistry of allosamidin and derivatives, including synthetic and X-ray crystallographic studies, have been reviewed by many researchers [14,15,16].

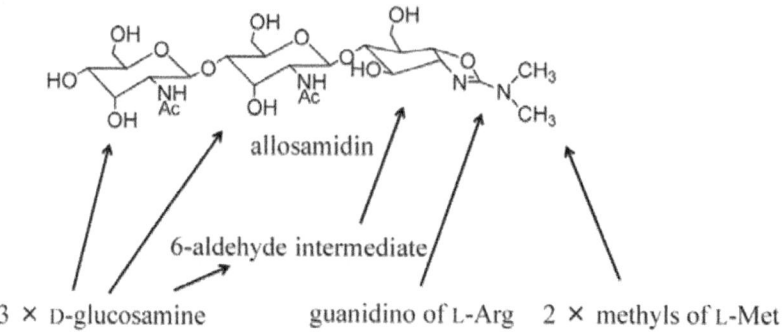

Figure 1: Structure and biosynthesis of allosamidin.

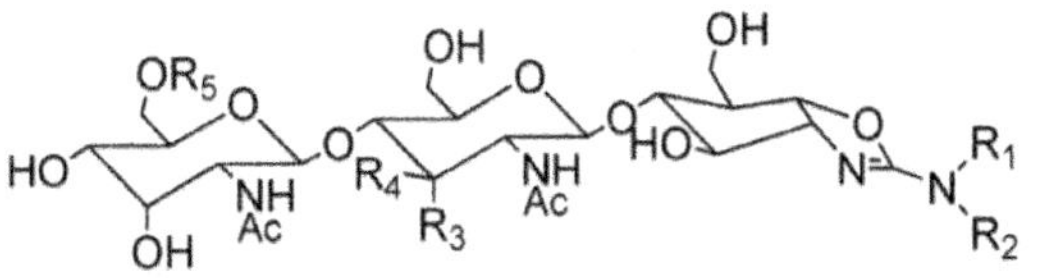

allosamidin $R_1 = R_2 = CH_3$, $R_3 = OH$, $R_4 = H$, $R_5 = H$
methylallosamidin $R_1 = R_2 = CH_3$, $R_3 = OH$, $R_4 = H$, $R_5 = CH_3$
demethylallosamidin $R_1 = CH_3$, $R_2 = H$, $R_3 = OH$, $R_4 = H$, $R_5 = H$
glucoallosamidin A $R_1 = R_2 = CH_3$, $R_3 = H$, $R_4 = OH$, $R_5 = CH_3$
glucoallosamidin B $R_1 = CH_3$, $R_2 = H$, $R_3 = H$, $R_4 = OH$, $R_5 = CH_3$
methyl-N-demethylallosamidin $R_1 = CH_3$, $R_2 = H$, $R_3 = OH$, $R_4 = H$, $R_5 = CH_3$
didemethylallosamidin $R_1 = R_2 = H$, $R_3 = OH$, $R_4 = H$, $R_5 = H$

Figure 2: Structures of natural allosamidins.

Allosamidin can inhibit all family 18 chitinases, but does not inhibit family 19 chitinases [17,18,19]. Family 18 chitinases cleave chitin to yield the β configuration at C1 by a mechanism that leads to retention of the anomeric configuration after hydrolysis. On the other hand, the reaction mediated by family 19 chitinases yields the α configuration at C1 with an inversion of the anomeric configuration. Family 18 chitinases are believed to catalyze chitin hydrolysis through a substrate-assisted mechanism in which an oxazolium ion intermediate is produced during the enzymatic reaction [20]. The allosamizoline moiety of allosamidin may bind to the active center as a mimic of the intermediate [21], leading to inhibition of the enzyme reaction (Scheme 1). Allosamidin binding with family 18 chitinases have been further examined by theoretical [22], NMR [23], X-ray crystallographic [24,25,26,27,28], and thermodynamic [29,30,31] studies.

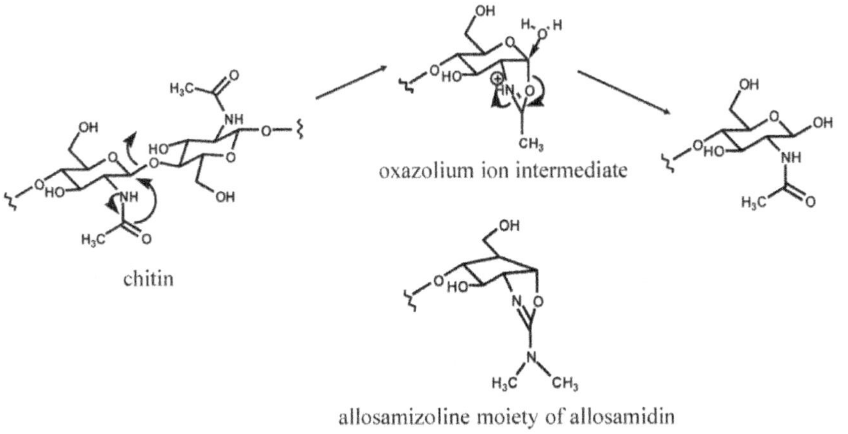

Scheme 1: Mechanism of inhibition of family 18 chitinase by allosamidin.

Allosamidins have been used to investigate physiological roles in a variety of organisms, which are summarized in Table 1. Allosamidins inhibit insect moulting [7,32,33,34] and the cell separation of yeast [35,36], indicating that these enzymes play an essential role in degrading the old cuticle and during fission of the septum, respectively. In fungi, it has been postulated that chitinase is necessary for fungal growth such as hyphal growth and branching [37], but a critical effect of allosamidins in fungal growth has not been observed [38,39], though a weak fungistatic effect of allosamidin on autolyzing *Penicillium* mycelia has been reported [40]. Therefore, it is still unclear if a chitinase inhibitor would be a potentially efficacious antifungal agent. Allosamidin inhibits encystment of a parasite [41] and the transmission of the malaria parasite in mosquito models [42,43,44,45], indicating that chitinases present in malaria ookinetes play a role for the penetration of ookinetes into the midgut epithelium of the mosquito by degrading the peritrophic matrix. In plants, one of the roles of chitinases may be a defensive function. It was found that allosamidin enhanced stress tolerance of plants [46], but molecular mechanism of allosamidin's effects on plants is unknown. Bacteria may produce chitinases to obtain nutrients through degradation of environmental chitin and to invade into host fungi [47,48]. We have shown the physiological role of allosamidin in its producing *Streptomyces*. Allosamidin promotes chitinase production and growth of its producer [49,50,51]. In mammals, the physiological role of chitinases remains unclear. It was reported that allosamidin shows anti-asthmatic activity in a mouse model of asthma [52] and reduces inflammatory signs observed in endotoxin-induced uveitis in rabbits [53]. In addition, it was reported that allosamidin promotes atherosclerosis in hyperlipidemic mice [54].

Table 1: Roles of chitinases and allosamidin's effects

Organisms	Roles of Chitinases	Allosamidin's Effects	Reference
insects	degradation of old cuticle	inhibition of moulting	[7,32-34]
yeasts	fission of septum	inhibition of cell separation	[35,36]
fungi	(hyphal growth and branching) [a]	no significant effect	
parasites	degradation of peritrophic matrix	inhibition of transmission of malaria ookinetes	[42-45]
		inhibition of encystment	[41]
plants	(defensive function) [a]	enhancement of stress tolerance	[46]
bacteria	obtaining nutrient	promotion of chitinase production of *Streptomyces*	[49-51]
mammals	(defensive function) [a]	anti-asthmatic activity	[52,68]

[a] Parentheses mean speculative roles.

In this article, recent works on these chitinase-production promoting and anti-asthmatic activities of allosamidins are reviewed.

PHYSIOLOGICAL ROLES OF ALLOSAMIDIN IN ITS PRODUCING *STREPTOMYCES*

Microbial Secondary Metabolites

Microorganisms can produce a variety of secondary metabolites, which have been used as important sources for developing useful drugs, including medicines, pesticides and perfumes. The novelty of the structures and biological activities of microbial secondary metabolites are of great interest in the basic sciences. Genome analyses have clarified that a strain of a microbe, such as *Streptomyces* sp. or *Aspergillus* sp., has the ability to produce dozens of secondary metabolites. Recent extensive information on biosynthesis of microbial secondary metabolites indicates that the substrate specificity of biosynthetic enzymes involved in secondary metabolism is high. These facts suggest that microorganisms have developed secondary metabolism mechanisms during evolution and that each secondary metabolite might have a physiological role in respective organism under certain circumstances. The role of antibiotic production is presumable but has not been proved. It is entirely unknown why microorganisms produce numerous other compounds without antibiotic activity, such as enzyme inhibitors.

Streptomyces Chitinases

Soil is rich in chitin originated from chitin-containing soil organisms, such as fungi and insects. Therefore, chitinases are important for soil bacteria in order to obtain chitin as a nutrient source. *Streptomyces* is thought to be a main microbe for the degradation of chitin in soil. A medium containing chitin as a sole carbon source can be used for the rough selection of *Streptomyces* [55]. There are many chitinase genes in a *Streptomyces* genome. For example, six family 18 chitinase genes and two family 19 chitinase genes are present in the *S. coelicolor* genome [56]. However, the regulatory mechanism of chitinase expression in *Streptomyces* has not been fully clarified yet. It is known that glucose suppresses the expression of chitinase genes with a common direct repeat sequence in the promoter region of the genes [57], and that N,N'-diacetylchitobiose [$(GlcNAc)_2$] induces chitinase expression by releasing the suppression [58]. These facts strongly suggest the presence of a suppressor protein that may bind to the direct repeat sequence in the presence of glucose and detach from it in the presence of $(GlcNAc)_2$, but the putative suppressor protein has not been identified. $(GlcNAc)_2$ is the main product of chitinase-mediated processing of chitin and a specific transporter for $(GlcNAc)_2$ has been identified in *Streptomyces* [59].

Chitinase Production Promoting Activity of Allosamidin in an Allosamidin-Producing Strain

Allosamidin is a typical secondary metabolite of *Streptomyces* sp. and allosamidin-producing strains are easily obtained from soil. Approximately 5% of randomly isolated *Streptomyces* strains produce allosamidin, suggesting that allosamidin might play a role in soil. To obtain a clue for investigating the physiological role of allosamidin, we first focused on the allosamidin-insensitive chitinase produced by an allosamidin-producing strain, *Streptomyces* sp. AJ9463, because no family 19 chitinase had been found in any bacteria other than strain AJ9463 at that time [60]. However, it was shown that *Streptomyces griseus* produced a family 19 chitinase [61] and an allosamidin-insensitive chitinase was not specially produced by allosamidin producers [62], but during the work for optimizing a medium suitable for chitinase production by strain AJ9463, we found that the addition of allosamidin into a chitin medium could enhance chitinase production of strain AJ9463 (Figure 3) [63]. This finding prompted us to elucidate the molecular mechanism of allosamidin's effect on chitinase production.

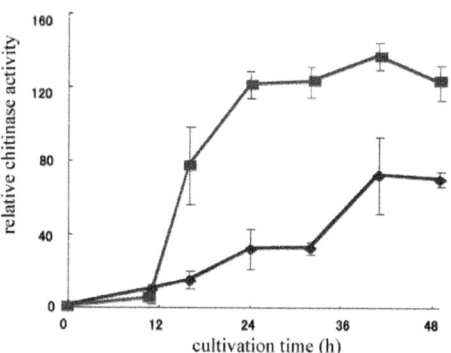

Figure 3: Chitinase production of *Streptomyces* sp. AJ9463 with (■) or without (◆) allosamidin (2 μM) in colloidal chitin medium.

Allosamidin enhanced chitinase activity of the culture filtrate of strain AJ9463 in a colloidal chitin medium at a concentration of 0.06–2.0 μM in a dose-depending manner [49]. Analysis of the chitinases in the culture filtrate by activity staining showed that two predominant chitinases (46 kDa and 105 kDa) were produced and that the amounts of both chitinases increased by addition of allosamidin in a dose-dependent manner. Since both chitinases were produced in the medium without allosamidin, chitinase production appeared to be promoted by allosamidin. The enzymatic activity of the chitinases was inhibited by allosamidin at a concentration of more than 10 μM, but inhibition was not observed at the concentration of 0.06–2.0 µM. Furthermore, the amount of allosamidin produced by allosamidin producers was less than 1.0 µM in the culture broth. These facts suggest that allosamidin is produced at a physiologically significant concentration.

Chitinase production may strongly affect growth when a bacterium is grown in a medium containing chitin as a sole carbon source. The addition of allosamidin clearly promoted mycelial growth in strain AJ9463, suggesting that allosamidin's effect on chitinase production may be physiologically important in its producer.

Molecular Mechanism of Allosamidin's Function on Chitinase Production

The 46 kDa chitinase, whose production is enhanced by allosamidin, was found to be encoded by the *chi65* gene. The amino acid sequence of Chi65 protein deduced from the nucleotide sequence of the gene showed that it contained chitin binding domain, fibronectin type III domain, and catalytic domain, but 46 kDa chitinase lacked the chitin binding and fibronectin type III domains (Figure 4) [50]. The 105 kDa chitinase was thought to be a dimer of the 46 kDa protein because the two chitinases had the same amino acid sequence [49]. A direct repeat sequence was present in the promoter region of *chi65*, and two genes (*chi65S* and *chi65R*) encoding a sensor histidine kinase and response regulator, respectively, were present at the 5'-upstream region of *chi65* (Figure 4) and can comprise a two-component regulatory system. This suggested that expression of Chi65 was regulated not only by the mechanism related to the direct repeat sequence and suppressor protein, but also by the regulation concerning a two-component system of Chi65S and Chi65R.

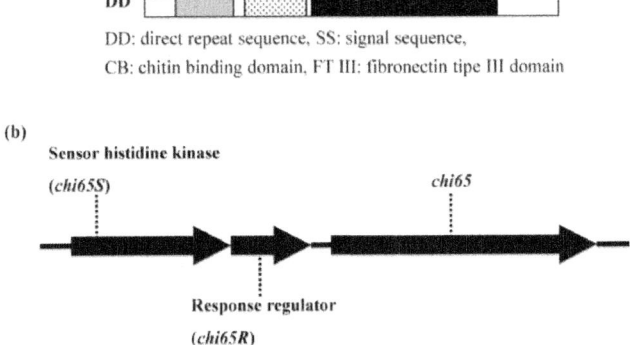

Figure 4: Schematic representation of *chi65* (a), *chi65R* and *chi65S* (b).

(GlcNAc)$_2$ is a key regulator for chitinase production in *Streptomyces* as described above. (GlcNAc)$_2$ enhanced production of 46 kDa and 105 kDa chitinases under the same conditions that were used to test the effect of as allosamidin, but its activity was much weaker than that of allosamidin. A mutant strain with disrupted *chi65R* and *chi65S* genes was found to be sensitive to (GlcNAc)$_2$, but it became insensitive to allosamidin. This results indicated that (GlcNAc)$_2$, but not allosamidin, can act on the putative suppressor and that allosamidin enhances Chi65 production through the two-component regulatory system. In the experiments with a medium containing only inorganic salts, allosamidin did not promote chitinase production but (GlcNAc)$_2$ did induce production of 46 kDa and 105 kDa chitinases. Furthermore, allosamidin enhanced chitinase production under the presence of (GlcNAc)$_2$ in the inorganic salts medium, indicating that (GlcNAc)$_2$ is necessary for the function of allosamidin. These findings suggested the following regulatory mechanism for Chi65 expression. First, (GlcNAc)$_2$ binds to the suppressor and expression of *chi65* starts. Allosamidin may subsequently act on the sensor region of Chi65S and strongly enhance *chi65* expression through the two-component regulatory system.

Localization of Allosamidin

Allosamidin localizes in the mycelia of strain AJ9463 when it is cultured in a medium without chitin. This localization creates a big problem if allosamidin is supposed to act on a membrane sensor from outside of the cells. However, we could find a phenomenon that allosamidin is released into a culture filtrate from the mycelia in a chitin medium. Chitin or its degradation product was hypothesized to act as a releasing factor during the process of allosamidin release, and the inducing factor was identified as (GlcNAc)$_2$ through the experiments with the inorganic salts medium (Suzuki *et al.*, unpublished data). (GlcNAc)$_2$ was found to induce allosamidin release from the strain AJ9463 cells at a concentration of several micromolar, which is a similar concentration that induced chitinase production. Therefore, it was speculated that chitinase production and allosamidin release were induced by (GlcNAc)$_2$ in a successive manner.

Generality of the Action of Allosamidin in Streptomyces

Allosamidin enhances production of the chitinase originated from *chi65h* of *Streptomyces halstedii* MF425, which is an allosamidin producer [51]. Chi65h is highly homologous to *chi65*, and the direct repeat sequence is present in the promoter region of *chi65h* and two genes homologous to *chi65S* and *chi65R* were present at the 5'-upstream region of *chi65h*. Allosamidin also enhanced chitinase production of allosamidin non-producers, *Streptomyces coelicolor* A3(2) and *Streptomyces griseus*. It promoted production of chitinases encoded by *chiC* of *S. coelicolor* A3(2) and *chiIII* of *S. griseus*, which have high homology to *chi65*. Two genes homologous to *chi65S* and *chi65R* were also present at their 5'-upstream regions. Furthermore, when allosamidin's effect was tested with six *Streptomyces* strains randomly isolated from soil, allosamidin enhanced chitinase production in all of the strains. All six strains possessed a set of three genes homologous to *chi65*, *chi65S*, and *chi65R*. Analysis of 16S rDNA indicated that allosamidin-sensitive strains are widely distributed in *Streptomyces*. These observations suggest that allosamidin can affect the common regulatory system for production of a chitinase with a two-component regulatory system in *Streptomyces*.

Role of Allosamidin in Nature

The 46 kDa chitinase was observed in the culture filtrate of strain AJ9463 when cultured in a chitin medium. The *chi65* gene was speculated to produce 65 kDa chitinase. The 65 kDa protein was not present in the culture filtrate, but was detectable in the membrane fraction of strain AJ9463. Based on these observations, we propose the following model of the function of allosamidin (Scheme 2). When strain AJ9463 present in soil comes in contact with chitin, a small amount of $(GlcNAc)_2$ is produced by the action of chitinase of the strain itself or other microbes, and it induces Chi65 production by binding to the suppressor (Scheme 2, step 1). Chi65 is secreted from the cells, but still located on the membrane as the 65 kDa protein (step 2). After proteolysis of the 65 kDa protein, 46 kDa chitinase is produced, which is secreted out of cells. The concentration of $(GlcNAc)_2$ produced by degradation of chitin increases rapidly by the action of the 65 kDa and 45 kDa chitinases (step 3). Allosamidin is transferred to outside of the cells by attaching Chi65 or by an unknown mechanism induced by $(GlcNAc)_2$ produced (step 4). Allosamidin or allosamidin-chitinase complex binds to the sensor moiety of Chi65S (step 5), leading to enhancement of Chi65 expression using the two-component regulatory system of Chi65S and Chi65R (step 6). The concentration of $(GlcNAc)_2$ further increases dramatically by the action of the 65 kDa and 45 kDa chitinases as well as other chitinases whose expression are induced by $(GlcNAc)_2$ (step 7). Since $(GlcNAc)_2$ is used as a nutrient, the growth of the bacterium is dramatically promoted. Allosamidin also affects the production and activity of chitinases of neighboring bacteria. It is hypothesized that allosamidin can enhance chitinase production of almost all *Streptomyces* strains and could inhibit the chitinase activities of other bacteria, leading to nutrient conditions suitable for the growth of *Streptomyces*. In our preliminary model experiment, colonies of *Streptomyces* clearly increased when soil was cultured in a chitin medium containing allosamidin. These findings suggest that allosamidin may act as an important signal molecule for chitin metabolism in an environment such as soil.

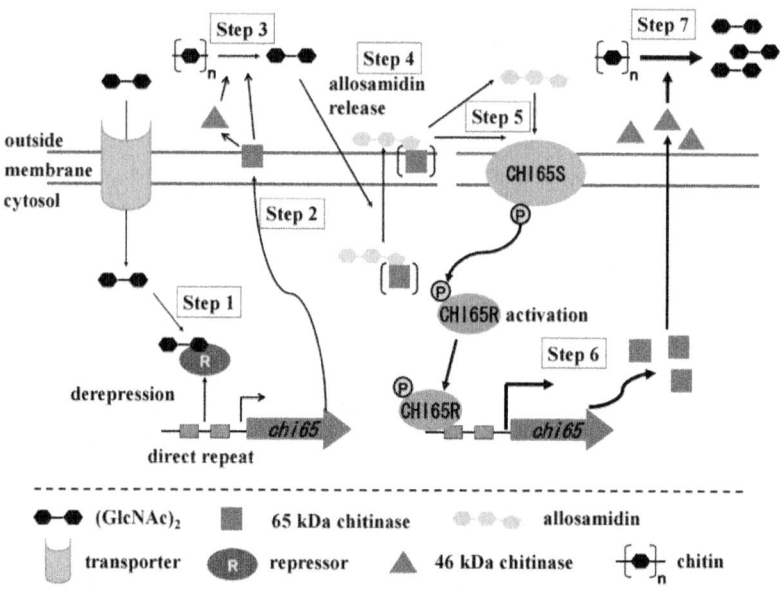

Scheme 2: Putative molecular mechanism of allosamidin's function on chitinase production in allosamidin-producing *Streptomyces*.

ANTI-ASTHMATIC ACTIVITY OF ALLOSAMIDINS

Asthma and Acidic Mammalian Chitinase

Bronchial asthma is a chronic inflammatory disease characterized by eosinophilic infiltration, airway hyperresponsiveness to non-specific stimuli, and remodeling of the airways. Today, asthma prevalence has reached 5% worldwide, and therefore more effective drugs are necessary to cure the disease, especially intractable asthma. T-helper-2 (Th2) cytokines are essential for generating asthmatic abnormalities. Among Th2 cytokines, IL-13 is now considered particularly critical (Figure 5) [64].

Although chitin is not present in mammals, two chitinases, chitotriosidase and acidic mammalian chitinase (AMCase), are present in human and mouse [65,66]. The physiological roles of the chitinases are not clear, but it is speculated that one of their roles is a defensive function against pathogens. AMCase has recently been associated with animal models of asthma. It was reported that AMCase expression is upregulated in the response to allergen exposure or IL-13-induced inflammation in the lung [52,67]. Inhibition of AMCase with anti-acidic mammalian chitinase sera leads to lower eosinophil counts and reduction in airway hyper-responsiveness in a murine model of asthma. Allosamidin was found to suppress allergen-induced airway eosinophilia in the asthma model [52] and inhibition of AMCase by allosamidin was reported [65]. Therefore, the effect of allosamidin on asthma may support the importance of AMCase in the mouse asthma. These observations suggest that AMCase acts as a proinflammatory mediator in IL-13 effector responses (Figure 5), and thus a compound with stronger AMCase inhibitory activity would be expected to show stronger anti-asthmatic activity than allosamidin.

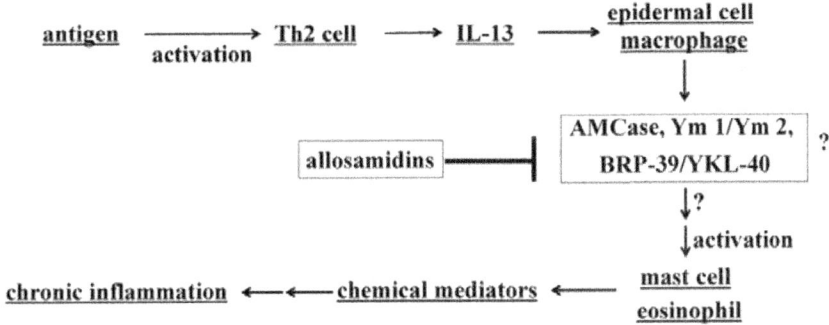

Figure 5: Asthma, and chitinase and chitinase-like proteins.

Demethylallosamidin is an allosamidin congener with a monomethylamino group (Figure 2) and has much stronger inhibitory activity toward yeast chitinases [35] and human chitotriosidase [27]. AMCase and chitotriosidase have more than 50% amino acid sequence similarity in both human and mouse enzymes. Therefore, we expected that demethylallosamidin inhibits AMCase more strongly than allosamidin and shows stronger anti-asthmatic activity.

Activities of Allosamidin and Demethylallosamidin on AMCase and Asthma

Recombinant mouse-AMCase expressed in COS-7 cells was used to test the inhibitory activity of allosamidins and demethylallosamidin on AMCase. The inhibitory activity of the two compounds did not differ in the pH range of 2 to 7.5 [68]. Both compounds inhibited AMCase more strongly at neutral pH than an acidic pH, similarly to the cases observed in other chitinases [69]. In contrast, the two compounds exhibited very different *in vivo* activities in the IL-13-induced asthmatic model of mice. IL-13 and allosamidin or demethylallosamidin were administered to mice intratracheally and intraperitoneally, respectively. IL-13 treatment induced eosinophilia counts and eotaxin concentration in bronchoalveolar lavage (BAL) fluid. Allosamidin or demethylallosamidin decreased the eosinophil counts and eotaxin concentration in a dose-depending manner. Very interestingly, demethylallosamidin was much more effective than allosamidin. Allosamidin and demethylallosamidin completely inhibited IL-13-induced eosinophilia and eotaxin at 10 and 1 mg/kg, respectively. IL-13 also induced airway hyperresposiveness to inhaled acetylcholine. Airway pressure was increased by administration of IL-13. 10 mg/kg allosamidin did not decrease the airway pressure to the control level, but 1 mg/kg demethylallosamidin did it. These results indicate only demethylallosamidin can suppress IL-13-induced hyperresponsiveness and has a much superior potential than allosamidin as an anti-asthmatic agent [68].

IL-13 enhanced the chitinase activity and AMC expression in BAL fluid. The AMCase expression enhanced by IL-13 was not inhibited by allosamidin or demethylallosamdin. The chitinase activity in BAL fluid increased by IL-13 was decreased by 10 mg/kg allosamidin or 1 mg/kg demethylallosamidin. The decreased levels of chitinase activity coincided with the amount of inhibitors, but did not coincide with the inhibitory activity on eosinophil counts or airway hyperresponsiveness of the two compounds. These results suggest that it may be necessary to consider other target molecules to explain the difference between anti-asthmatic activities of allosamidin and demethylallosamidin.

Targets of Allosamidins for Their Anti-asthmatic Activity

Photoaffinity probes of allosamidin and demethylallosamidin were prepared to investigate their binding proteins (Figure 6) [70]. The photoaffinity probes possess photoactive aryl azido moiety and biotin moiety. They maintain strong inhibitory activities toward *Trichoderma* chitinase. By the experiments with these probes, Ym1 was identified as a possible allosamidin-binding protein present in the BAL fluid of IL-13 induced asthmatic mice. Ym1 belongs to chitinase-like proteins that have structures homologous to chitinases but do not have chitinase activity due to a lack of essential amino acid residue(s) commonly present at the active site of family 18 chitinases [71,72]. Chitinase-like proteins are present in a wide range of organisms including mammals, insects, and plants. However, little is known about their physiological roles at the molecular level. The chitinase-like proteins in mouse include Ym 1, Ym 2 (Ym 1 homolog), BRP-39, and oviductin, and those in human include YKL-40 (human homolog of BRP-39), YKL-39, and oviductin. Some of them can bind to poly and/or oligosaccharide. Ym 1 has been reported to bind to chitin [73], and YKL-40 and YKL-39 can bind to *N*-acetylglucosamine oligomers [74,75]. The carbohydrate-binding properties of chitinase-like proteins are thought to be involved in their biological activity. Therefore, allosamidins may have the potential to inhibit the function of these proteins by binding to them.

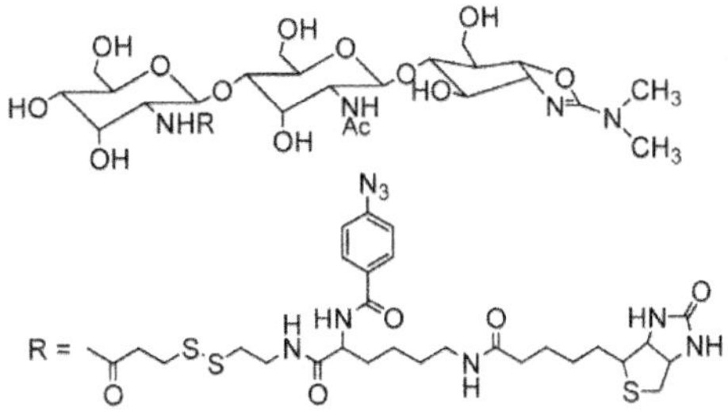

Figure 6: Photoaffinity probe of allosamidin.

It has been suggested that Ym 1 and BRP-39/YKL-40 have an important role in asthma, similarly to AMCase [71,72,76]. The expression of Ym 1 and BRP-39 was upregulated in the lung of mouse asthma model [77,78]. Moreover, Ym 1 induced eosinophil chemotaxis and anti-sense Ym1 RNA suppressed asthmatic responses in the model [77]. It was suggested that Ym1/Ym 2 inhibited 12/15(*S*)-lipoxygenase and promoted Th2 cytokine production [79]. The knockout of BRP-39 attenuated asthmatic responses in mice [78], and thus BRP-39 may function as an effector molecule of IL-13. BRP-39 and AMCase were shown to suppress apoptosis and a mutant AMCase lacking chitinase activity also showed apoptosis suppression activity [79]. A strong correlation between YKL-40 expression levels in lung and human asthma has been demonstrated [78]. These reports may suggest that all of Ym 1, Ym 2, BRP-39, and AMCase have a potential as targets of allosamidins in mice (Figure 5). Further studies are needed to investigate the action of allosamidin and demethylallosamidin on each of these proteins in detail to "understand why demethylallosamidin shows much stronger anti-asthmatic activity than allosamidin".

CONCLUSIONS

Allosamidin has been used as a probe in basic research for studying chitinases since it was first discovered in 1986. The finding of the chitinase-production promoting activity of allosamidin may provide a new angle for the field of research studying microbial secondary metabolites. Studies on the physiological roles of secondary metabolites may be very important not only as basic researches but also as their application to agriculture kind for environment. The anti-asthmatic activity of allosamidins indicates their potential used not only as lead compounds for developing effective anti-asthmatic drugs but also as probes for investigating physiological roles of chitinase-like proteins.

REFERENCES

1. Muzzarelli, R.A.A. *Chitin*; Pergamon Press: Oxford, UK, 1977.
2. Kramer, K.J.; Koga, D. Insect chitin; physical state, synthesis, degradation and metabolic regulation. *Insect biochem.* 1986, *16*, 851–877.
3. Cohen, E. Chitin biochemistry; Synthesis and inhibition. *Ann. Rev. Entomol.* 1987, *32*, 71–93. [CrossRef]
4. Cohen, E.; Casida, J.E. Properties and inhibition of insect integumental chitin synthase. *Pestic. Biochem. Physiol.* 1982, *17*, 301–306.
5. Sakuda, S. The biochemical significance of allosamidins as chitinase inhibitors. In *Binomium Chitin-Chitinase: Resent Issues*; Musumeci, S., Paoletti, M.G., Eds.; Nova Science: Hauppauge, NY, USA, 2009.
6. Sakuda, S.; Isogai, A.; Matsumoto, S.; Suzuki, A.; Koseki, K. Structure of allosamidin, a novel insect chitinase inhibitor, Produced by *Streptomyces* sp. *Tetrahedron Lett.* 1986, *27*, 2475–2478.
7. Sakuda, S.; Isogai, A.; Matsumoto, S.; Suzuki, A. Search for microbial insect growth regulators II. Allosamidin. A novel insect chitinase inhibitor. *J. Antibiot.* 1987, *40*, 296–300. [PubMed]
8. Sakuda, S.; Isogai, A.; Makita, T.; Matsumoto, S.; Koseki, K.; Kodama, H.; Suzuki, A. Structures of allosamidins, Novel insect chitinase inhibitors, Produced by actinomycetes. *Agric. Biol. Chem.* 1987, *51*, 3251–3259.
9. Sakuda, S.; Isogai, A.; Matsumoto, S.; Suzuki, A.; Koseki, K.; Kodama, H.; Yamada, Y. Absolute configuration of allosamizoline, an aminocyclitol derivative of the chitinase inhibitor allosamidin. *Agric. Biol. Chem.* 1988, *52*, 1615–1617.
10. Zhou, Z.Y.; Sakuda, S.; Yamada, Y. Biosynthetic studies on the chitinase inhibitor, allosamidin. Origin of the carbon and nitrogen atoms. *J. Chem. Soc. Perkin Trans. I* 1992, 1649–1652.
11. Zhou, Z.Y.; Sakuda, S.; Kinoshita, M.; Yamada, Y. Biosynthetic studies of allosamidin 2. Isolation of didemethylallosamidin, and conversion experiments of ^{14}C-labeled demethylallosamidin, didemethylallosamidin and their related compounds. *J. Antibiot.* 1993, *46*, 1582–1588.
12. Sakuda, S.; Sugiyama, Y.; Zhou, Z.Y.; Takao, H.; Ikeda, H.; Kakinuma, K.; Yamada, Y.; Nagasawa, H. Biosynthetic studies on the cyclopentane ring formation of allosamizoline, an aminocyclitol component of the chitinase inhibitor allosamidin. *J. Org. Chem.* 2001, *66*, 3356–3361.

13. Nishimoto, Y.; Sakuda, S.; Takayama, S.; Yamada, Y. Isolation and characterization of new allosamidins. *J. Antibiot.* 1991, *44*, 716–722.
14. Berecibar, A.; Grandjean, C.; Siriwardena, A. Synthesis and biological activity of natural aminocyclopentitol glycosidase inhibitors: Mannostatins, Trehazoline, Allosamidins, And their analogues. *Chem. Rev.* 1999, *99*, 779–844.
15. Anderses, O.A.; Dixon, M.J.; Eggleston, I.M.; van Aalten, D.M.F. Natural product family 18 chitinase inhibitors. *Nat. Prod. Rep.* 1005, *22*, 563–579.
16. Gangliang, H. Recent progress on synthesis and activities of allosamidin and its analogues. *Med. Chem.* 2012, *12*, 665–670.
17. Sakuda, S.; Isogai, A.; Suzuki, A.; Yamada, Y. Chemistry and biochemistry of the chitinase inhibitors, allosamidins.*Actinomycetologica* 1993, *7*, 50–57.
18. Henrissat, B.A. Classification of glycosyl hydrolases based on amino acid sequence similarities. *Biochem. J.* 1991, *280*, 309–316.
19. Spindler, K.D.; Spindler-Barth, M. Inhibitor of chitinases. In *Chitin and Chitinases*; Jolles, P., Muzzarelli, R.A.A., Eds.; Birkhauser Verlag.: Basel, Switzerland, 1999; pp. 201–209.
20. Tews, I.; van Scheltinga, A.C.T.; Perrakis, A.; Wilson, K.S.; Dijkstra, B.W. Substrate-assisted catalysis unifies two families of chitinolytic enzymes. *J. Am. Chem. Soc.* 1997, *119*, 7954–7959.
21. van Scheltinga, A.C.T.; Armand, S.; Kalk, K.H.; Isogai, A.; Henrissat, B.; Dijkstra, B.W. Stereochemistry of chitin hydrolysis by a plant chitinase/lysozyme and X-ray structure of a complex with allosamidin: Evidence for substrate assisted catalysis. *Biochemstry* 1995, *34*, 15619–15623.
22. Brameld, K.A.; Shrader, W.D.; Imperiali, B.; Gddard, W.A., III. Substrate assistance in the mechanism of family 18 chitinases: Theoretical studies of potential intermediates and inhibitors. *J. Mol. Biol.* 1998, *280*, 913–923.
23. Germer, A.; Klod, S.; Peter, M.G.; Kleinpeter, E. NMR spectroscopic and theoretical study of the complexation of the inhibitor allosamidin in the binding pocket of the plant chitinase hevamine. *Mol. Model.* 2002, *8*, 231–236.
24. Papanikolau, Y.; Tavlas, G.; Vorgias, C.E.; Petratos, K. De novo purification scheme and crystallization conditions yield high-resolution structures of chitinase A and its complex with the inhibitor allosamidin. *Acta Cryst.* 2003, *D59*, 400–403.

25. van Aalten, D.M.F.; Komander, D.; Synstad, B.; Gaseidnes, S.; Peter, M.G.; Eijsink, V.G.H. Structural insights into the catalytic mechanism of a family 18 exo-chitinase. *Proc. Natl. Acad. Sci. USA* 2001, *98*, 8979–8984.
26. Bortone, K.; Monzingo, A.F.; Ernst, S.; Robertus, J.D. The structure of an allosamidin complex with the *Coccidioides immitis* chtinase defines a role for a second acid residue in substrate-assisted mechanism. *J. Mol. Biol.* 2002, *320*, 293–302.
27. Rao, F.V.; Houston, D.R.; Boot, R.G.; Aerts, J.M.F.; Sakuda, S.; van Aalten, M.F. Crystal structures of allosamidin derivatives in complex with human macrophage chitinase. *J. Biol. Chem.* 2003, *278*, 20110–20116.
28. Zhao, Y.S.; Zheng, Q.C.; Zhang, H.X.; Chu, H.Y.; Sun, C.C. Analysis of a three-dimensional structure of human acidic mammalian chitinase obtained by homology modeling and ligand binding studies. *J. Mol. Model.* 2009, *15*, 499–505.
29. Cederkvist, F.H.; Saua, S.F.; Karlsen, V.; Sakuda, S.; Eijsink, V.G.H.; Sorlie, M. Thermodynamic analysis of allosamidin binding to a family 18 chitinase. *Biochemistry* 2007, *46*, 12347–12354.
30. Zakariassen, H.; Klemetsen, L.; Sakuda, S.; Vaaje-Kolstad, G.; Varum, KM.; Sorlie, M.; Eijsink, V.G.H. Effect of enzyme processivity on the efficacy of a competitive chitinase inhibitor. *Carbohydr. Polym.* 2010, *82*, 779–785.
31. Baban, J.; Fjeld, S.; Sakuda, S.; Eijsink, V.G.H.; Sorlie, M. The roles of three *Serratia marcescens* chitinases in chitin conversion are reflected in different thermodynamic signatures of allosamidin binding. *J. Phys. Chem. B* 2010, *114*, 6144–6149.
32. Blattner, R.; Gerard, P.J.; Spindler-Barth, M. Synthesis and biological activity of allosamidin and allosamidin analogues. *Pestic. Sci.* 1997, *50*, 312–318.
33. Somers, P.J.B.; Yao, R.C.; Doolin, L.E.; McGowan, M.J.; Fukuda, D.S.; Mynderse, J.S. Methods for detection and quantitation of chitinase inhibitors in fermentation broths; isolation and insect life cycle effect of A82516. *J. Antibiot.* 1987, *40*, 1751–1756.
34. Filho, B.P.D.; Lemos, F.J.A.; Secundino, N.F.C.; Pascoa, V.; Pereira, S.T.; Pimenta, P.F.P. Presence of chitinase and beta-*N*-acetyl glucosaminidase in the *Aedes aegypti* a chitinolytic system involving peritrophic matrix formation and degradation. *Insect Biochem. Mol. Biol.* 2002, *32*, 1723–1729.

35. Sakuda, S.; Nishimoto, Y.; Ohi, M.; Watanabe, M.; Takayama, S.; Isogai, A.; Yamada, Y. Effects of demethylallosamidin, a potent yeast chitinase inhibitor, on the cell division of yeast. *Agric. Biol. Chem.* 1990, *54*, 1333–1335.
36. Yamanaka, S.; Tsuyoshi, N.; Kikuchi, R.; Takayama, S.; Sakuda, S.; Yamada, Y. Effect of demethylallosamidin, a chitinase inhibitor, on morphology of fungus *Geotrichum candidum. J. Gen. Appl. Microbiol.* 1994, *40*, 171–174.
37. Adams, D.J. Fungal cell wall chitinases and glucanases. *Microbiology* 2004, *150*, 2029–2035.
38. Dickinson, K.; Keer, V.; Hitchcock, C.A.; Adams, D.J. Chtinase activity from *Candida albicans* and its inhibition by allosamidin. *J. Gen. Microbiol.* 1989, *135*, 1417–1421.
39. Yamazaki, H.; Yamazaki, D.; Takaya, N.; Takagi, M.; Ohta, M.; Horiuchi, H. A chitinase gene, chiB, involved in the autolysis process of *Aspergillus nidulans. Curr. Genet.* 2007, *51*, 89–98.
40. Sami, L.; Pusztahelyi, T.; Emri, T.; Varecza, Z.; Fekete, A.; Grallert, A.; Karanyi, Z.; Kiss, L.; Pocsi, I. Autolysis and aging of *Penicillium chrysogenum* cultures under carbon starvation: Chitinase production and antifungal effect of allosamidin. *J. Gen. Appl. Microbiol.* 2001, *47*, 201–211.
41. Villagomez-Castro, J.C.; Calvo-Mendez, C.; Lopez-Romero, E. Chtinase activity in encysting *Entamoeba invadens* and its inhibition by allosamidin. *Mol. Biochem. Parasitol.* 1992, *52*, 53–62.
42. Shahabuddin, M.; Toyoshima, T.; Aikawa, M.; Kaslow, D.C. Transmission-blocking activity of a chitinase inhibitor and activation of malarial parasite chitinase by mosquito protease. *Proc. Natl. Acad. Sci. USA* 1993, *90*, 4266–4270.
43. Vinetz, J.M.; Dave, S.K.; Specht, C.A.; Brameld, K.A.; Xu, B.; Hayward, R.; Fidock, D.A. The chitinase PfCHT1 from the human malaria parasite *Plasmodium falciparum* lacks proenzyme and chitin-binding domains and displays unique substrate preferences. *Proc. Natl. Acad. Sci. USA* 1999, *96*, 14061–14066.
44. Takeo, S.; Hisamori, D.; Matsuda, S.; Vinetz, J.; Sattabongkot, J.; Tsuboi, T. Enzymatic characterization of the Plasmodium vivax chitinase, a potential malaria transmission-blocking target. *Parasitol. Int.* 2009, *58*, 243–248.
45. Wu, Y.; Egerton, G.; Underwood, A.P.; Sakuda, S.; Bianco, A.E. Expression and secretion of a larval-specific chtinase (family 18 glycosyl

hydrolase) by the infective stages of the parasitic nematode, *Onchocerca volvulus. J. Biol. Chem.*2001, *276*, 42557–42564.
46. Takenaka, Y.; Nakano, S.; Tamoi, M.; Sakuda, S.; Fukamizo, T. Chitinase gene expression in response to environmental stresses in *Arabidopsis thaliana*: Chitinase inhibitor allosamidin enhances the stress. *Biosci. Biotechnol. Biochem.* 2009, *73*, 1066–1071.
47. Sampson, M.N.; Gooday, G.W. Involvement of chitinases of *Bacillus thuringiensis* during pathogenesis in insects.*Microbiology* 1998, *144*, 2189–2194.
48. Boer, W.D.; Gunnewiek, P.J.A.K.; Kowalchuk, G.A.; van Veen, J.A. Growth of chitinolytic dune soil β-subclass*Proteobacteria* in response to invading fungal hyphae. *Appl. Environ. Microbiol.* 2001, *67*, 3358–3362.
49. Suzuki, S.; Nakanishi, E.; Ohira, T.; Kawachi, R.; Nagasawa, H.; Sakuda, S. Chitinase inhibitor allosamidin is a signal molecule for chitinase production in its producing *Streptomyces*. I. Analysis of the chitinase whose production is promoted by allosamidin and growth accelerating activity of allosamidin. *J. Antibiot.* 2006, *59*, 402–409.
50. Suzuki, S.; Nakanishi, E.; Ohira, T.; Kawachi, R.; Ohnishi, Y.; Horinouchi, S.; Nagasawa, H.; Sakuda, S. Chitinase inhibitor allosamidin is a signal molecule for chitinase production in its producing *Streptomyces*. II. Mechanism for regulation of chitinase production by allosamidin through a two-component regulatory system. *J. Antibiot.* 2006, *59*, 410–417.
51. Suzuki, S.; Nakanishi, E.; Furihata, K.; Miyamoto, K.; Tsujibo, H.; Watanabe, T.; Ohnishi, Y.; Horinouchi, S.; Nagasawa, H.; Sakuda, S. Chtinase inhibitor allosamidin promotes chitinase production of *Streptomyces* generally. *Int. J. Biol. Macromol.* 2008, *43*, 13–19.
52. Zhu, Z.; Zheng, T.; Homer, R.J.; Kim, Y.K.; Chen, N.Y.; Cohn, L.; Hamid, Q.; Elias, J.A. Acidic mammalian chitinase in asthmatic Th2 inflammation and IL-13 pathway activation. *Science* 2004, *304*, 1678–1682.
53. Bucolo, C.; Musumeci, M.; Maltese, A.; Drago, F.; Musumeci, S. Effect of chitinase inhibitors on endotoxin-induced uveitis (EIU) in rabbits. *Pharmacol. Res.* 2008, *57*, 247–252.
54. Kitamoto, S.; Egashira, K.; Ichiki, T.; Han, X.; McCurdy, S.; Sakuda, S.; Sunagawa, K.; Boisvert, W.A. Chitinase inhibition promotes atherosclerosis in hyperlipidemic mice. *Am. J. Pathol.* 2013. in press.
55. Lingappa, Y.; Lockwood, J.L. A chitin medium for isolation, growth and maintenance of actinomycetes. *Nature,* 1961,*189*, 158–159.

56. Bentley, S.D.; Chater, K.F.; Cerdeño-Tárraga, A.M.; Challis, G.L.; Thomson, N.R.; James, K.D.; Harris, D.E.; Quail, M.A.; Kieser, H.; Harper, D.; et al. Complete genome sequence of the model actinomycete Streptomyces coelicolor A3(2). Nature 2002, 417, 141–147.
57. Ni, X.; Westpheling, J. Direct repeat sequences in the Streptomyces chitinase-63 promoter direct both glucose repression and chitin induction. Proc. Natl. Acad. Sci. USA 1997, 94, 13116–13121.
58. Saito, A.; Ishizaka, M.; Francisco, P.B., Jr.; Fujii, T.; Miyashita, K. Transcriptional co-regulation of five chitinase genes scattered on the Streptomyces coelicolor A3(2) chromosome. Microbiology 2000, 146, 2937–2946.
59. Saito, A.; Schrempf, H. Mutational analysis of the binding affinity and transport activity for N-acetylglucosamine of the novel ABC transporter Ngc in the chitin-degrader Streptomyces olivaceoviridis. Mol. Genet. Gen. 2004, 271, 545–553.
60. Wang, Q.; Zhou, Z.Y.; Sakuda, S.; Yamada, Y. Purification of allosamidin-sensitive and -insensitive chitinases produced by allosamidin-producing Streptomyces. Biosci. Biotech. Biochem. 1993, 57, 467–470.
61. Ohno, T.; Armand, S.; Hara, T.; Nikaidou, N.; Henrissat, B.; Mitsutomi, M.; Watanabe, T. A modular family 19 chitinase found in the prokaryotic organism Streptomyces griseus HUT 6037. J. Bacteriol. 1996, 178, 5065–5070.
62. Matsuura, H.; Okamoto, S.; Anamnart, S.; Wang, Q.; Zhou, Z.Y.; Nihira, T.; Yamada, Y.; Kuzuyama, T.; Seto, H.; Nakayama, J.; Suzuki, A.; Nagasawa, H.; Sakuda, S. Nucleotide sequences of genes encoding allosamidin-sensitive and -insensitive chitinases produced by allosamidin-producing Streptomyces. Biosci. Biotech. Biochem. 2003, 67, 2002–2005.
63. Nakanishi, E.; Okamoto, S.; Matsuura, H.; Nagasawa, H.; Sakuda, S. Allosamidin, a chitinase inhibitor produced by Streptomyces, acts as an inducer of chitinase production in its producing strain. Proc. Japan Academy Ser. B 2001, 77, 79–82.
64. Wills-Karp, M.; Luyimbazi, J.; Xu, X.; Schofield, B.; Neben, T.Y.; Karp, C.L.; Donaldson, D.D. Interleukin-13: Central mediator of allergic asthma. Science 1998, 282, 2258–2261.
65. Boot, R.G.; Blommaart, E.F.C.; Swart, E.; Ghauharali-van der Vlugt, K.; Bijl, N.; Moe, C.; Place, A.; Aerts, M.F. Identification of a novel acidic mammalian chitinase distinct from chitotriosodase. J. Biol. Chem. 2001, 276, 6770–6778.
66. Boot, R.G.; Bussink, A.P.; Verhoek, M.; de Boer, P.A.J.; Moorman,

A.F.M.; Aerts, J.M.F.G. Marked diffrences in tissue-specific expression of chitinases in mouse and man. *J. Histochem. Cytochem.* 2005, *53*, 1283–1292.
67. Homer, R.J.; Zhu, Z.; Cohn, L.; Lee, C.G.; White, W.I.; Chen, S.; Elias, J.A. Differential expression of chitinases identify subsets of murine airway epithelial cells in allergic inflammation. *Am. J. Physiol. Lung Cell Mol. Physiol.* 2006, *291*, L502–L511.
68. Matsumoto, T.; Inoue, H.; Sato, Y.; Kita, Y.; Nakano, T.; Noda, N.; Eguchi-Tsuda, M.; Moriwaki, A.; Kan-o, K.; Matsumoto, K.; Shimizu, T.; Nagasawa, H.; Sakuda, S.; Nakanishi, Y. demethylallosamidin, a chitinase inhibitor, suppresses airway inflammation and hyperresponsiveness. *Biochem. Biophys. Res. Commun.* 2009, *390*, 103–108.
69. Karasuda, S.; Yamamoto, K.; Kono, M.; Sakuda, S.; Koga, D. Kinetics analysis of a chitinase from red sea bream,*Pagrus major*. *Biosci. Biotech. Biochem.* 2004, *68*, 1338–1344.
70. Sato, Y.; Suzuki, S.; Muraoka, S.; Kikuchi, N.; Noda, N.; Matsumoto, T.; Inoue, H.; Nagasawa, H.; Sakuda, S. Preparation of allosamidin and demethylallosamidin photoaffinity probes and analysis of allosamidin-binding proteins in asthmatic mice. *Bioorg. Med. Chem.* 2011, *19*, 3054–3059.
71. Ober, C.; Chupp, G.L. The chitinase and chitinase-like proteins: A review of genetic and functional stidies in asthma and immune-mediated diseases. *Curr. Opin. Allergy Clin. Immunol.* 2009, *9*, 401–408.
72. Sutherland, T.E.; Maizels, R.M.; Allen, J.E. Chitinases and chitinase-like proteins: Potential therapeutic targets for the treatment of T-helper type 2 allergies. *Clin. Experi. Allergy* 2009, *39*, 943–955.
73. Owashi, M.; Arita, H.; Hayai, N. Identification of a novel eosinophil chemotactic cytokine (ECF-L) as a chitinase family protein. *J. Biol. Chem.* 2000, *275*, 1279–1286.
74. Houston, D.R.; Recklies, A.D.; Krupa, J.C.; van Aalten, D.M.F. Structure and ligand-induced conformational change of the 39-kDa glycoprotein from human articular chondrocytes. *J. Biol. Chem.* 2003, *278*, 30206–30212.
75. Schimpl, M.; Rush, C.L.; Betou, M.; Eggleston, I.M.; Recklies, A.D.; van Aalten, D.M.F. Human YKL-39 is apseudo-chitinase with retained chitooligosaccharide-binding properties. *Biochem. J.* 2012, *146*, 149–157.
76. Lee, C.G.; Elias, J.A. Role of breast regression protein-39/YKL-40 in

asthma and allergic responses. *Allergy Asthma Immunol. Res.* 2010, *2*, 20–27.

77. Iwashita, H.; Morita, S.; Sagiya, Y.; Nakanishi, A. Role of eosinophil chemotactic factor by T lymphocytes on airway hyperresponsiveness in a murine model of allergic asthma. *Am. J. Respir. Cell Mol. Biol.* 2006, *35*, 103–109.

78. Lee, C.G.; Hartl, D.; Lee, G.R.; Koller, B.; Matsuura, H.; Da Silva, C.A.; Sohn, M.H.; Cohn, L.; Homer, R.J.; Kozhich, A.A.; *et al.* Role of breast regression protein 39 (BRP-39)/chitinase 3-like-1 in Th2 and IL-13-induced tissue responses and apoptosis. *J. Exp. Med.* 2009, *206*, 1149–1166.

79. Cai, Y.; Kumar, R.K.; Zhou, J.; Foster, P.S.; Webb, D.C. Ym1/2 promotes Th2 cytokine expression by inhibiting 12/15(S)-lipoxygenase: Identification of a novel pathway for regulating allergic inflammation. *J. Immunol.* 2009, *182*, 5393–5399.

Chapter 4

BIOLOGICAL ACTIVITIES OF HYDRAZONE DERIVATIVES

Sevim Rollas and Ş. Güniz Küçükgüzel

Marmara University, Faculty of Pharmacy, Department of Pharmaceutical Chemistry, 34668, Turkey

ABSTRACT

There has been considerable interest in the development of novel compounds with anticonvulsant, antidepressant, analgesic, antiinflammatory, antiplatelet, anti- malarial, antimicrobial, antimycobacterial, antitumoral, vasodilator, antiviral and antischistosomiasis activities. Hydrazones possessing an azometine -NHN=CH- proton constitute an important class of compounds for new drug development. Therefore, many researchers have synthesized these compounds as target structures and evaluated their biological activities. These observations have been guiding for the development of new hydrazones that possess varied biological activities.

INTRODUCTION

Hydrazones have been demonstrated to possess, among other, antimicrobial, anticonvulsant, analgesic, antiinflammatory, antiplatelet, antitubercular and antitumoral activities. For example, isonicotinoyl hydrazones are antitubercular; 4-hydroxybenzoic acid[(5-nitro-2-furyl)methylene]-hydrazide (nifuroxazide) is an intestinal antiseptic; 4-fluorobenzoic acid[(5-nitro-2-furyl)methylene]-hydrazide [1] and 2,3,4-pentanetrione-3-[4-[[(5-nitro-2-furyl)methylene] hydrazino]carbonyl]phenyl]-hydrazone [2], which were synthesized in our Department, have antibacterial activity against both *Staphylococcus aureus* ATCC 29213 and *Mycobacterium tuberculosis* H37Rv at a concentration of 3.13 µg/mL. N^1-(4-Methoxybenzamido)benzoyl]-N^2-[(5-nitro-2-furyl) methylene]hydrazine, which was also synthesized in our Department [3], demonstrated antibacterial activity. In addition, some of the new hydrazide-hydrazones that we have recently synthesized were active against the same strain of *M. tuberculosis* H37Rv between the concentrations of 0.78-6.25 µg/mL [4].

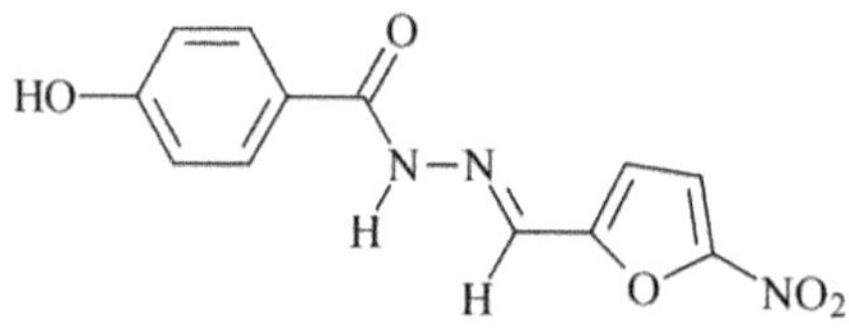

Nifuroxazide

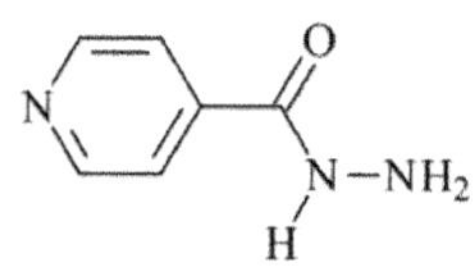

Isoniazid

Isonicotinic acid hydrazide (isoniazid, INH) has very high *in vivo* inhibitory activity towards *M. tuberculosis* H37Rv. Sah and Peoples synthesized INH hydrazide-hydrazones 1 by reacting INH with various aldehydes and ketones. These compounds were reported to have inhibitory activity in mice infected with various strains of *M. tuberculosis* [5]. They also showed less toxicity in these mice than INH [5,6] Buu-Hoi *et al.* synthesized some hydrazide-hydrazones that were reported to have lower toxicity than hydrazides because of the blockage of $-NH_2$ group. These findings further support the growing importance of the synthesis of hydrazide-hydrazones compound [7].

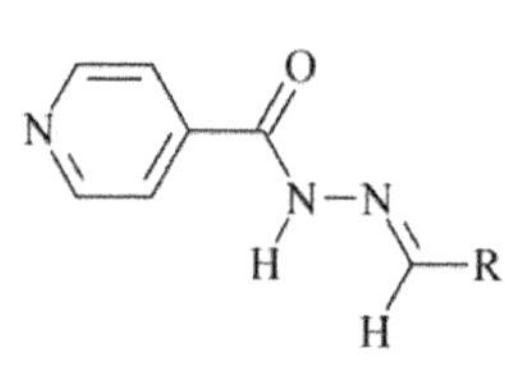

1

Iron is necessary for the biochemical reactions of living organisms. Desferrioxamine is an agent which is used for the treatment of a complication called "Iron Overload Disease". Researchers have synthesized hydrazones of INH by using various aldehydes and their iron complexes and evaluated these complexes for their antitumoral activity. The mechanism of antitumoral activity of iron complexes is the inhibition of ribonucleotide reductase, which is an important enzyme for conversion of ribonucleotides to deoxyribonucleotides. Copper complexes of INH that facilitate the intercellular transport of INH were synthesized and evaluated for their antitubercular activity.

Hydrazones containing an azometine -NHN=C$\underline{H}$- proton are synthesized by heating the appropriate substituted hydrazines/hydrazides with aldehydes and ketones in solvents like ethanol, methanol, tetrahydrofuran, butanol, glacial acetic acid, ethanol-glacial acetic acid. Another synthetic route for the synthesis of hydrazones is the coupling of aryldiazonium salts with active hydrogen compounds. In addition, 4-acetylphenazone isonicotinoylhydrazones was prepared by Amal and Ergenç [8] by exposing an alcohol solution of 4-acetylphenazone and INH to sunlight or by mixing them with a mortar in the absence of the solvent.

Hydrazide-hydrazones compounds are not only intermediates but they are also very effective organic compounds in their own right. When they are used as intermediates, coupling products can be synthesized by using the active hydrogen component of –CONHN=CH- azometine group [9]. *N*-Alkyl hydrazides can be synthesized by reduction of hydrazones with NaBH$_4$ [10], substituted 1,3,4-oxadiazolines can be synthesized when hydrazones are heated in the presence of acetic anhydride [1,11,12]. 2-Azetidinones can be synthesized when hydrazones react with trietylamine chloro acetylchloride[13]. 4-Thiazolidinones are synthesized when hydrazones react with thioglycolic acid/ thiolactic acid [3,14] (Scheme 1).

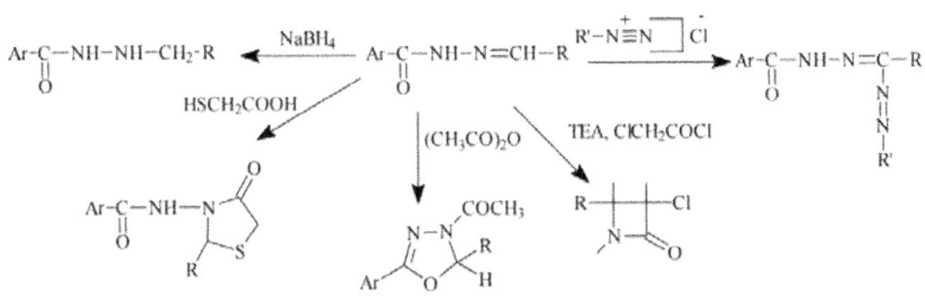

Scheme 1.

Many effective compounds, such as iproniazide and isocarboxazide, are synthesized by reduction of hydrazide-hydrazones. Iproniazide, like INH, is used in the treatment of tuberculosis. It has also displays an antidepressant effect and patients appear to have a better mood during the treatment. Another clinically effective hydrazide-hydrazones is nifuroxazide, which is used as an intestinal antiseptic.

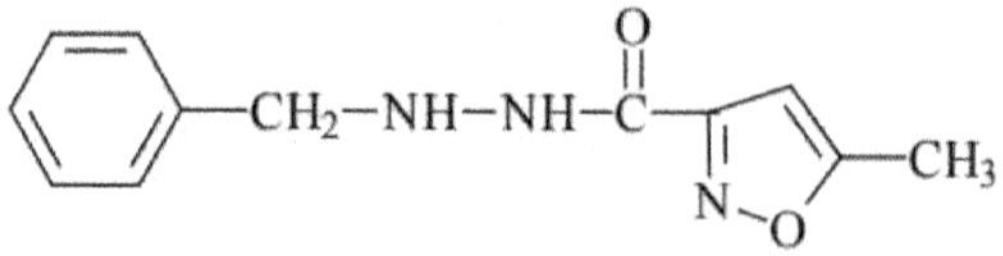

Isocarboxazide

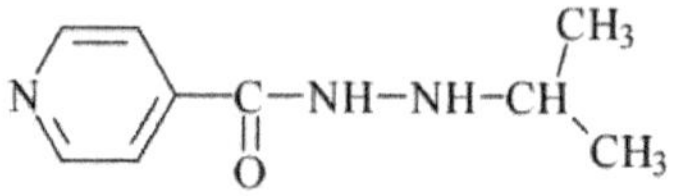

Iproniazide

A number of studies have investigated the *in-vitro* and *in-vivo* metabolism of hydrazide-hydrazones. In *in-vitro* metabolism studies, it has been found that hydrazide-hydrazones undergo hydrolytic reactions and aromatic rings undergo aromatic hydroxylation reactions [15,16] (Scheme 2).

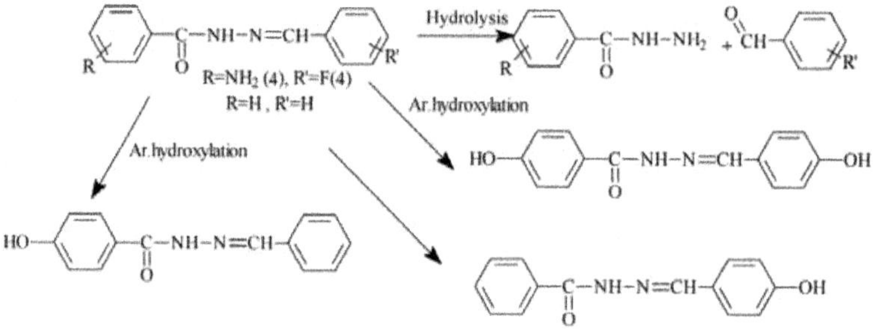

Scheme 2.

Gülerman *et al.* investigated the *in vivo* metabolism of 4-fluorobenzoic acid ((5-nitro-2-furyl)-methylene-hydrazide, a hydrazide derivative that is effective against *S. aureus* ATCC 29213. They confirmed the presence of the substrate and 4-fluorobenzoic acid metabolite in blood and blood cells [17] (Scheme 3).

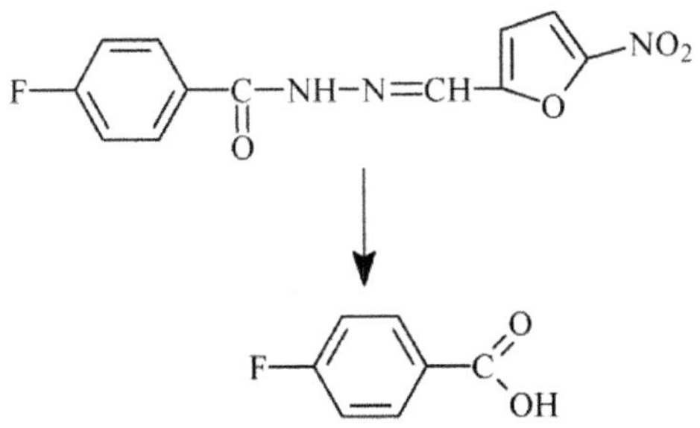

Scheme 3.

Küçükgüzel et al. studied the *in vitro* hepatic microsomal metabolism of N-(4-chlorobenzyl)-N-benzoylhydrazine (CBBAH). The corresponding hydrazone, namely benzoic acid (4-chlorophenyl)-methylenehydrazide was detected as the major *in vitro* metabolic product [18] (Scheme 4).

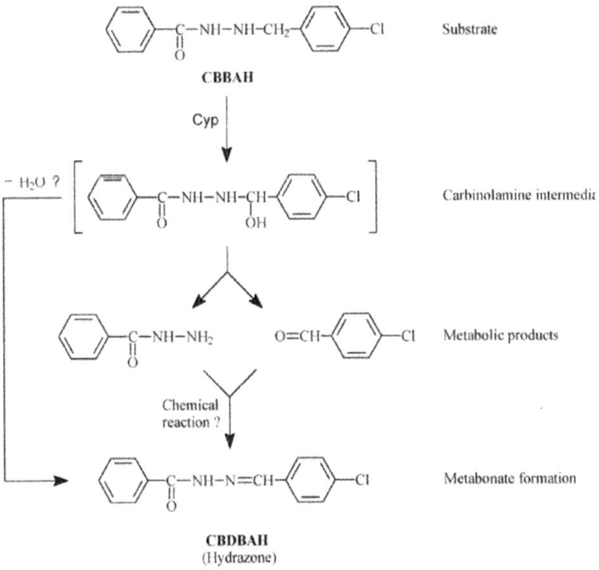

Scheme 4.

It has been known that the hydrazides (like INH) form α-ketoglutaric acid and form hydrazones with vitamin B6 and pyruvic acid. It is clinically important that when tuberculosis patients are treated with INH, reaction of INH with vitamin B6 leads to formation of a hydrazone and development of vitamin B6 deficiency, therefore, patients who are treated with INH should be administered vitamin B6 (Scheme 5).

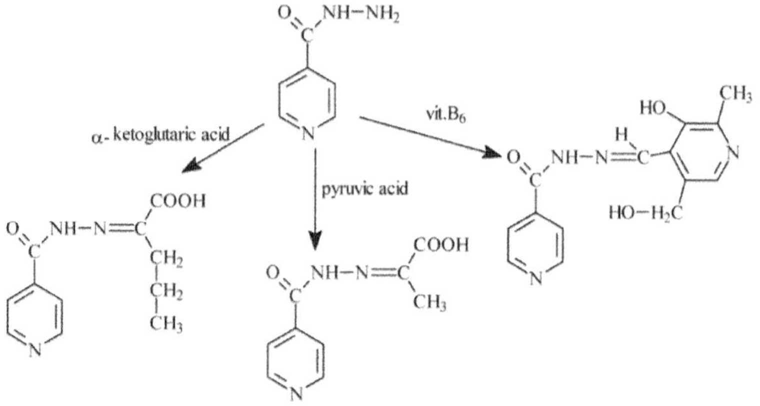

Scheme 5.

This review critically evaluates the pharmacological activity of the hydrazones that were reported in the past ten years.

BIOLOGICAL ACTIVITY

Anticonvulsant Activity

Epilepsy is a common neurological disorder and a collective term given to a group of syndromes that involve spontaneous, intermittent, abnormal electrical activity in the brain. The pharmacotherapy of epilepsy has been archieved during the last decade. Furthermore, although for the last twenty years new antiepileptic drugs have been introduced into clinical practice, the maximal electroshock (MES) test and the subcutaneous pentylenetetrazole (scPTZ) test are the most widely used animal models of epilepsy to characterize the anticonvulsant activity.

The biological results revealed that in general, the acetylhydrazones **2** provided good protection against convulsions while the oxamoylhydrazones **3** were significantly less active. [19].

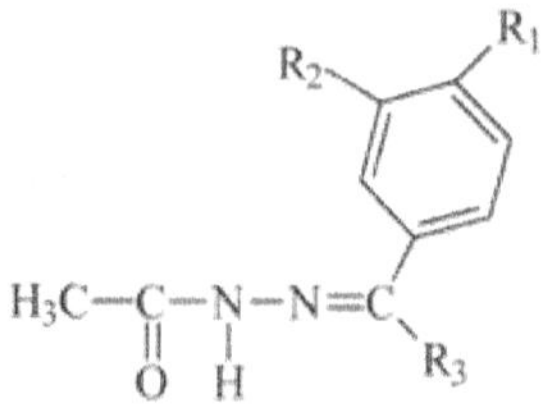

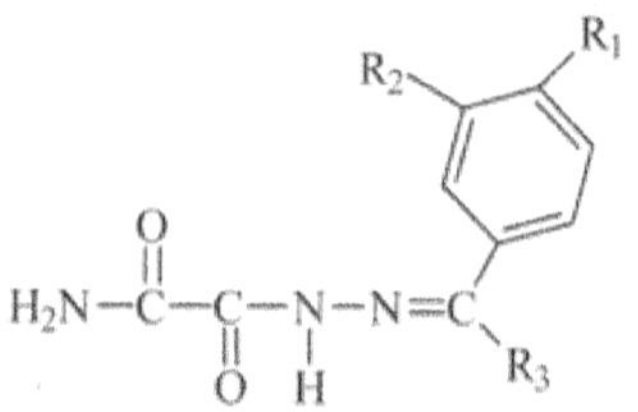

Fifteen new hydrazones of (2-oxobenzoxazoline-3-yl) acetohydrazide 4 were synthesised and their antiepileptic activity was tested in scPTZ test. The 4-fluoro derivative was found to be more active than the others [20].

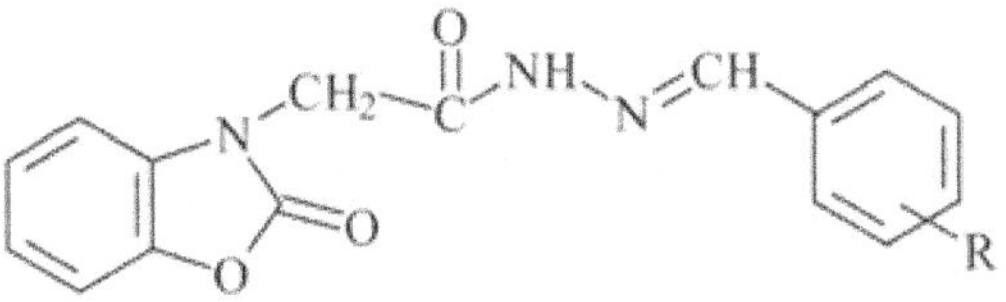

4

4-Aminobutyric acid (GABA) is the principal inhibitory neurotransmitter in the mammalian brain. GABA hydrazones 5 were designed and synthesized and evaluated for their anticonvulsant properties in different animal models of epilepsy such as MES, scPTZ, subcutaneous strychine (scSTY) and intraperitonal picrotoxin (ipPIC) induced seizure tests. Some of the compounds were effective in these models [21].

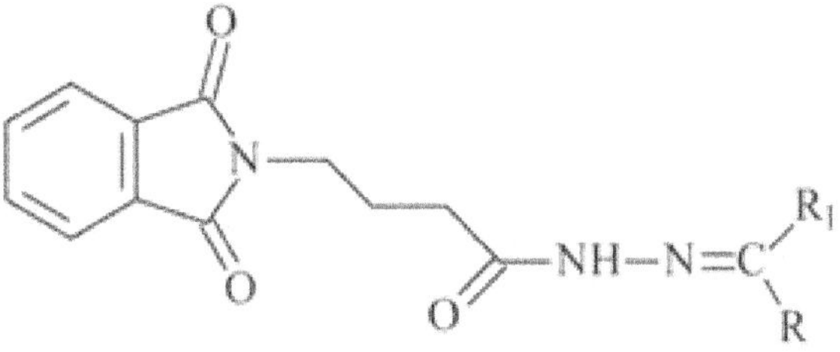

5

Antidepressant Activity

Iproniazide, isocarboxazide and nialamide, which are hydrazide derivatives, exert their action by inhibiting the enzyme monoamine oxidase (MAO). Inhibition results in increased levels of norepinephrine, dopamine, tyramine and serotonin in brain neurons and in various other tissues. There have been many reports on the antidepresant / MAO-inhibiting the activity of hydrazones derived from substituted hydrazides and reduction products.

Ten new arylidenehydrazides 6 which were synthesized by reacting 3-phenyl-5-sulfonamidoindole-2-carboxylic acid hydrazide with various aldehydes, evaluated for their antidepresant activity. 3-Phenyl-5-sulfonamidoindole-2-carboxylic acid 3,4-methylenedioxy / 4-methyl / 4-nitrobenzylidene-hydrazide showed antidepresant activity at 100 mg/kg [10].

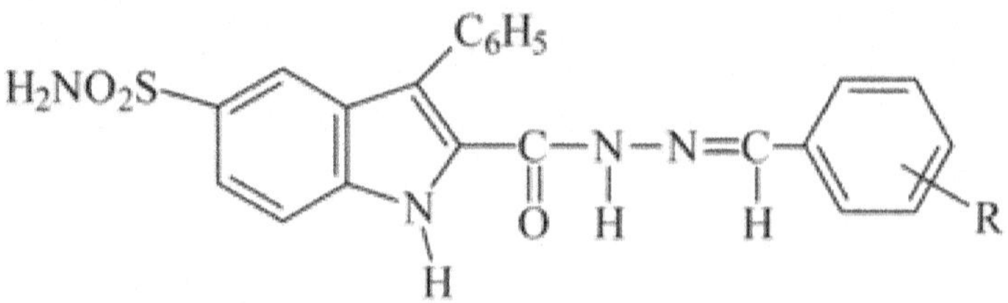

6

Analgesic, Antiinflammatory and Antiplatelet Activity

Non-steroidal anti-inflammatory drugs (NSAIDs) have a wide clinical use for the treatment of inflammatory and painful conditions including rheumatoid arthritis, soft tissue and oral cavity lesions, respiratory tract infections and fever. The two isoforms of cyclooxygenase (COX) are poorly distinguishable by most of the classical NSAIDs and these agents actually inhibit COX-1 extensively, besides COX-2, leading to gastrointestinal injury, suppression of TXA2 formation and platelet aggregation. The combination of these interactions is probably the reason for gastrointestinal bleeding as the most serious complication of these drugs. Some evidences suggest that the hydrazone moiety present in some compounds possess a pharmacophoric character for the inhibition of COX.

The most important antiinflammatory derivative 2-(2-formylfuryl) pyridylhydrazone **7** presented a 79 % inhibition of pleurisy at a dose of 80.1 µmol/kg. The authors also described the results concerning the mechanism of the action of these series of *N*-heterocyclic derivatives in platelet aggregation that suggests a Ca^{2+} scavenger mechanism. Compound **7** was able to complex Ca^{2+} in in-vitro experiments at 100 µM concentration, indicating that these series of compounds can act as Ca^{2+} scavenger depending on the nature of the aryl moiety present at the imine subunit [22].

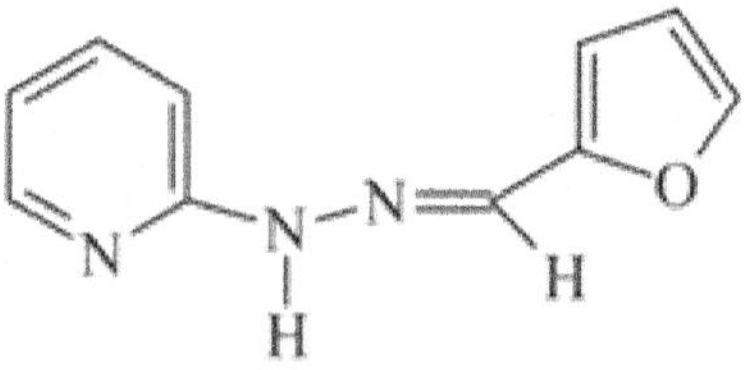

7

A new series of antinociceptive compounds that belong to the N-acylarylhydrazone class were synthesized from natural safrole. [(4'-N,N-Dimethylaminobenzylidene-3-(3',4'-methylenedioxyphenyl) propionylhydrazine] 8 was more potent than dipyrone and indomethacine, are used as standard antiinflammatory/antinociceptive drugs [23].

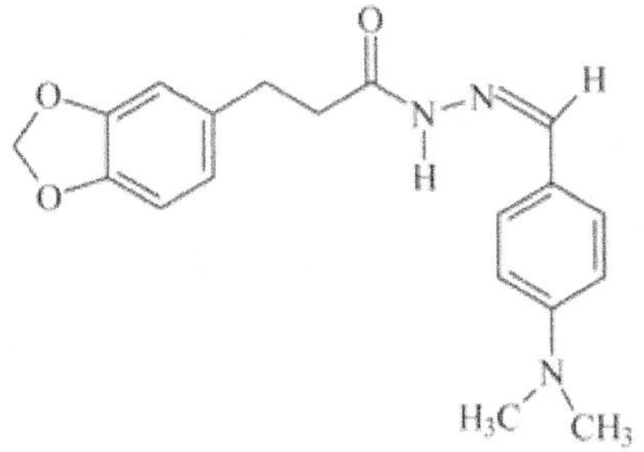

8

The antiplatelet activity of novel tricyclic acylhydrazone derivatives 9 was evaluated by their ability to inhibit platelet aggregation of rabbit platelet-rich plasma induced by platelet-activating factor (PAF) at 50 nM. Benzylidene- / 4'-bromobenzylidene 3-hydroxy-8-methyl-6-phenylpyrazolo[3,4-b]thieno-[2,3-d]pyridine-2-carbohydrazide were evaluated at 10 μM, presenting, respectively, 10.4 and 13.6% of inhibition of the PAF-induced platelet aggregation [24].

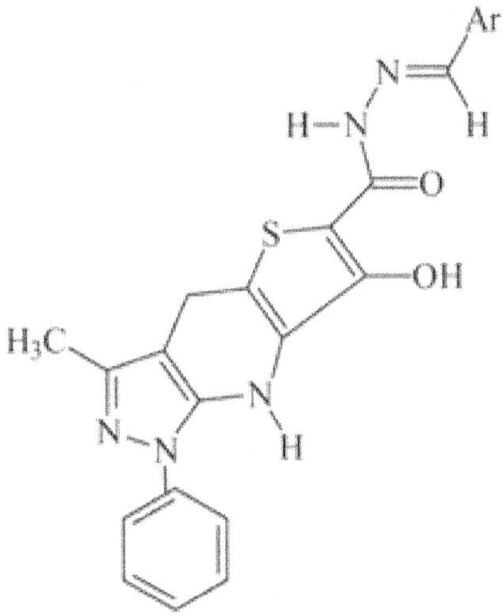

9

The evaluation of platelet antiaggregating profile let to identification of a new potent prototype of antiplatelet derivative, that is benzylidene 10H-phenothiazine-1-carbohydrazide (IC_{50}=2.3 μM) 10, which acts in the arachidonic acid pathway probably by inhibition of platelet COX-1 enzyme.

Additionally, the change in *para*-substituent group of acylhydrazone framework permitted to identify a hydrophilic carboxylate derivative and a hydrophobic bromo derivative as two new analgesics that are more potent than dipyrone, which is the standard, possessing selective peripheral or central mechanism of action [25].

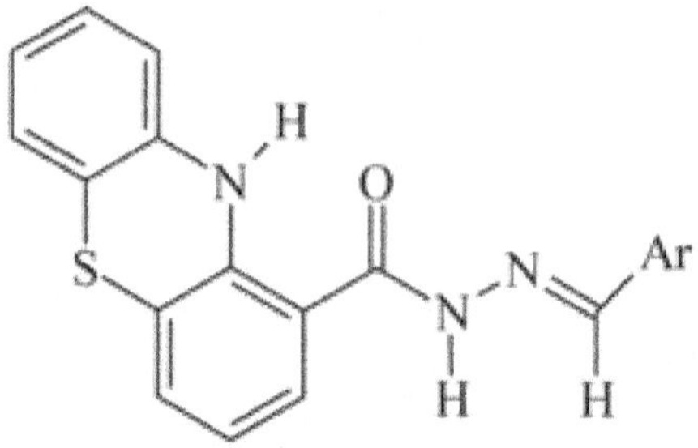

10

Gökhan-Kelekçi *et al.* synthesized hydrazones containing 5-methyl-2-benzoxazoline. The analgesic effects of 2-[2-(5-methyl-2-benzoxazoline-3-yl)acetyl]-4-chloro- / 4-methyl benzylidene hydrazine 11c and 11d were found to be higher than those of morphine and aspirin. In addition, 2-[2-(5-methyl-2-benzoxazoline-3-yl)acetyl]-4- methoxybenzylidene hydrazine11e at 200 mg/kg dose possessed the most antiinflammatory activity [26].

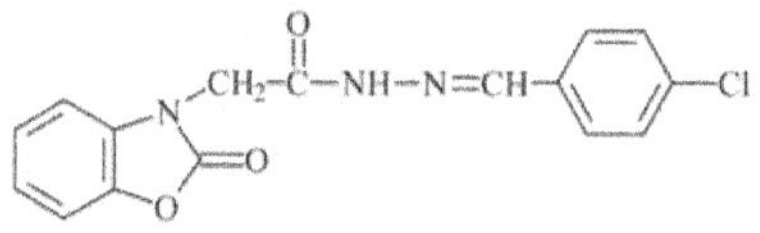

11c 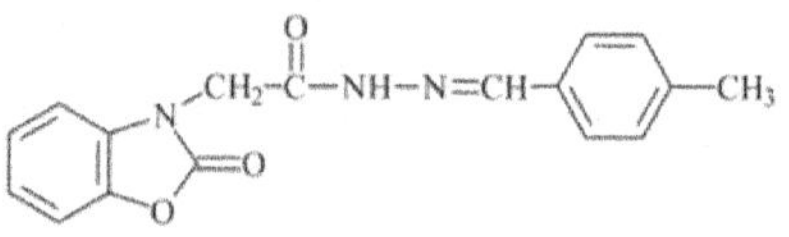   **11d**

Duarte *et al.* have described *N'*-(3,5-Di-tert-butyl-4-hydroxybenzylidene)-6-nitro-1,3-benzodioxole-5-carbohydrazine 12c as a novel antiinflammatory compound [27].

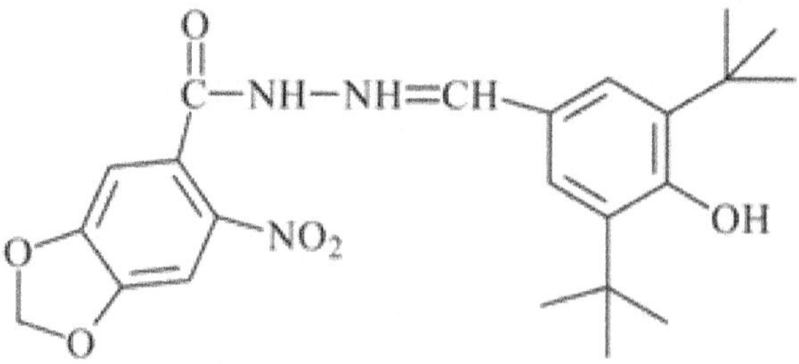

12c

Antimalarial Activity

Malaria is a disease caused by parasitic protozoa of the genus *Plasmodium* which afflicts more than 500 million people worldwide and causes approximately 2 million deaths each year. The spread of multidrug-resistant *Plasmodium falciparum* has highlighted the urgent need to discover new antimalarial drugs.

The aroylhydrazone chelator 2-hydroxy-1-naphthylaldehyde isonicotinoyl hydrazone 13 showed greater antimalarial agent activity than desferrioxamine against chloroquine-resistant and -sensitive parasites [28].

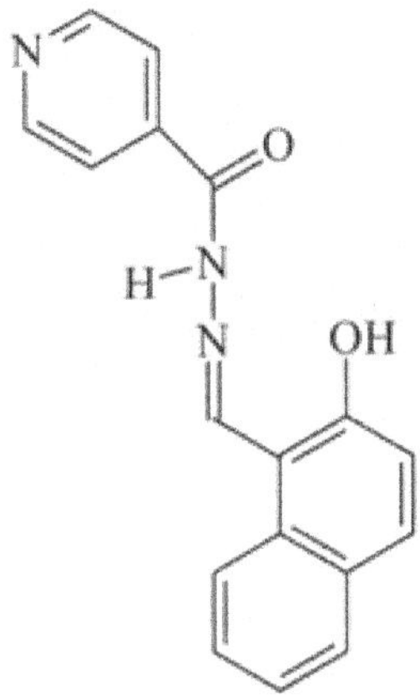

13

A series of N^1-arylidene-N^2-quinolyl- 14 and N^2-acrydinylhydrazones- 15 were synthesized and tested for their antimalarial properties. The new synthesized compounds, including 14d-g and 15a-c showed an antiplasmodial activity against the chloroquine-sensitive D10 strain in the same range of chloroquine (CQ). Similarly, 14f and 14g displayed the same activity as CQ against chloroquine-sensitive 3D-7 strain, while compound 15b was 10 times more potent than CQ. Two analogues 15b and 15c, were more active against W2 CQ-resistant than D10 CQ-sensitive strains [29].

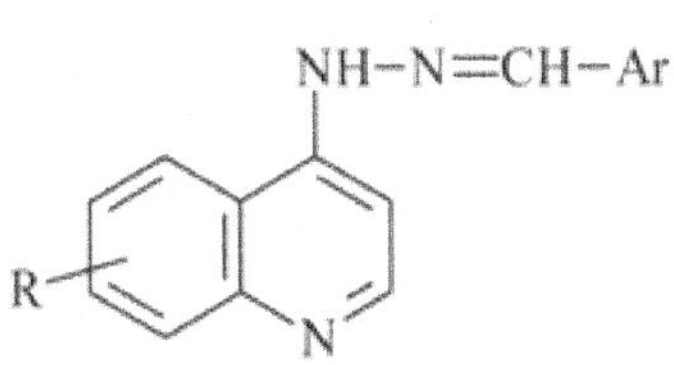

14d-g 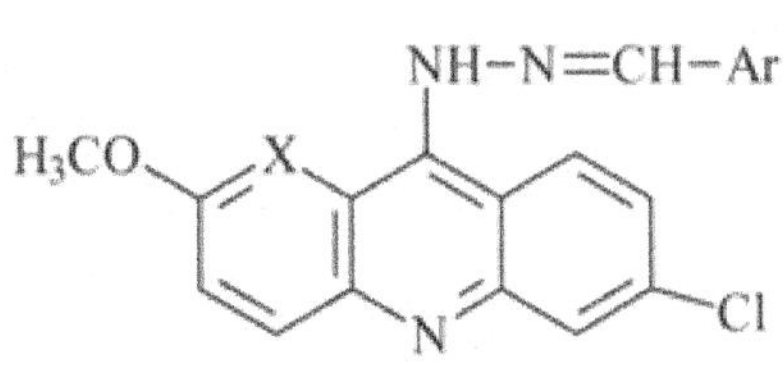   **15a-c**

1-Substituted phenyl-*N'*-[(substitutedphenyl)methylene]-1*H*-pyrazole-4-carbohydrazides 16 were synthesized and their leishmanicidal and cytotoxic effects were compared to the prototype drugs (ketoconazole, benznidazole, allopurinol and pentamidine) *in vitro*. The 1*H*-pyrazole-4-carbohydrazide derivatives with X = Br, Y = NO_2 and X = NO_2, Y = Cl demonstrated the highest activity and they were more effective on promastigotes forms of *L. amazonensis* than on *L. chagasi* and *L. braziliensis* species [30].

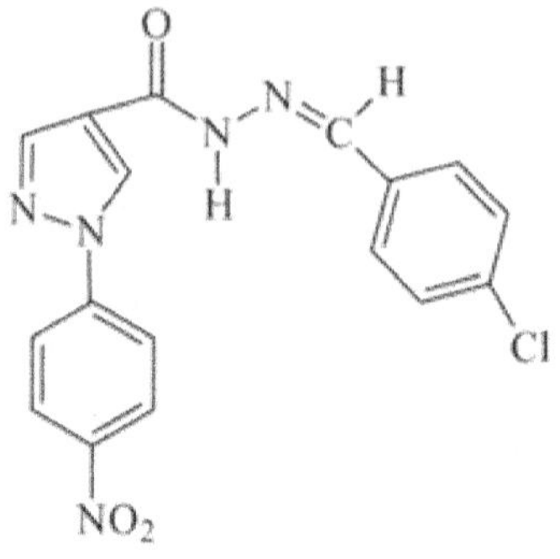

16

Antimicrobial Activity

The dramatically rising prevalence of multi-drug resistant microbial infections in the past few decades has become a serious health care problem. The search for new antimicrobial agents will consequently always remain as an important and challenging task for medicinal chemists.

Ethyl 2-arylhydrazono-3-oxobutyrates 17 were synthesized in order to determine their antimicrobial properties. Compound 17d showed significant activity against *S. aureus* whereas the others had no remarkable activity on this strain. Compound 17e was found to be more active than the others against *Mycobacterium fortuitum* at a MIC value of 32 µg/ml [31].

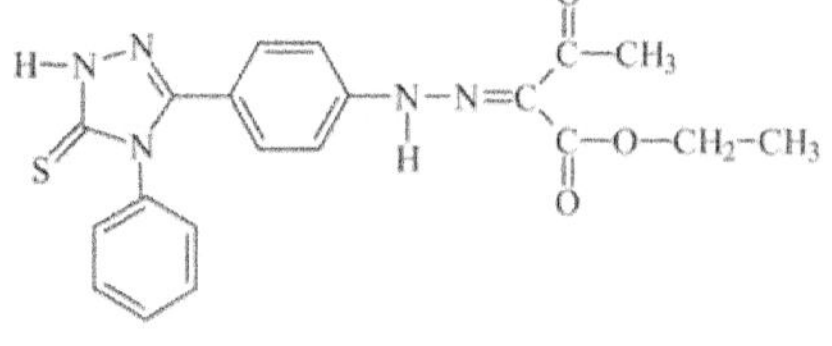

17d 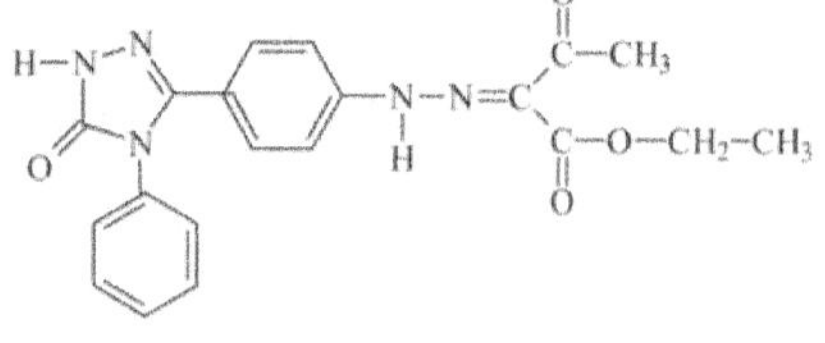   **17e**

N^1-(4-methoxybenzamido)benzoyl]-N^2-[(5-nitro-2-furyl)methylene] hydrazine 18 inhibited the growth of several bacteria and fungi [3].

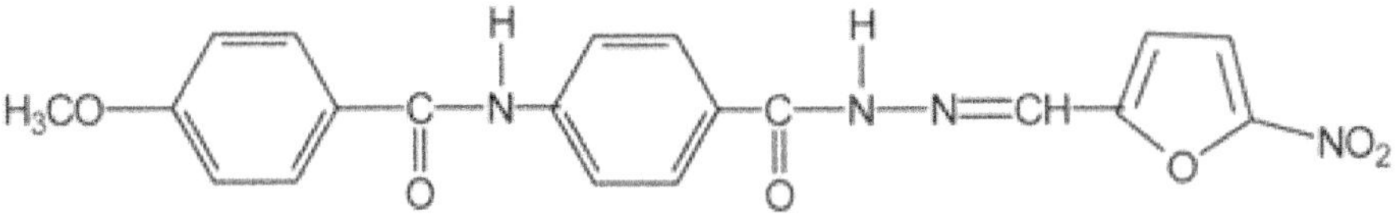

18

Nifuroxazide and six analogs 19 were synthesized by varying the substituent at the *p*-position of the benzene ring and the heteroatom of the heterocyclic ring. These compounds were evaluated for their antimicrobial activity against *S. aureus* ATCC 25923 and found to be active at concentration 0.16-63.00 µg/mL [32].

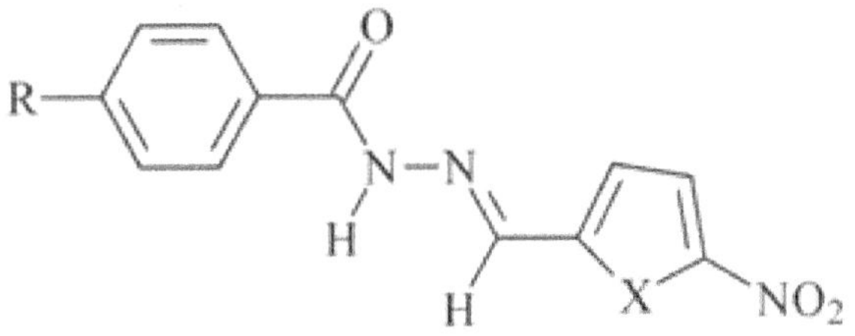

19

N^2-Substituted alkylidene/arylidene-6-phenylimidazo[2,1-b]thiazole-3-acetic acid hydrazides 20 were synthesized and evaluated for their *in vitro* antimicrobial activity. Some compounds showed antimicrobial activity against *S. aureus* ATCC 6538, *S. epidermidis* ATCC 12228, *T. mentagrophytes var. Erinacei* NCPF-375, *T. rubrum* and *M. audounii* (MIC 25-0.24 µg/mL) [33].

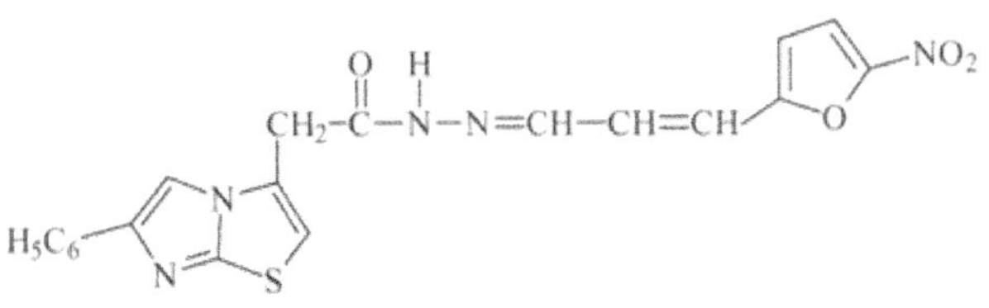

20j

Turan-Zitouni *et al.* found 5-bromoimidazo[1,2-a]pyridine-2-carboxylic acid benzylidene-hydrazide 21 and 5-bromoimidazo[1,2-a]pyridine-2-carboxylic acid 4-methoxybenzylidenehydrazide 22 to possess antimicrobial activity at 3.9 µg/mL against *E. fecalis* and *S. epidermis* [34].

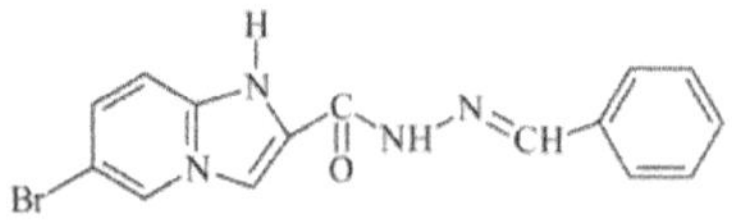

21 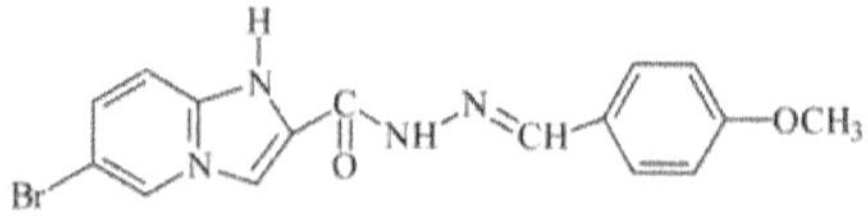　　　　　　　　　　　　　**22**

A series of hydrazones derived from 1,2-benzisothiazole hydrazides (R_1=H) 23-27 as well as the parent cyclic (23 and 26) and acyclic (24, 25 and 27) 1,2-benzisothiazole hydrazides, were synthesized and evaluated as antibacterial and antifungal agents. All of the 2-amino-1,2-benzisothiazole-3(2H)-one derivatives, belonging to series 23 and 26 showed good antibacterial activity against Gram positive bacteria. Most of them were also active against yeasts, too [35].

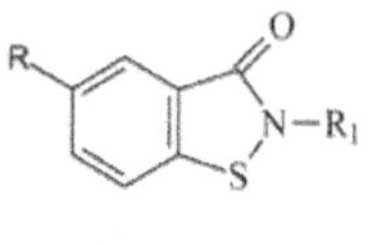

23 (R=H), 26 (R=CH$_3$)　　　　**24** 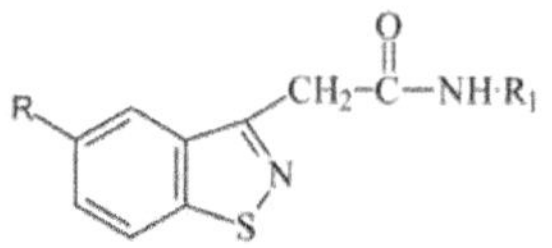　　　　**25(R=H), 27 (R=CH$_3$)**

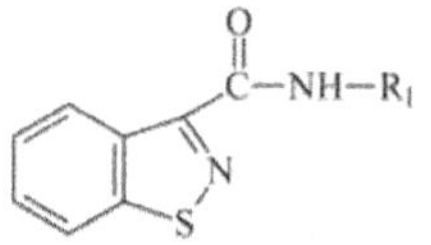

Rollas *et al.* synthesized a series of hydrazide hydrazones 28 and 1,3,4-oxadiazolines of 4-fluoro-benzoic acid hydrazide as potential antimicrobial agents and tested these compounds for their antibacterial and antifungal activities against *S. aureus*, *E. coli*, *P. aeruginosa* and *C. albicans*. From these compounds, 4-fluorobenzoic acid[(5-nitro-2-furyl)methylene] hydrazide (28a) showed equal activity as ceftriaxone against *S. aureus*. In addition, the MIC values of compounds 28c and 28d for the same strain were in the range of those reported for ceftriaxone according to NCCLS 1997 [1].

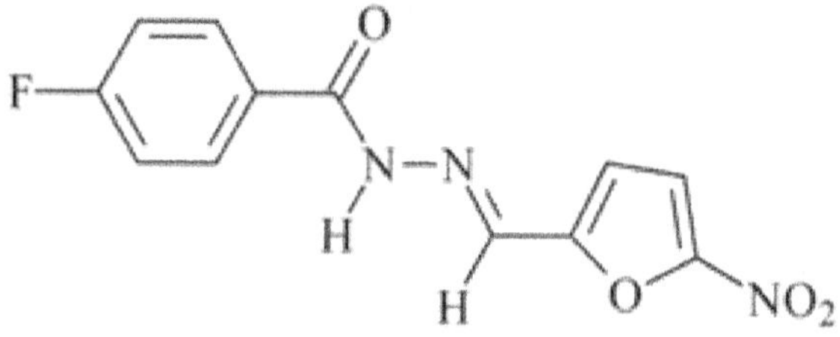

28a

Küçükgüzel *et al.* synthesized diflunisal hydrazide-hydrazone derivatives. 2',4'-Difluoro-4-hydroxybiphenyl-3-carboxylic acid [(5-nitro-2-furyl) methylene] hydrazide (29a) has shown activity against *S. epidermis* HE-5 and *S.*

aureus HE-9 at 18.75 µg/mL and 37.5 µg/mL, respectively. 2',4'-Difluoro-4-hydroxybiphenyl-3-carboxylic acid [(2,4,6-trimethylphenyl)methylene] hydrazide (29e) has exhibited activity against *Acinetobacter calcoaceticus* IO-16 at a concentration of 37.5 µg/mL, whereas Cefepime, the drug used as standard, has been found to be less active against the same microorganisms [36].

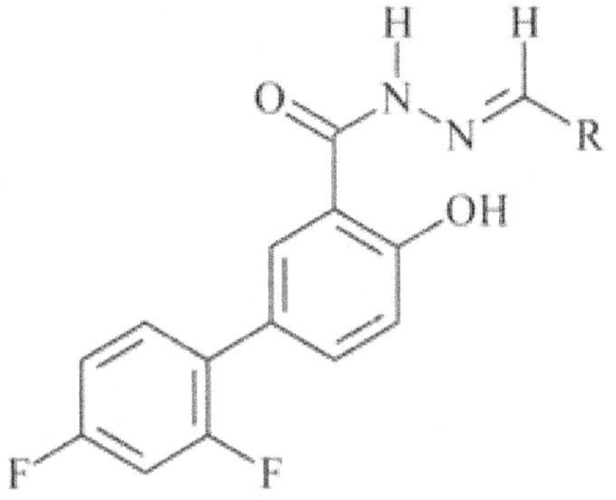

29

A series of hydrazones synthesized from various cholesterol derivatives **30** were evaluated for their *in vitro* antimicrobial properties against human pathogens. The activity was highly dependent on the structure of the different compounds involved. The best results have been obtained with tosylhydrazone cholesterol derivatives exhibiting activities against *C. albicans* (CIP 1663-80) at a concentration of 1.5 µg/mL [37].

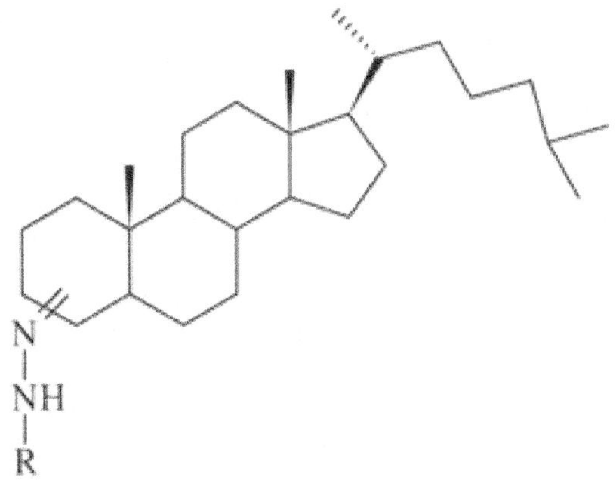

30

4-Substituted benzoic acid [(5-nitro-thiophene-2-yl)methylene] hydrazides 31 were synthesized as potential bacteriostatic activity and some of them indeed showed bactericidal activity [38].

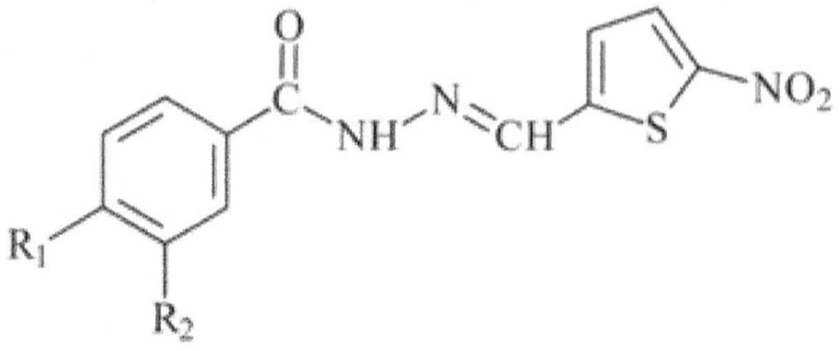

31

Antimycobacterial Activity

Tuberculosis is a serious health problem that causes the death of some three million people every year worldwide [39]. In addition to this, the increase in *M. tuberculosis* strains resistant to front-line antimycobacterial drugs such as rifampin and INH has further complicated the problem, which clearly indicates the need for more effective drugs for the efficient management of tuberculosis. Meyer and Mally prepared new hydrazones by reacting isoniazid (INH) with benzaldehyde, o-chlorobenzaldehyde and vanilin [5]. Shchukina *et al.* prepared INH hydrazide-hydrazones 1 by reacting INH with various aldehydes and ketones; the compounds were reported to have activity in mice which had been infected with various strains of *M. tuberculosis*, and also indicated lower toxicity than INH [5,6].

The reaction of 1-methyl-1H-2-imidazo[4,5-b]pyridinecarboxylic acid hydrazide with substituted aldehydes yielded the corresponding hydrazide-hydrazones. Compound 32 exhibited antimycobacterial activity against *M. tuberculosis H37 Rv, M. tuberculosis 192, M. tuberculosis 210*, isolated from patients and resistant against INH, ethambutol, rifampicine at 31.2 μg/mL[40].

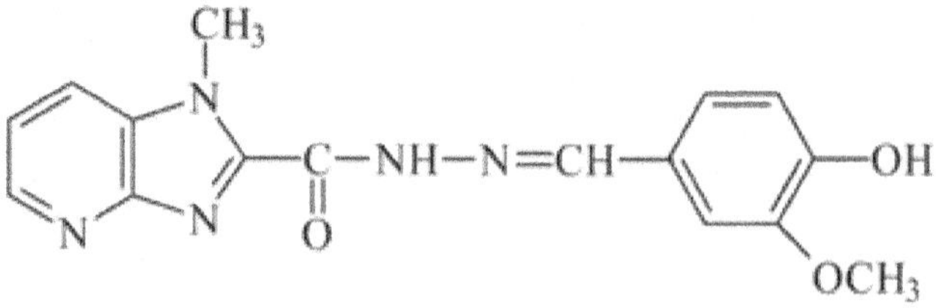

32

Various different 2,3,4-pentanetrione-3-[4-[[(5-nitro-2-furyl/ pyridyl/substituted-phenyl)-methylene]hydrazino]carbonyl]phenyl] hydrazones 33 were synthesized. All the synthesized compounds were evaluated for their antimycobacterial activity against *M. fortuitum* ATCC 6841 and *M. tuberculosis* H37Rv. Of the compounds screened, 33e and 33g were found to be active against *M. fortuitum* at an MIC value of 32 μg/mL. Compound 33a, which exhibited > 90% inhibition in the primary screen at 12.5 μg/mL against *M. tuberculosis* H37Rv, was the most promising derivative for antituberculosis activity. Results obtained from the level II screening showed that the actual MIC and IC_{50} values of 32a were 3.13 and 0.32 μg/mL, respectively. The same compound was also tested against *Mycobacterium avium*, which was observed not to be susceptible to 33a [2].

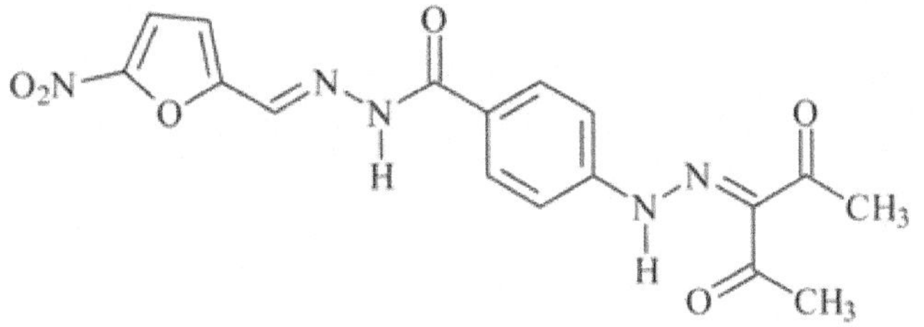

33a

Isonicotinoylhydrazones have been further reacted with pyridinecarboxaldehydes to give the corresponding pyridylmethyleneamino derivatives 34. The new synthesized hydrazones and their pyridylmethyleneamino derivatives were tested for their activity against mycobacteria, Gram-positive and Gram-negative bacteria. The cytotoxicity was also tested. Several compounds showed a good activity against *M. tuberculosis* H37Rv and some isonicotinoylhydrazones showed a moderate activity against a clinically isolated *M. tuberculosis* (6.25-50 μg/mL) which was INH resistant [41].

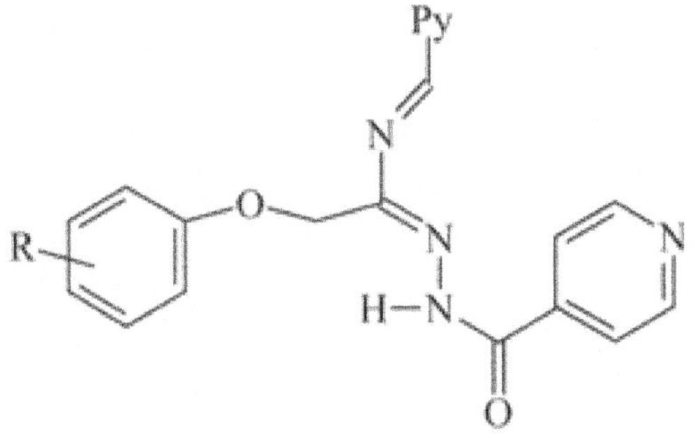

34

The reaction of 2-acetylimidazo[4,5-b]pyridine with INH yielded the corresponding hydrazide-hydrazones 35. This compound exhibited activity against *M. tuberculosis H37 Rv* , *M. tuberculosis 192*, *M. tuberculosis 210*, isolated from patients and resistant against INH, ethambutol, rifampicine at 3.13 µg/mL [42].

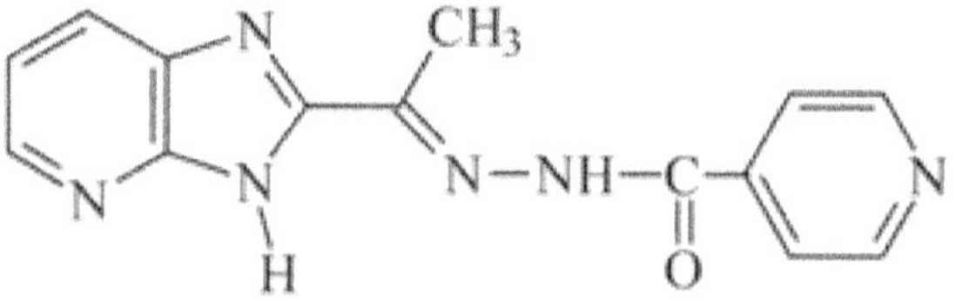

35

N^2-Substitutedalkylidene/arylidene-6-phenylimidazo[2,1-b]thiazole-3-acetic acid hydrazides 20 were synthesized and evaluated for *in vitro* antimycobacterial activity. The compounds exhibited different degrees of inhibition (17-98 %) against *M. tuberculosis H37 Rv* [33].

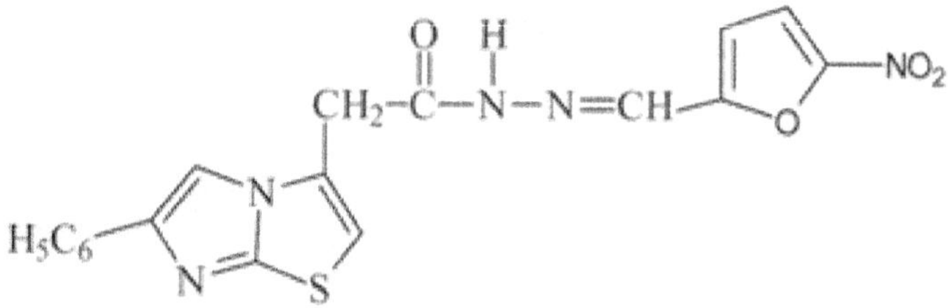

20i

[5-(Pyridine-2-yl)-1,3,4-thiadiazole-2-yl]thio]acetic acid arylidenehydrazide derivatives 36 were synthesized and tested for their in vitro antimycobacterial activity. Some compounds showed activity at 20 µg/mL against *M. tuberculosis* and at 40 µg/mL against *M. avium* [43].

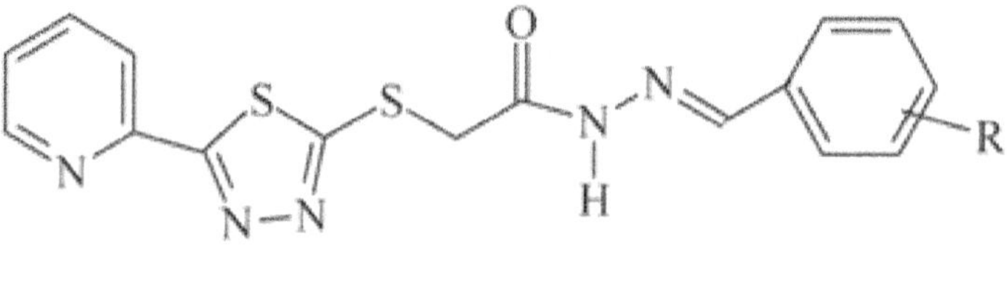

36

N-Alkylidene/arylidene-5-(2-furyl)-4-ethyl-1,2,4-triazole-3-mercaptoacetic acid hydrazides 37 were synthesized and evaluated for in vitro antimycobacterial activity. The compounds exhibited different degrees of inhibition (3-61%) against *M. tuberculosis* H37 Rv at 6.25 μg/mL [44].

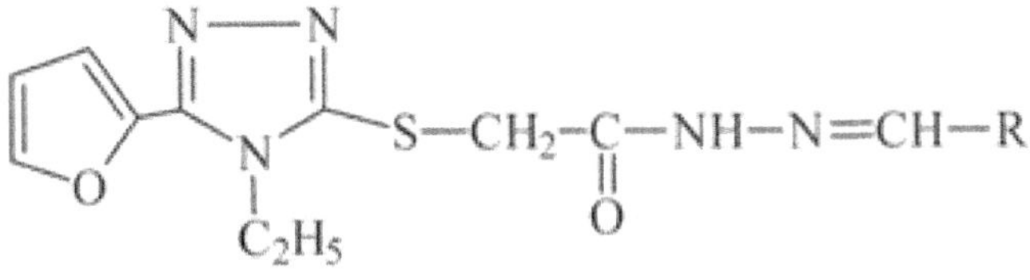

37

A series of 4-quinolylhydrazones 38 were synthesized and tested against *M. tuberculosis* H37Rv. Preparation of the title compounds was achieved by reaction of 4-quinolylhydrazine and aryl- or heteroarylcarboxaldehydes. Most of the derivates had antitubercular properties; two compounds were identified with the highest activity and they were tested also against *M. avium* [45].

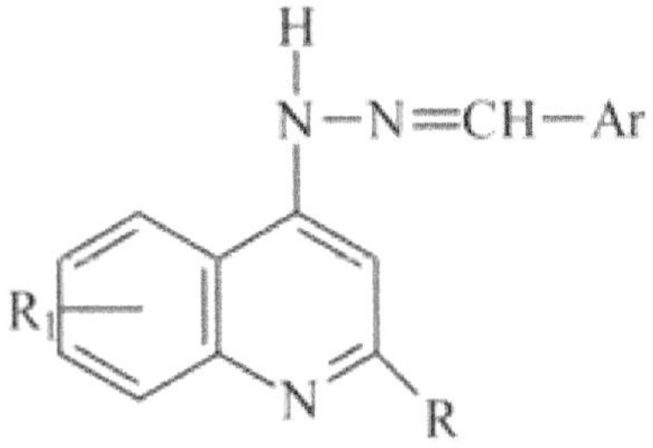

38

Benzoic acid [(5-nitro-thiophene-2-yl)methylene] hydrazide series 39 were synthesized and tested against *M. tuberculosis*H37Rv. Rando and co-workwers have applied *Topliss* methodolgy to a set of nitrogen analogues. 4-Methoxybenzoic acid[(5-nitrothiophene-2-yl)methylene] hydrazide (39a) was demonstrated as being the most active, with a MIC value of 2.0 μg/mL [46].

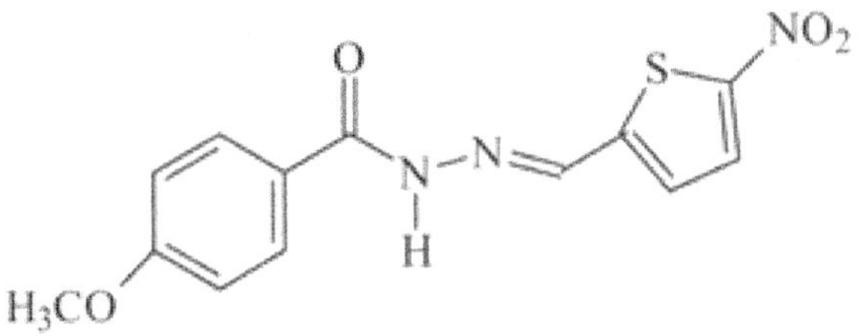

39a

Both hydrazone products, ethyl 2-[(3,5-dimethylpyrazole-4-yl)hydrazono]-3-oxobutyrate 40d and methyl 2-[(3,5-dimethylpyrazole-4-yl)hydrazono]-4-methoxy-3-oxobutyrate 40e showed 29 and 28% inhibition against *M. tuberculosis* H37Rv, respectively [47].

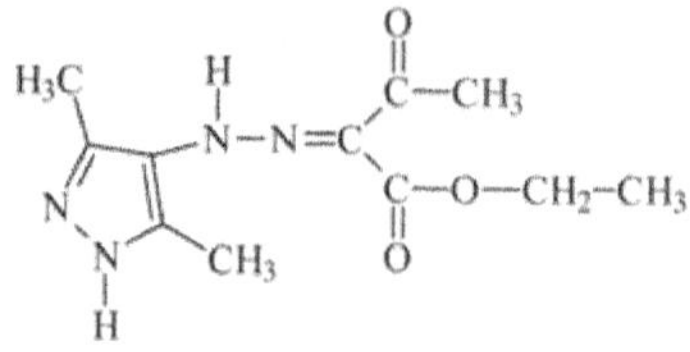

40d 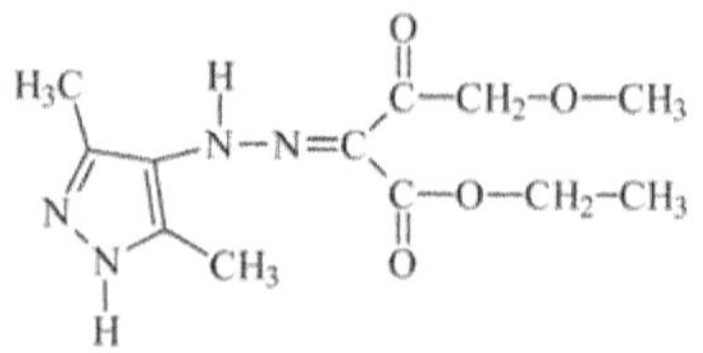   **40e**

Novel coupling products 41 were synthesized and evaluated for their antimycobacterial activity against *M. tuberculosis* H37Rv and *M. avium*. Compound 41b was found to be the most potent derivatives of these series with the MIC value of 6.25 µg/mL against *M. tuberculosis* H37Rv [48].

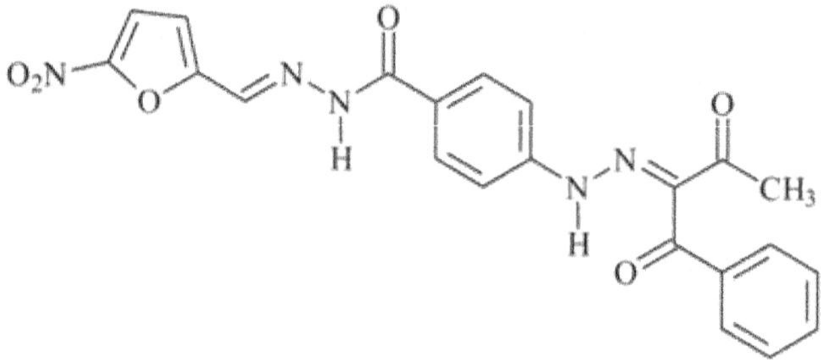

41b

[5-(Pyridine-2-yl)-1,3,4-thidiazole-2-yl]acetic acid (3,4-diaryl-3H-thiazole-2-ylidene)hydrazide derivatives 42 were synthesized and tested for their *in vitro* antimycobacterial activity towards three strains. Compound 42s was exhibited at 20 µg/mL against *M. tuberculosis* 190, isolated from bronchial aspirates [49].

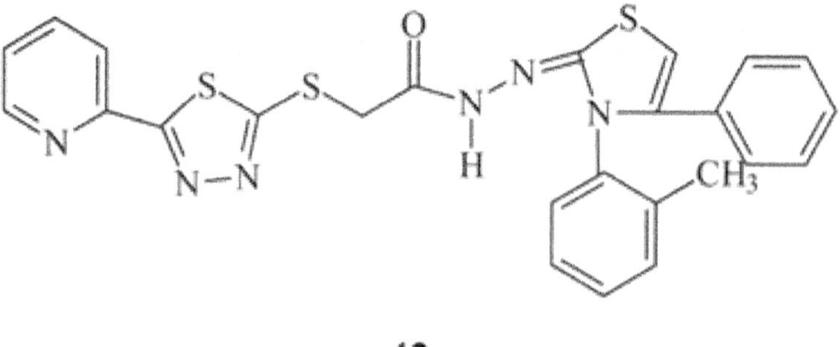

42s

N'-{1-[2-hydroxy-3-(piperazine-1-yl-methyl)phenyl]ethylidene} isonicotinohydrazide 43 was found to be the most active compound with the MIC of 0.56 µM, and it was more potent than INH (MIC of 2.04 µM). After 10 days of treatment, same compound decreased the bacterial load in murine lung tissue by 3.7-log10 as compared to controls, which was equipotent to INH [50].

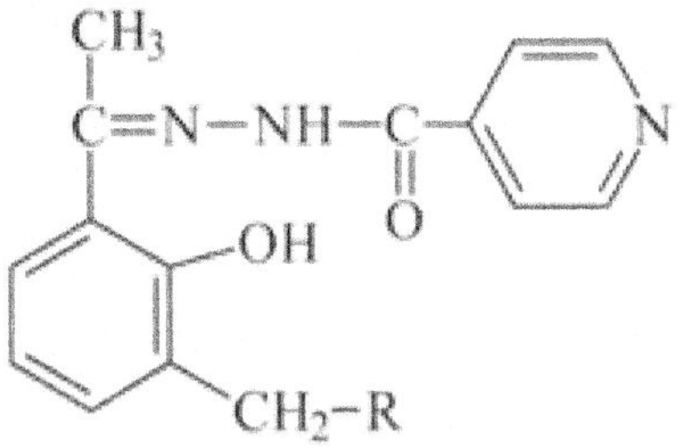

43

As a part of an ongoing search for the new isoniazid-related isonicotinoylhydrazones (ISNEs), 2'-monosubstituted isonicotinohydrazides and cyanoboranes 44-48 were studied and evaluated *in vitro* advanced antimycobacterial screening. Some of tested compounds displayed excellent (MICs ranging from 0.025 to 0.2 µg/mL) to moderate (6.25 to 12.5 µg/mL) MICs against ethambutol- and rifampin-resistant strains [51].

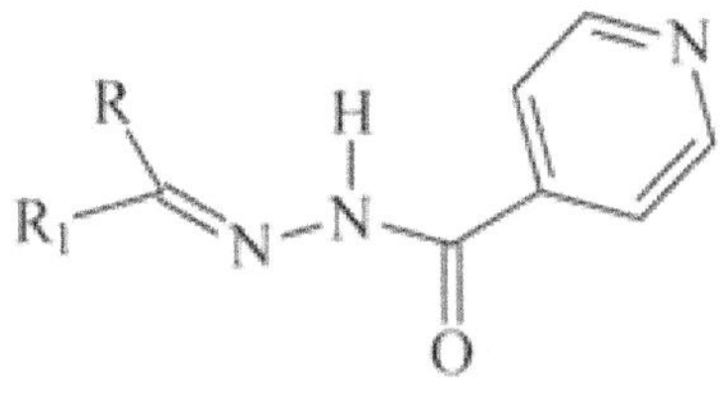

44-48

Novel fluoroquinolones 49 containing a hydrazone structure were synthesized and evaluated *in vivo* against *M. tuberculosis* H37Rv in Swiss albino mice by Shindikar *et al*. Results of the study indicate the potent antitubercular activity of the test compouds [52].

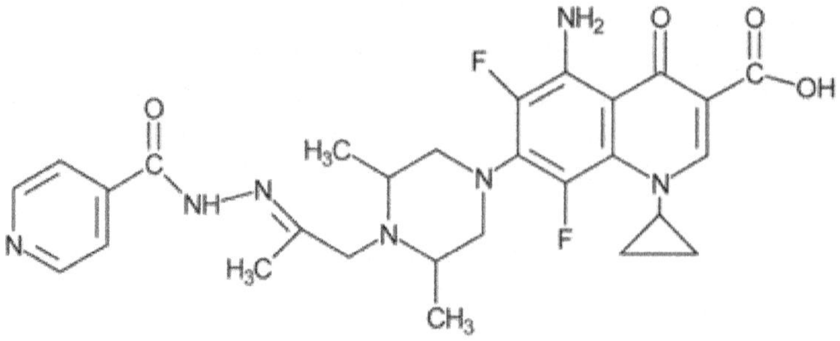

49

N'-Arylidene-*N*-[2-oxo-2-(4-aryl-piperazin-1-yl)ethyl]hydrazide derivatives 50 containing INH hydrazide-hydrazones were synthesized and evaluated antimycobacterial activity against *M. tuberculosis* H37Rv ATCC 27294 and *M. tuberculosis* clinical isolates. Compound 50h showed *in vitro* activity against *M. tuberculosis* H37Rv ATCC 27294 (at 1 μg/mL) and clinical isolates (sensitive and resistant at 0.25-0.5, 2-4 μg/mL, respectively) [53].

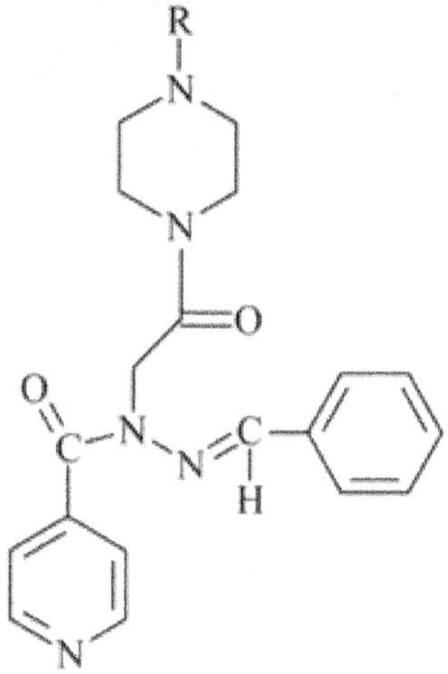

50

Sriram *et al.* synthesized a new series of antimycobacterial agents 51 containing INH hydrazide-hydrazones. 1-(4-Fluorophenyl)-3-(4-{1-[pyridine-4-carbonyl)hydrazono]ethyl}phenyl)thiourea 51d was found to be most potent compound, with MIC of 0.49 μM against *M. tuberculosis* H37Rv and INH-resistant *M. tuberculosis* [54].

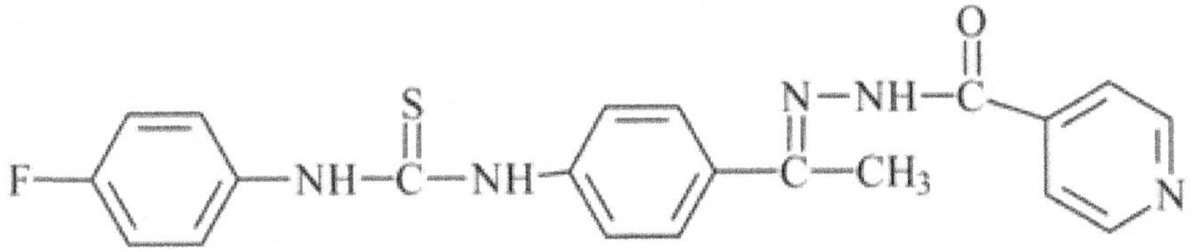

51d

In 2006 Nayyar *et al.* found that the most active compounds of type 52, *N*-(2-fluorophenyl)-*N'*-quinoline-2-yl-methylenehydrazine, *N*-(2-adamantan-1-yl)-*N'*-quinoline-4-yl-methylene)-N'-4-fluoro-phenyl)hydrazine and *N*-(2-cyc-lohexyl)-*N'*-quinoline-4-yl-methylene)-(2-fluorophenyl)hydrazine exhibited 99% inhibition at the lowest tested concentration of 3.125 µg/mL against drug-sensitive *M. tuberculosis* H37 strain [55].

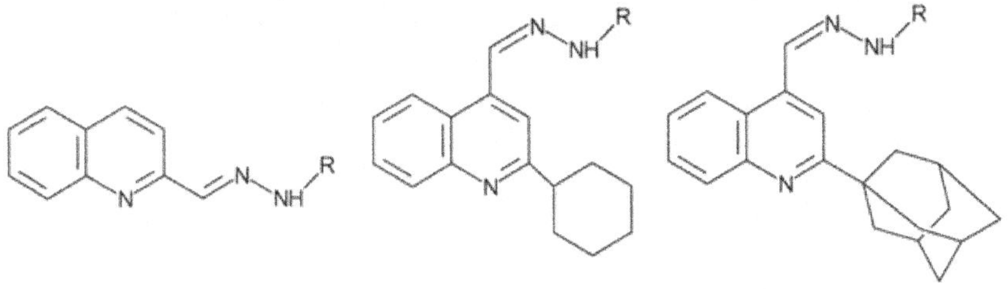

52

Various diclofenac acid hydrazones 53 were synthesized and evaluated for their antimycobacterial activities against *M. tuberculosis in vitro* and *in vivo*. Preliminary results indicated that most of the compounds demonstrated better *in vitro* antimycobacterial activity at concentrations ranging from 0.0383 to 7.53 µM [56].

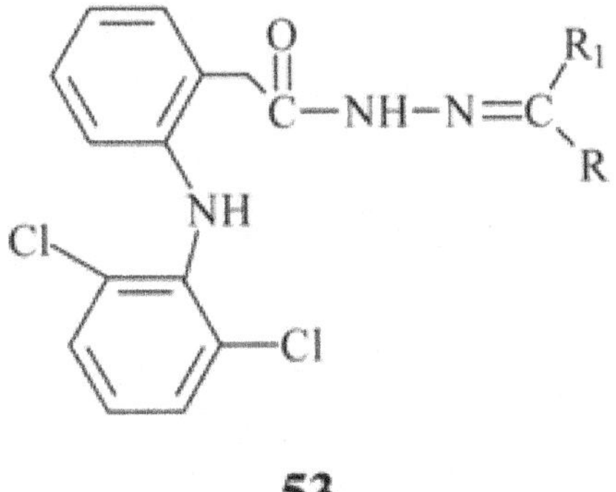

53

Hydrazide-hydrazones 54, based on series of 4-substituted benzoic acid were synthesized and screened for antituberculosis activity. 4-Fluorobenzoic [(5-nitrothiophene-2-yl)methylene]hydrazide 54a showed the highest inhibition (99%) at a constant concentration level (6.25 μg/mL) [4].

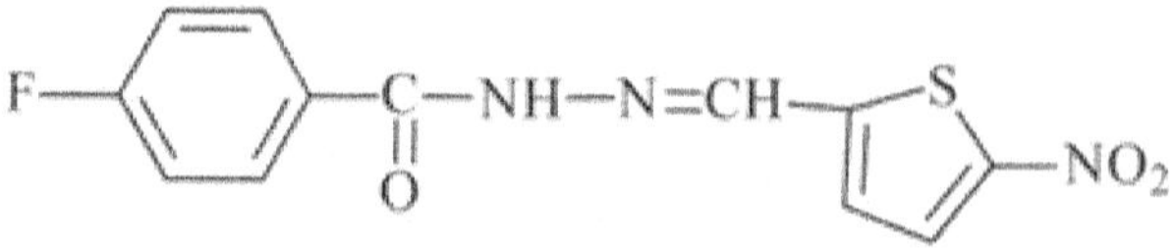

54a

Sixteen new hydrazones 55 containing a pyrrole ring were synthesized as potential tuberculostatics and nine showed 92-100% inhibition of *M. tuberculosis* H37Rv at 6.25 μg/mL. Two leads exhibited low minimum inhibitory concentrations (MIC) and excellent selectivity indexes (SI) [57].

55

N-[3-({2-[(2*E*)-2-Benzylidenehydrazino]-2-oxoethyl}sulfanyl)-5-(-({2[(acetyl)amino]-1,3-thiazol-4-yl}methyl)-4H-1,2,4-triazol-4-yl]-3-nitrobenzamide 56f and *N*-[3-({2-[(2*E*)-2-benzylidene-hydrazino]-2-oxoethyl}sulfanyl)-5-(-({2[(benzoyl)amino]-1,3-thiazol-4-yl}methyl)-4H-1,2,4-triazol-4-yl]-3-nitrobenzamide 56g have been proven to be most active, with MIC values ranging from 0.39 to 0.78μM [58].

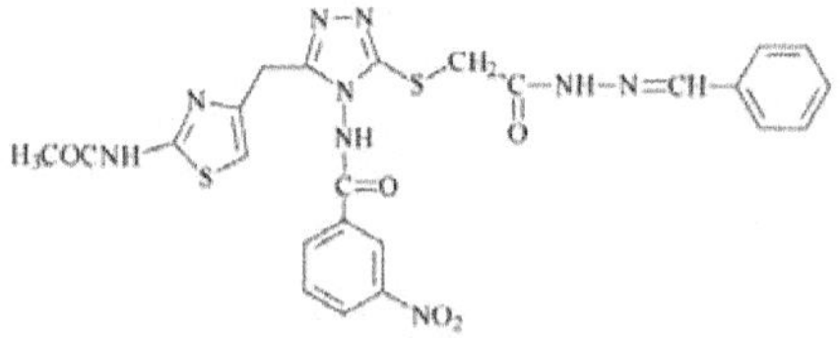

56f 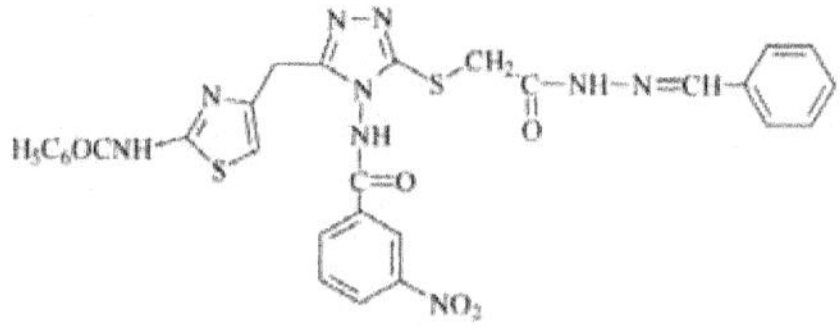   **56g**

A series of hydrazones 57 were synthesized from INH, pyrazineamide, *p*-aminosalicylic acid (PAS), ethambutol and ciprofloxacin. 2-Hydroxy-4-{[(isonicotinoylhydrazono)methyl]amino}benzoic acid **57d** showed the highest inhibition (96%) of *M. tuberculosis* H37Rv at 0.39 μg/mL [59].

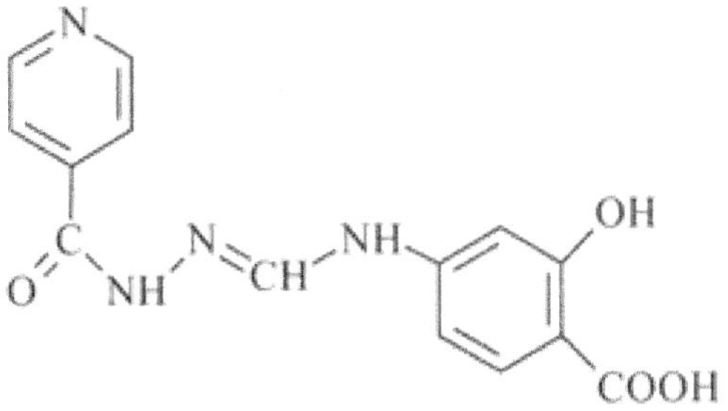

57d

Antitumoral Activity

A variety of antitumoral drugs are currently in clinical use. The search for antitumoral drugs led to the discovery of several hydrazones having antitumoral activity. Some of diphenolic hydrazones showed maximum uterotrophic inhibition of 70%, whereas compound 58 exhibited cytotoxicity in the range of 50-70% against MCF-7 and ZR-75-1 human malignant breast cell lines [60].

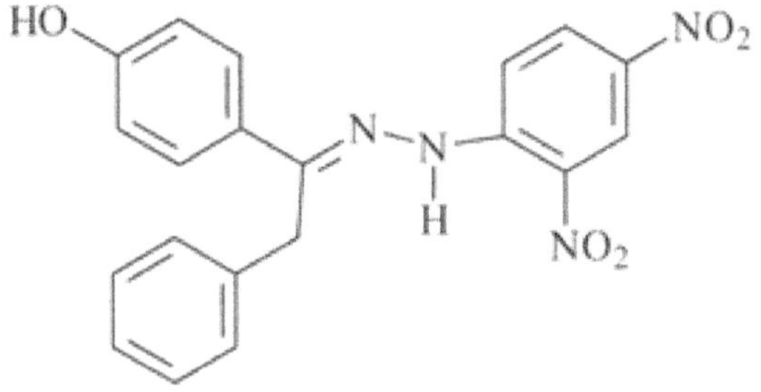

58

N'-(1-{1-[4-nitrophenyl-3-phenyl-1H-pyrazole-4-yl}methylene)-2-chlorobenzohydrazide 59 was found to be the most active, with full panel median growth inhibition, total growth concentration and median lethal concentration mean graph mid-point of 3.79, 12.5 and 51.5 µM, respectively [61].

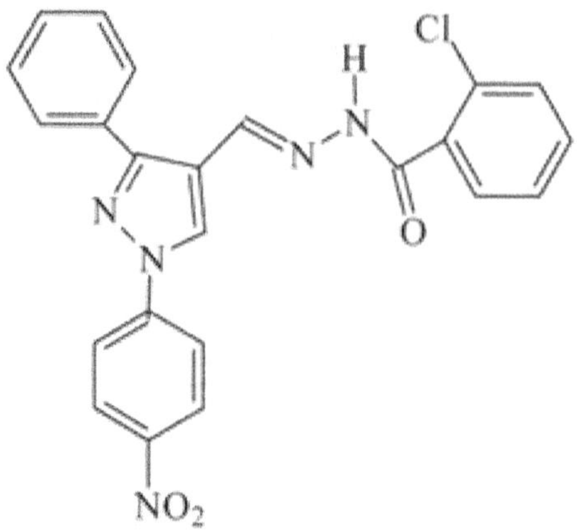

59

Some novel 2,6-dimethyl-N'-substituted-phenylmethyleneimidazo[2,1-*b*][1,3,4]thiadiazole-5-carbohydrazides 60 were synthesized. (2,6-Dimethyl-N'-(2-hydroxyphenylmethylidene)imidazo[2,1-*b*][1,3,4]thiadiazole-5-carbohydrazide 60c showed the most favorable cytotoxicity. In the *in vitro* screening of National Cancer Institute's 60 human tumor cell lines, this compound demonstrated the most marked effects on the ovarian cancer cell line (OVCAR $\log_{10} GI_{50}$ value −5.51) [62].

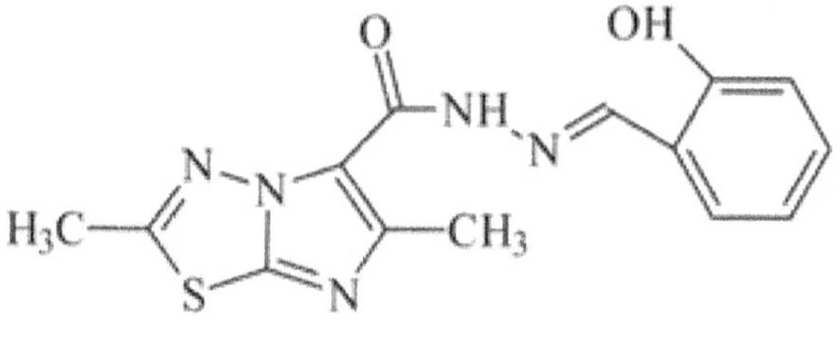

60c

3-[[(6-Chloro-3-phenyl-4(3*H*)-quinazolinone-2-yl)mercaptoacetyl]hydrazono]-5-fluoro-1*H*-2-indolinone 61o showed the most favourable cytotoxicity against the renal cancer cell line UO-31 ($\log_{10} GI_{50}$ value −6.68) [63].

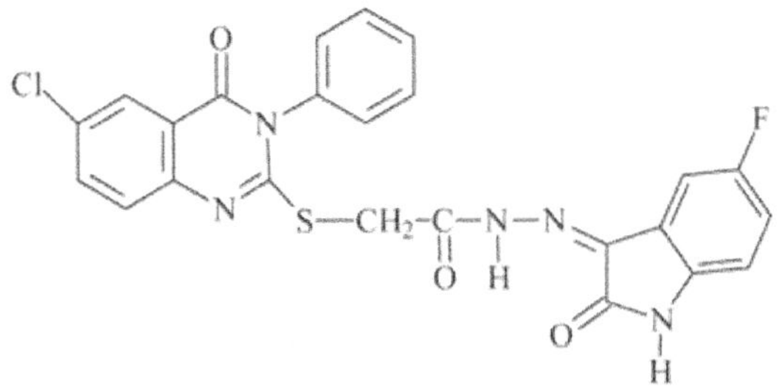

61o

Some recently synthesized compounds were found to possess antiproliferative properties. The most active compound of the series was the 3- and 5-methylthiophene-2-carboxaldehyde α-(N)-heterocyclichydrazones derivatives 62, which exhibited tumor growth inhibition activity against all cell lines at GI_{50} values between 1.63 and 26.5 µM [64].

62

5-Chloro-3-methylindole-2-carboxylic acid(4-nitrobenzylidene)hydrazide 63a was found to arrest T47D cells in G_2/Mphase of the cell cycle and to induce apoptosis as measured by the flow cytometry analysis. A 20-fold increase of apoptotic activity was achieved from the screening hit to 5-methyl-3-phenylindole-2-carboxylic acid(4-methylbenzylidene) hydrazide 64a and 5-chloro-3-phenylindole-2-carboxylic acid(4-nitrobenzylidene)hydrazide 64b, with EC_{50} values of 0.1 µM in the caspase activation assay in T47D breast cancer cells. Compound 64b also was found to be highly active in a standard growth inhibition assay with a GI_{50} value of 0.9 µM in T47D cells. Compound 63a and its analogs were found to inhibit tubulin polymerization, which is the most probable primary mechanism of the action of these compounds [65].

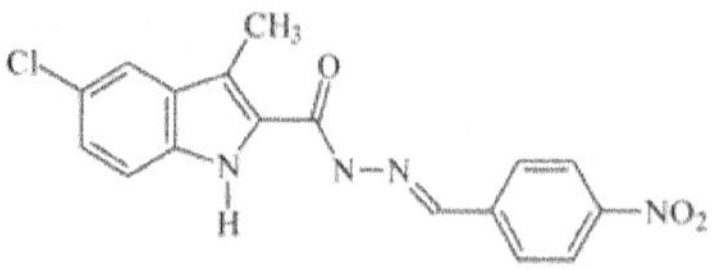

63a

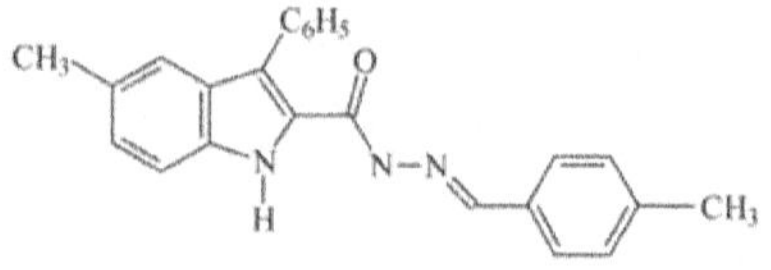

64a

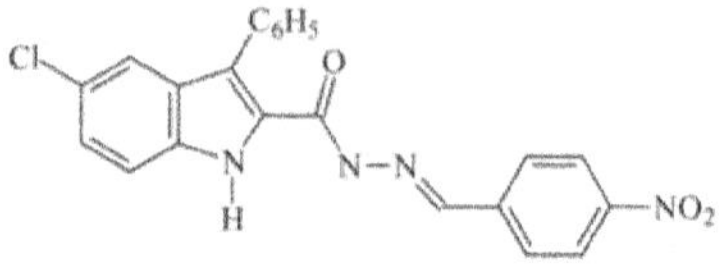

64b

Demirbaş et al. synthesized the new hydrazide-hydrazones containing 65 5-oxo-[1,2,4]triazole ring. Some of these compounds had inhibitory effect on mycelial growth whereas Compounds 65c and 65f were found to possess antitumor activity in breast cancer. [66].

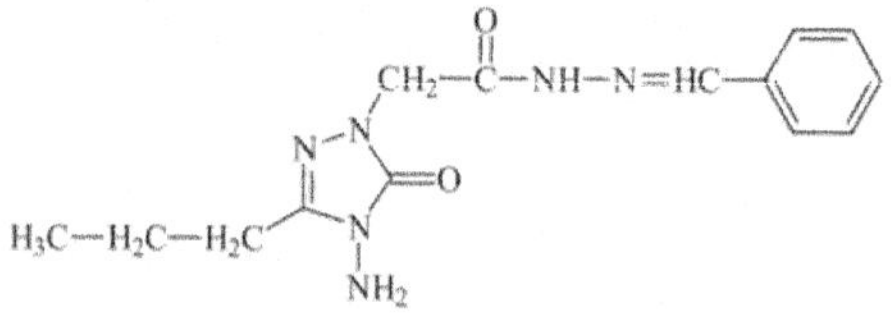

65c

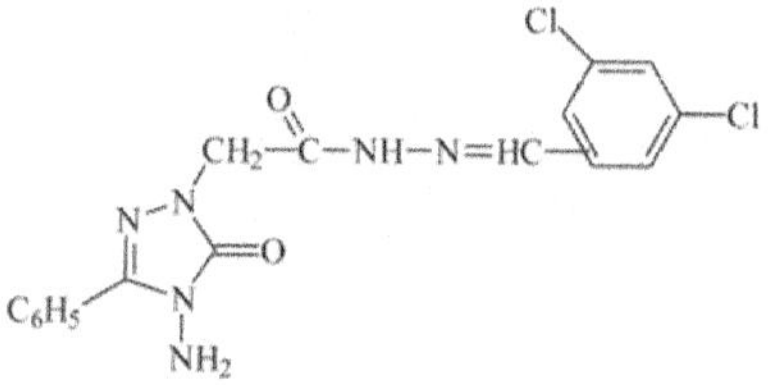

65f

Hydrazinopyrimidine derivatives 66 were evaluated for their *in vitro* antitumoral activity in nine different types of human cancers. Some of the newly prepared compounds demonstrated inhibitory effects on the growth of a wide range of cancer cell lines generally at 10^{-5} M to 10^{-7} M concentrations [67].

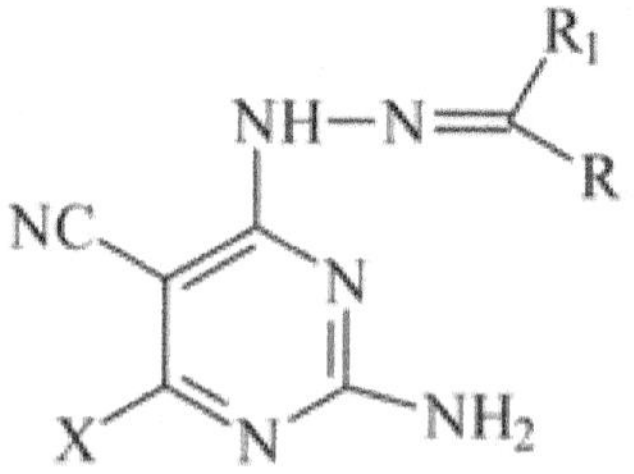

66

Several benzo[*d*]isothiazole hydrazones have been tested for antitumoral activity. Compound 67h, bearing a hydroxy group at *o*-position of the benzylidene moiety, was the most potent, with the IC50 against the various cell lines ranging between 0.5 and 8.0 µM, thus acting equally potent as 6-mercaptopurine against the haematological tumors [68].

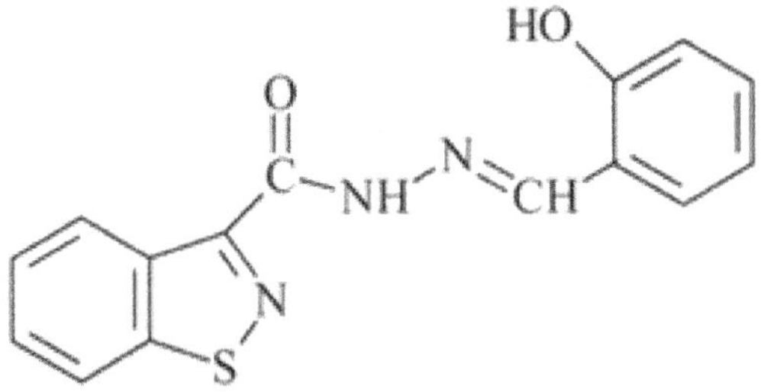

67h

N'-Substituted-benzylidene-3,4,5-trimethoxybenzohydrazide 68 were synthesized and evaluated for their antitumoral activity against some cancer cells. Many hydrazone compounds containing the active moiety (-CONH-N=CH-) showed good antitumor activity. Compounds 68a-c and 68f were highly effective against PC3 cells and 68a, 68c and 68f showed moderate activities against Bcap37 and BGC823 cells [69].

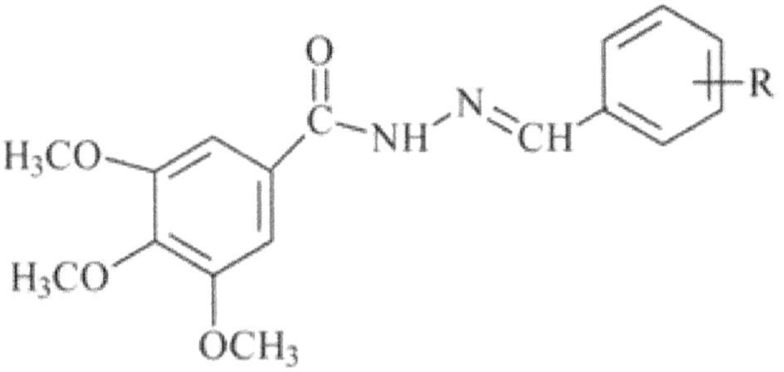

68

6-Amino-4-aryl-2-oxo-1-(1-pyrid-3-yl- or 4-yl-ethylidene-amino)-1,2-dihydropyridine-3,5-dicarbo-nitrile series 69exhibited a high percentage of tumor growth inhibition at concentrations of 10^{-5} to 10^{-7} M in all cancer cell lines [70].

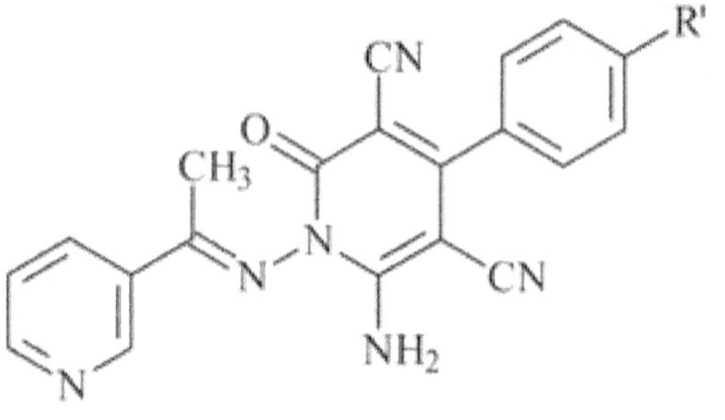

69

Duarte *et al.* described *N'*-(3,5-Di-tert-butyl-4-hydroxybenzylidene)-6-nitro-1,3-benzodioxole-5-carbohydrazine 12c as a novel antiproliferative compound. They observed that 12c was able to inhibit T-cell proliferation (66 % at 10μM) [27].

A series of arylidenehydrazides 70 were synthesized and evaluated in the National Cancer Institue's against the full panel of 60 human tumour cell lines. Compound 70c demonstrated the most effect on prostate cancer cell line [71].

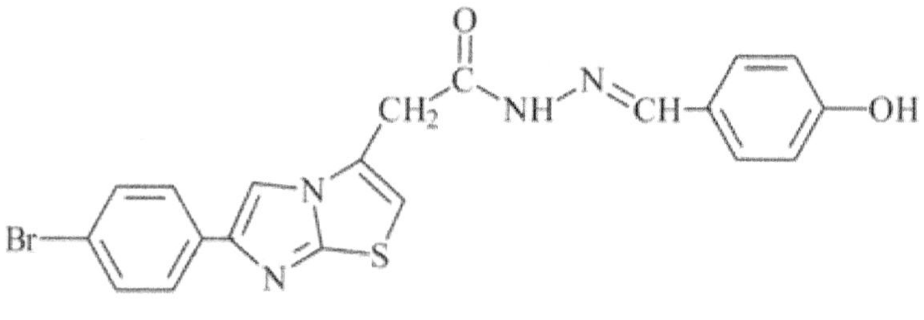

70c

Vasodilator Activity

Conventional therapy to treat hypertension often involves arterial vasodilation. It is important to find new vasodilators with a potential for clinical use.

A new bioactive compound of the *N*-acylhydrazone class, 3,4-methylenedioxybenzoyl-2-thienyl hydrazone 71, named LASSBio-294, was shown to have inotropic and vasodilatory effects. New derivatives of LASSBio-294 were designed and tested on the contractile responses of rat vascular smooth muscle *in vitro*. Phenylephrine-induced contractions of aorta was inhibited by the derivatives *N*-methyl-2-thienylidene-3,4-methylenedioxy-benzoyl hydrazine, named LASSBio-785 and *N*-allyl-2-thienylidene-3,4-methylenedioxy-benzoyl hydrazine, named LASSBio-788. The concentrations necessary to cause 50% reduction in maximum contractions (IC50) were 10.2 +/- 0.5 and 67.9 +/- 6.5 µM. Vasodilation induced by both derivatives is likely to be mediated by a direct effect on smooth muscle because it was not dependent on the integrity of vascular endothelium. LASSBio-785 was seven times more potent than the reference compound LASSBio-294 (IC50 = 74 µM) in producing an endothelium-independent vasodilator effect [72].

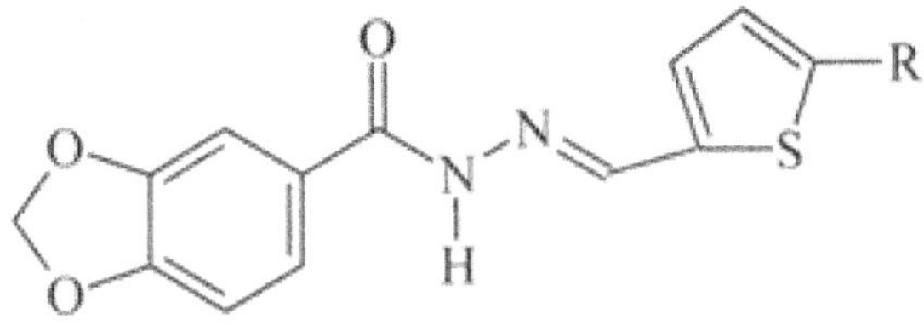

71

Antiviral Activity

HIV infection and AIDS represent one of the first diseases for which the discovery of drugs was performed entirely via a rational drug design approach. Current treatment regimens are based on the use of two or more drugs that belong to group of inhibitors termed as highly active antiretroviral therapy (HAART). Some thiourea compounds were reported to be non-nucleoside inhibitors (NNIs) of the reverse transcriptase (RT) enzyme of the human immunodeficiency virus (HIV). Such hydrazones have been reported to be the potent inhibitors of ribonucleotide reductase activity.

N-Arylaminoacetylhydrazones and *O*-acetylated derivatives of sugar *N*-arylaminoacetyl hydrazones were synthesized and evaluated for their antiviral activity against *Herpes simplex* virus-1 (HSV-1) and hepatitis-A virus (HAV). Some compounds revealed the highest antiviral activity against HAV-27 and HSV-1 [73].

3.0. Schistosomiasis

Schistosomiasis or bilharzia is a parasitic disease caused by several species of flatform. Currently, schistosomiasis affects roughly 200 million people in tropical countries, and in certain African communities the process of overcoming schistosomiasis is an important rite of passage. Schistosomiasis causes debilitating nutritional, hematologic and cognitive deficits, with substantial morbidity and mortality in populations. There are five species of flatworms that cause schistosomiasis. *Schistosoma mansoni, S. intercalatum, S. haematobium, S. japonicum* and *S. mekongi. Schistosoma mansoni* and *S.*

intercalatum, *S. japonicum* and *S. mekongi* cause intestinal and Asian intestinal schistosomiasis, respectively. *S. haematobium*resides in the venous plexus, which causes urinary schistosomiasis [74].

9-Acridanone hydrazones have been developed by Hoffmann-La Roche (Basel-Switzerland). One of these compounds (RO 15-5458/000) was administered at two dose levels 25 mg and 15 mg/kg body-weight to *S. mansoni* infected vervet-monkeys [75]. In addition, same compounds were found to be effective against *S. mansoni* in mice, killing almost all the skin schistosomules, when administered at the dose of 100mg/kg. In experiments carried out with Cebus monkeys, the compound RO 15-5458 / 000 was shown to be fully effective at 25 mg/kg [76].

REVIEW ARTICLES RELATED WITH HYDRAZONES

There critical reviews have been published recently, and they may give an outlook on the latest research developments on antimycobacterial substances [77,78,79].

REFERENCES AND NOTES

1. Rollas, S.; Gülerman, N.; Erdeniz, H. Synthesis and antimicrobial activity of some new hydrazones of 4-fluorobenzoic acid hydrazide and 3-acetyl-2,5-disubstituted-1,3,4-oxadiazolines. *Farmaco* 2002, *57*, 171–174.
2. Küçükgüzel, Ş.G.; Rollas, S.; Küçükgüzel, İ; Kiraz, M. Synthesis and Antimycobacterial activity of some coupling products from 4-aminobenzoic acid hydrazones. *Eur. J. Med. Chem.* 1999, *34*, 1093–1100.
3. Küçükgüzel, Ş.G.; Oruç, E.E.; Rollas, S.; Şahin, F.; Özbek, A. Synthesis, Characterization and biological activity of novel 4-thiazolidinones, 1,3,4-oxadiazoles and some related compounds. *Eur. J. Med. Chem.* 2002, *37*, 197–206.
4. Kaymakçıoğlu, K.B.; Oruç, E.E.; Unsalan, S.; Kandemirli, F.; Shvets, N.; Rollas, S.; Anatholy, D. Synthesis and characterization of novel hydrazide-hydrazones and the study of their structure-antituberculosis activity. *Eur. J. Med. Chem* 2006, *41*, 1253–1261.
5. Sah, P.P.T.; Peoples, S.A. Isonicotinyl hydrazones as antitubercular agents and derivatives for idendification of aldehydes and ketones. *J. Am. Pharm. Assoc.* 1954, *43*, 513–524. [CrossRef]
6. Bavin, E.M.; Drain, D.J.; Seiler, M.; Seymour, D.E. Some further studies on tuberculostatic compounds. *J. Pharm. Pharmacol.* 1954, *4*, 844–855.

7. Buu-Hoi, P.H.; Xuong, D.; Nam, H.; Binon, F.; Royer, R. Tuberculostatic hydrazides and their derivatives . *J. Chem. Soc.* 1953, 1358–1364.
8. Amâl, H.; Ergenç, N. Some isonicoticoyl-hydrazones. .*Ü. Fen Fakültesi Mecmuası.* 1957, *22*, 390–392.
9. Singh, V.; Srivastava, V.K.; Palit, G.; Shanker, K. Coumarin congeners as antidepressants. *Arzneim-Forsch. Drug. Res.*1992, *42*, 993–996.
10. Ergenç, N.; Günay, N.S. Synthesis and antidepressant evaluation of new 3-phenyl-5-sulfonamidoindole derivatives.*Eur. J. Med. Chem.* 1998, *33*, 143–148.
11. Durgun, B.B.; Çapan, G.; Ergenç, N.; Rollas, S. Synthesis, characterisation and biological evaluation of new benylidenebenzohydrazides and 2,5-disubstituted-2,3-dihydro-1,3,4-oxadiazoles. *Pharmazie* 1993, *48*, 942–943.
12. Doğan, H.N.; Duran, A.; Rollas, S.; Şener, G.; Armutak, Y.; Keyer-Uysal, M. Synthesis and structure elucidation of some new hydrazones and oxadiazolines : anticonlsant activitites of 2-(3-acetyloxy-2-naphtyl)-4-acetyl-5-substituted-1,3,4-oxadiazolines. *Med. Sci. Res.* 1998, *26*, 755–758.
13. Kalsi, R.; Shrimali, M.; Bhalla, T.N.; Barthwal, J.P. Synthesis and anti-inflammatory activity of indolyl azetidinones.*Indian J. Pharm. Sci.* 2006, *41*, 353–359.
14. Küçükgüzel, Ş.G.; Kocatepe, A.; De Clercq, E.; Şahin, F.; Güllüce, M. Synthesis and biological activity of 4-thiazolidinones, thiosemicarbazides derived from diflunisal hydrazide. *Eur. J. Med. Chem.* 2006, *41*, 353–359.
15. Kömürcü, Ş.G.; Rollas, S.; Ülgen, M.; Gorrod, J.W.; Çevikbaş, A. Evaluation of Some Arylhydrazones of p-Aminobenzoic acid hydrazide as Antimicrobial Agents and Their *in-vitro* Hepatic Microsomal Metabolism. *Boll. Chim. Farm.* 1995, *134*, 375–379.
16. Ülgen, M.; Durgun, B.B.; Rollas, S.; Gorrod, J.W. The in vitro hepatic microsomal metabolism of benzoic acid benzylidenehydrazide. *Drug Metab. Interact.* 1997, *13*, 285–294.
17. Gülerman, N.N.; Oruç, E.E.; Kartal, F.; Rollas, S. In vivo metabolism of 4-fluorobenzoic acid [(5-nitro-2-furanyl)methylene] hydrazide in rats. *Eur. J .Drug Metab. Pharmacokinet.* 2000, *25*, 103–108.
18. Küçükgüzel, Ş.G.; Küçükgüzel, İ.; Ülgen, M. Metabolic and Chemical Studies on N-(4-chlorobenzyl)-N'-benzoylhydrazine. *Farmaco* 2000, *55*, 624–630.
19. Dimmock, J.R.; Vashishtha, S.C.; Stables, J.P. Anticonvulsant properties

of various acetylhydrazones, oxamoylhydrazones and semicarbazones derived from aromatic and unsaturated carbonyl compounds. *Eur. J. Med. Chem.* 2000, *35*, 241–248.

20. Çakır, B.; Dağ, Ö.; Yıldırım, E.; Erol, K.; Şahin, M.F. Synthesis and anticonvulsant activity of some hydrazones of 2-[(3H)-oxobenzoxazolin-3-yl-aceto]hydrazide. *J. Fac. Pharm. Gazi.* 2001, *18*, 99–106.

21. Ragavendran, J.; Sriram, D.; Patel, S.; Reddy, I.; Bharathwajan, N.; Stables, J.; Yogeeswari, P. Design and synthesis of anticonvulsants from a combined phthalimide-GABA-anilide and hydrazone pharmacophore. *Eur. J. Med. Chem.* 2007,*42*, 146–151.

22. Todeschini, A.R.; Miranda, A.L.; Silva, C.M.; Parrini, S.C.; Barreiro, E.J. Synthesis and evaluation of analgesic, antiinflammatory and antiplatelet properties of new 2-pyridylarylhydrazone derivatives. *Eur. J. Med. Chem.* 1998, *33*, 189–199.

23. Lima, P.C.; Lima, L.M.; Silva, K.C.; Leda, P.H.; Miranda, A.L.P.; Fraga, C.A.M; Barreiro, E.J. Synthesis and analgesic activity of novel N-acylhydrazones and isosters, derived from natural safrole. *Eur. J. Med. Chem.* 2000, *35*, 187–203.

24. Fraga, A.G.M.; Rodrigues, C.R.; Miranda, A.L.P.; Barreiro, E.J.; Fraga, C.A.M. Synthesis and pharmacological evaluation of novel heterocyclic acylhydrazone derivatives, designed as PAF antagonists. *Eur. J. Pharm. Sci.* 2000, *11*, 285–290.

25. Silva, G.A.; Costa, L.M.M.; Brito, F.C.F.; Miranda, A.L.P.; Barreiro, E.J.; Fraga, C.A.M. New class of potent antinociceptive and antiplatelet 10H-phenothiazine-1-acylhydrazone derivatives. *Bioorg. Med. Chem* 2004, *12*, 3149–3158.

26. Salgın-Gökşen, U.; Gökhan-Kelekçi, N.; Göktaş, Ö.; Köysal, Y.; Kılıç, E.; Işık, Ş.; Aktay, G.; Özalp, M. 1-Acylthiosemicarbazides, 1,2,4-triazole-5(4H)-thiones, 1,3,4-thiadiazoles and hydrazones containing 5-methyl-2-benzoxazolinones: Synthesis, analgesic-anti-inflammatory and antimicrobial activities. *Bioorg. Med. Chem.* 2007, *15*, 5738–5751.

27. Duarte, C.D.; Tributino, J.L.M.; Lacerda, D.I.; Martins, M.V.; Alexandre-Moreira, M.S.; Dutra, F.; Bechara, E.J.H.; De-Paula, F.S.; Goulart, M.O.F.; Ferreira, J.; Calixto, J.B.; Nunes, M.P.; Bertho, A.L.; Miranda, A.L.P.; Barreiro, E.J.; Fraga, C.A.M. Synthesis, pharmacological evaluation and electrochemical studies of novel 6-nitro-3,4-methylenedioxyphenyl-*N*-acylhydrazone derivatives : Discovery of LASSBio-881, a new ligand of cannabinoid receptors. *Bioorg. Med. Chem.* 2007, *15*, 2421–2433.

28. Walcourt, A.; Loyevsky, M.; Lovejoy, D.B.; Gordeuk, V.R.; Richardson,

D.R. Novel aroylhydrazone and thiosemicarbazone iron chelators with anti-malarial activity against chloroquine-resistant and -sensitive parasites. *Int. J. Biochem. Cell Biol.* 2004, *36*, 401–407.

29. Gemma, S.; Kukreja, G.; Fattorusso, C.; Persico, M.; Romano, M.; Altarelli, M.; Savini, L.; Campiani, G.; Fattorusso, E.; Basilico, N. Synthesis of N1-arylidene-N2-quinolyl- and N2-acrydinylhydrazones as potent antimalarial agents active against CQ-resistant P. falciparum strains. *Bioorg. Med. Chem. Lett.* 2006, *16*, 5384–5388.

30. Bernardino, A.; Gomes, A.; Charret, K.; Freitas, A.; Machado, G.; Canto-Cavalheiro, M.; Leon, L.; Amaral, V. Synthesis and leishmanicidal activities of 1-(4-X-phenyl)-N'-[(4-Y-phenyl)methylene]-1H-pyrazole-4-carbohydrazides. *Eur. J. Med. Chem.* 2006, *41*, 80–87.

31. Küçükgüzel, Ş.G.; Rollas, S.; Erdeniz, H.; Kiraz, M. Synthesis, Characterization and Antimicrobial Evaluation of Ethyl 2-Arylhydrazono-3-oxobutyrates. *Eur. J. Med. Chem.* 1999, *34*, 153–160.

32. Tavares, L.C.; Chiste, J.J.; Santos, M.G.B.; Penna, T.C.V. Synthesis and biological activity of nifuroxazide and analogs. II . *Boll. Chim. Farm.* 1999, *138*, 432–436.

33. Ulusoy, N.; Çapan, G.; Ötük, G.; Kiraz, M. Synthesis and antimicrobial activity of new 6- phenylimidazo[2,1-b]thiazole derivatives. *Boll. Chim. Farm.* 2000, *139*, 167–172.

34. Turan-Zitouni, G.; Blache, Y.; Güven, K. Synthesis and antimicrobial activity of some imidazo-[1,2-a]pyridine-2-carboxylic acid arylidenehydrazide derivatives. *Boll. Chim. Farm.* 2001, *140*, 397–400.

35. Vicini, P.; Zani, F.; Cozzini, P.; Doytchinova, I. Hydrazones of 1,2-benzisothiazole hydrazides: synthesis, antimicrobial activity and QSAR investigations. *Eur. J. Med. Chem.* 2002, *37*, 553–564.

36. Küçükgüzel, Ş.G.; Mazi, A.; Şahin, F.; Öztürk, S.; Stables, J. P. Synthesis and biological activities of diflunisal hydrazide-hydrazones. *Eur. J. Med. Chem.* 2003, *38*, 1005–1009.

37. Loncle, C.; Brunel, J.; Vidal, N.; Dherbomez, M.; Letourneux, Y. Synthesis and antifungal activity of cholesterol-hydrazone derivatives. *Eur. J. Med. Chem.* 2004, *39*, 1067–1071.

38. Masunari, A.; Tavares, L.C. A new class of nifuroxazide analogues : Synthesis of 5-nitrophene derivatives with antimicrobial activity against multidrug-resistant *Staphylococcus aureus*. *Bioorg. Med. Chem.* 2007, *15*, 4229–4236.

39. www.TAACF.org

40. Bukowski, L.; Janowiec, M. 1-Methyl-1H-2-imidazo[4,5-b] pyridinecarboxylic acid and some of derivatives with suspected antituberculotic activity. *Pharmazie* 1996, *51*, 27–30.
41. Cocco, M.T.; Congiu, C.; Onnis, V.; Pusceddo, M.C.; Schivo, M.L.; De Logu, A. Synthesis and antimycobacterial activity of some isonicotinoylhydrazones. *Eur. J. Med. Chem.* 1999, *34*, 1071–1076.
42. Bukowski, L.; Janowiec, M.; Zwolska-Kwiek, Z.; Andrzejczyk, Z. Synthesis and some reactions of 2- acetylimidazo[4,5-b]pyridine. Antituberculotic activity of the obtained compounds. *Pharmazie* 1999, *54*, 651–654.
43. MamoloM., G.; Falagiani, V.; Zampieri, D.; Vio, L.; Banfi, E. Synthesis and antimycobacterial activity of [5-(pyridin-2-yl)-1,3,4-thiadiazole-2-ylthio]acetic acid arylidene-hydrazide derivatives. *Farmaco* 2001, *56*, 587–592.
44. Ulusoy, N.; Gürsoy, A.; Ötük, G.; Kiraz, M. Synthesis and antimicrobial activity of some 1,2,4-triazole-3-mercaptoacetic acid derivatives. *Farmaco* 2001, *56*, 947–952.
45. Savini, L.; Chiasserini, L.; Gaeta, A.; Pellerano, C. Synthesis and Anti-tubercular Evaluation of Quinolylhydrazones.*Bioorg. Med. Chem.* 2002, *10*, 2193–2198.
46. Rando, D.; Sato, D.N.; Siqueira, L.; Malvezzi, A.; Leite, C.Q.F.; Amaral, A.T.; Ferreira, E.I.; Tavares, L.C. Potential tuberculostatic agents. Topliss application on benzoic acid [(5-Nitro-thiophene-2-yl)methylene] hydrazide series.*Bioorg. Med. Chem.* 2002, *10*, 557–560.
47. Kaymakçıoğlu, K.B.; Rollas, S. Synthesis, characterization and evaluation of antituberculosis activity of some hydrazones. *Farmaco* 2002, *57*, 595–599.
48. Küçükgüzel, Ş.G.; Rollas, S. Synthesis, Characterization of Novel Coupling Products and 4-Arylhydrazono-2-pyrazoline-5-ones as Potential Antimycobacterial Agents. *Farmaco* 2002, *57*, 583–587.
49. Mamolo, M.G.; Falagiani, V.; Zampieri, D.; Vio, L.; Banfi, E.; Scialino, G. Synthesis and antimycobacterial activity of (3,4-diaryl-3H-thiazole-2-ylidene)hydrazide derivatives. *Farmaco* 2003, *58*, 631–637.
50. Sriram, D.; Yogeeswari, P.; Madhu, K. Synthesis and in vitro and in vivo antimycobacterial activity of isonicotinoyl hydrazones. *Bioorg. Med. Chem. Lett.* 2005, *15*, 4502–4505.
51. Maccari, R.; Ottana, R.; Vigorita, M.G. In vitro advanced antimycobacterial screening of isoniazid-related hydrazones, hydrazides and cyanoboranes:

Part 14. *Bioorg. Med. Chem. Lett.* 2005, *15*, 2509–2513.
52. Shindikar, A.V.; Viswanathan, C.L. Novel fluoroquinolones : design, synthesis, and in vivo activity in mice against*Mycobacterium tuberculosis* H37 Rv. *Bioorg. Med. ChemLett.* 2005, *15*, 1803–1806.
53. Sinha, N.; Jain, S.; Tilekar, A.; Upadhayaya, R.S.; Kishore, N.; Jana, G.H.; Arora, S.K. Synthesis of isonicotinic acid N'-Arylidene-N-[2-oxo-2-(4-aryl-piperazin-1-yl)ethyl]hydrazides as antituberculosis agents. *Bioorg. Med. Chem Lett.* 2005,*15*, 1573–1576.
54. Sriram, D.; Yogeeswari, P.; Madhu, K. Synthesis and in vitro antitubercular activity of some 1-[(4-sub)phenyl]-3-(4-{1-[(pyridine-4-carbonyl) hydrazono]ethyl} phenyl)thiourea. *Bioorg. Med. Chem.* 2006, *14*, 876–878.
55. Nayyar, A.; Malde, A.; Coutinho, E.; Jain, R. Synthesis, antituberculosis activity, and 3D-QSAR study of ring-substituted-2/4-quinolinecarbaldehyde derivatives. *Bioorg. Med. Chem.* 2006, *14*, 7302–7310.
56. Sriram, D.; Yogeeswari, P.; Devakaram, R.V. Synthesis , in vitro and in vivo antimycobacterial activities of diclofenac acid hydrazones and amides. *Bioorg. Med. Chem.* 2006, *14*, 3113–3118.
57. Bijev, A. New heterocyclic hydrazones in the search for antitubercular agents: Synthesis and in vitro evaluations.*Lett.Drug Des. Discov.* 2006, *3*, 506–512.
58. Shiradkar, M.R.; Murahari, K.K.; Gangadasu, H.R.; Suresh, T.; Kalyan, C.C.; Panchal, D.; Kaur, R.; Burange, P.; Ghogare, J.; Mokale, V.; Raut, M. Synthesis of new S-derivatives of clubbed triazolyl thiazole as anti-*Mycobacterium tuberculosis* agents. *Bioorg. Med. Chem.* 2007, *15*, 3997–4008.
59. Imramovský; Polanc, S.; Vinšová, J.; Kočevar, M.; Jampílek, J.; Rečková, Z.; Kaustová, J. A new modification of anti-tubercular active molecules. *Bioorg. Med. Chem.* 2007, *15*, 2551–2559.
60. Pandey, J.; Pal, R.; Dwivedi, A.; Hajela, K. Synthesis of some new diaryl and triaryl hydrazone derivatives as possible estrogen receptor modulators. *Arzneimittelforschung.* 2002, *52*, 39–44. [PubMed]
61. Abadi, A.H.; Eissa, A.A.H.; Hassan, G.S. Synthesis of novel 1,3,4-trisubstituted pyrazole derivatives and their evaluation as antitumor and antiangiogenic agents. *Chem. Pharm. Bull.* 2003, *51*, 838–844.
62. Terzioğlu, N.; Gürsoy, A. Synthesis and anticancer evaluation of some new hydrazone derivatives of 2,6-dimethylimidazo[2,1-b]-[1,3,4]

thiadiazole-5-carbohydrazide. *Eur. J. Med. Chem.* 2003, *38*, 781–786.
63. Gürsoy, A.; Karali, N. Synthesis and primary cytotoxicity evaluation of 3-[[(3-phenyl-4(3H)-quinazolinone-2-yl)mercaptoacetyl]hydrazono]-1H-2-indolinones. *Eur. J. Med. Chem.* 2003, *38*, 633–643.
64. Savini, L.; Chiasserini, L.; Travagli, V.; Pellerano, C.; Novellino, E.; Cosentino, S.; Pisano, M.B. New α-heterocyclichydrazones : evaluation of anticancer, anti-HIV and antimicrobial activity. *Eur.J.Med. Chem.* 2004, *39*, 113–122.
65. Zhang, H.; Drewe, J.; Tseng, B.; Kasibhatla, S.; Cai, S.X. Discovery and SAR of indole-2-carboxylic acid benzylidenehydrazides as a new series of potent apoptosis inducers using a cell-based HTS assay. *Bioorg. Med. Chem.*2004, *12*, 3649–3655.
66. Demirbas, N.; Karaoglu, S.; Demirbas, A.; Sancak, K. Synthesis and antimicrobial activities of some new 1-(5-phenylamino-[1,3,4]thiadiazol-2-yl)methyl-5-oxo-[1,2,4]triazole and 1-(4-phenyl-5-thioxo-[1,2,4]triazol-3-yl)methyl-5-oxo-[1,2,4]triazole derivatives. *Eur. J. Med. Chem.* 2004, *39*, 793–804.
67. Cocco, M. T.; Congiu, C.; Lilliu, V.; Onnis, V. Synthesis and in vitro antitumoral activity of new hydrazinopyrimidine-5-carbonitrile derivatives. *Bioorg. Med. Chem.* 2005, *14*, 366–372.
68. Vicini, P.; Incerti, M.; Doytchinova, I.; La Colla, P.; Busonera, B.; Loddo, R. Synthesis and antiproliferative activity of benzo[d]isothiazole hydrazones. *Eur. J.Med. Chem.* 2006, *41*, 624–632.
69. Jin, L.; Chen, J.; Song, B.; Chen, Z.; Yang, S.; Li, Q.; Hu, D.; Xu, R. Synthesis, structure, and bioactivity of N′-substitutedbenzylidene-3,4,5-trimethoxybenzohydrazide and 3-acetyl-2-substituted phenyl-5-(3,4,5-trimethoxyphenyl)-2,3-dihydro-1,3,4-oxadiazole derivatives. *Bioorg. Med. Chem.Lett.* 2006, *16*, 5036–5040.
70. Gürsoy, E.; Güzeldemirci-Ulusoy, N. Synthesis and primary cytotoxicity evaluation of new imidazo[2,1-b]thiazole derivatives. *Eur. J. Med. Chem.* 2007, *42*, 320–326.
71. El-Hawash, S.A.M.; Abdel Wahab, A.E.; El-Dewellawy, M.A. Cyanoacetic acid hydrazones of 3-(and 4-) acetylpyridine and some derived ring systems as potential antitumor and anti-HCV agents. *Arch. Pharm. Chem. Life Sci.* 2006, *339*, 14–23.
72. Silva, A.G.; Zapata-Suto, G.; Kummerle, A.E.; Fraga, C.A.M.; Barreiro, E.J.; Sudo, R.T. Synthesis and vasodilatory activity of new N-acylhydrazone derivatives, designed as LASSBio-294 analogues. *Bioorg. Med. Chem.* 2005, *13*, 3431–3437.

73. Abdel-Aal, M.T.; El-Sayed, W.A.; El-Ashry, E.H. Synthesis and antriviral evaluation of some sugar arylglycinoylhydrazones and their oxadiazoline derivatives. *Arch. Pharm. Chem. Life Sci.* 2006, *339*, 656–663.
74. Friedman, J.F.; Mital, P.; Kanzaria, H.K.; Olds, G.R.; Kurtis, J.D. Schistosomiasis and pregnancy. *Trends Parasitol.* 2007, *23*, 159–164.
75. Sulaiman, S.M.; Ali, H.M.; Homeida, M.M.; Bennett, J.L. Efficacy of a new Hoffmann-La Roche compound (RO-15-5458) against Schistosoma mansoni (Gezira strain, Sudan) in vervet monkeys (Cercopithecus aethiops). *Trop. Med. Parasitol.* 1989, *40*, 335–336.
76. Pereira, L.H.; Coelho, P.M.; Costa, J.O.; de Mello, R.T. Activity of 9-acridanone-hydrazone drugs detected at the prepostural phase, in the experimental schistosomiasis mansoni. *Mem. Inst. Oswaldo Cruz.* 1995, *90*, 425–428.
77. Nayyar, A.; Jain, R. Recent advances in new structural classes of anti-tuberculosis agents. *Curr. Med. Chem.* 2005, *12*, 1873–1886.
78. Scior, T.; Garcés-Eisele, S.J. Isoniazid is not a lead compound for its pyridyl ring derivatives, isonicotinoyl amides, hydrazides and hydrazones: A critical review. *Curr. Med. Chem.* 2006, *13*, 2205–2219.
79. Janin, Y. Antituberculosis drugs: Ten years of research. *Bioorg. Med. Chem.* 2007, *15*, 2479–2513.

Chapter 5

SYNTHESIS OF N-(6-ARYLBENZO[D] THIAZOLE-2-ACETAMIDE DERIVATIVES AND THEIR BIOLOGICAL ACTIVITIES: AN EXPERIMENTAL AND COMPUTATIONAL APPROACH

Yasmeen Gull[1,2], Nasir Rasool[1], Mnaza Noreen[1], Ataf Ali Altaf[3], Syed Ghulam Musharraf[4], Muhammad Zubair[1], Faiz-Ul-Hassan Nasim[5], Asma Yaqoob[5], Vincenzo DeFeo[6] and Muhammad Zia-Ul-Haq[7]

[1]Department of Chemistry, Faculty of Science and Technology, Government College University Faisalabad, Faisalabad 38000, Pakistan

[2]Department of Chemistry, Faculty of Science, University of Sargodhah, Bhakkar Campus, Bhakkar 30000, Pakistan

[3]Department of Chemistry, Faculty of Science, University of Gujrat, Hafiz Hayat Campus, Gujrat 50700, Pakistan

[4]International Center for Chemical and Biological Sciences, Hussain Ebrahim Jamal Research Institute of Chemistry, University of Karachi, Karachi 75270, Pakistan

[5]Department of Chemistry, Faculty of Science, Islamia University of Bahawalpur, Bahawalpur 63000, Pakistan

[6]Department of Pharmaceutical and Biomedical Sciences, University of Salerno, Via Ponte don Melillo, Fisciano (Salerno) I-84084, Italy

[7]Offices of Research, Innovation and Commercialization, Lahore College for Women University, Lahore 54600, Pakistan

ABSTRACT

A new series of *N*-(6-arylbenzo[d]thiazol-2-yl)acetamides were synthesized by C-C coupling methodology in the presence of Pd(0) using various aryl boronic pinacol ester/acids. The newly synthesized compounds were evaluated for various biological activities like antioxidant, haemolytic, antibacterial and urease inhibition. In bioassays these compounds were found to have moderate to good activities. Among the tested biological activities screened these compounds displayed the most significant activity for urease inhibition. In urease inhibition, all compounds were found more active than the standard used. The compound N-(6-(p-tolyl)benzo[d]thiazol-2-yl)acetamide was found to be the most active. To understand this urease inhibition, molecular docking studies were performed. The *in silico* studies showed that these acetamide derivatives bind to the non-metallic active site of the urease enzyme. Structure-activity studies revealed that H-bonding of compounds with the enzyme is important for its inhibition.

INTRODUCTION

Benzothiazoles consist of a benzene ring fused with a thiazole ring. Various benzothiazole derivatives serve as drugs, dyes and industrial chemicals [1,2,3]. Benzothiazoleand its derivatives such as esters have also been reported as active against Gram-positive and Gram-negative bacteria such as *Staphylococcus epidermidis, Escherichia coli, Enterobacter* and yeast (*Candida albicans*) [4]. Benzothiazole derivatives have also found to possess anticancer, antifungal and antibacterial activities [5,6]. 2-Aminobenzothiazole and a number of other aminobenzothiazole derivatives have been reported as muscle relaxants [7,8]. A literature survey performed for the current study revealed that 6-substituted-2-aminobenzothiazole derivatives such as 6-methyl, 6-methoxy, 6-ethoxy and 6-isopropoxy show antibacterial, anti-inflammatory and analgesic properties [9]. Various other derivatives are found to be cytotoxic against various tumors [10,11].

Suzuki cross coupling reactions are remarkable methods for C-C bond formation, utilized for the synthesis of agrochemicals, advanced materials and pharmaceuticals at both the industrial and laboratory scale [12,13,14,15,16,17].

The purpose of this study was to synthesize novel N-(6-arylbenzo[d] thiazol-2-yl)acetamides employing the Pd(0)-catalyzed Suzuki cross coupling methodology. This article describes our optimized experiments for the synthesis of N-protected 6-bromobenzothiazoles. The biological activities of these newly synthesized molecules were studied with the intention to explore their potential as future drugs. We investigated urease inhibition, nitric oxide scavenging, haemolytic and antibacterial activities. Molecular docking studies were performed to determine how they bind to the urease enzyme. To the best of our knowledge, all the studies reported in the current manuscript have not been reported in the literature to date.

RESULTS AND DISCUSSION

Chemistry

Majo *et al.* reported low to moderate yielding one-step Suzuki cross coupling reactions using various boronic acids/ester with 2-bromobenzothiazole under thermal conditions [18]. We have previously reported Pd(0)-catalyzed reactions of 2-amino-6-bromobenzothiazole with different arylboronic pinacol esters/ arylboronic acids using Suzuki cross coupling methodology with moderate yields [19]. We have not been able to achieve better yields as the amino moiety present in the benzothiazole molecule is basic and nucleophilic. In the current study, we have tried to enhance the yield of the synthesized molecules. In order to achieve high yields, the amino group was protected via acylation, which led to substantially enhanced yields of the products 3a–3h (Figure 1) compared to unprotected benzothiazole derivatives as reported in literature [19].

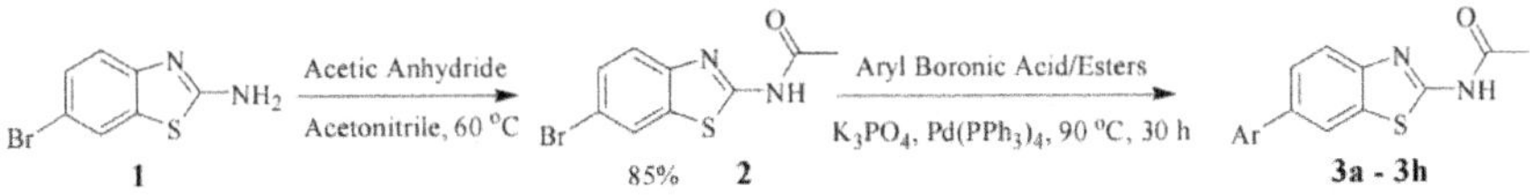

Figure 1: Synthesis of *N*-(6-bromobenzo[d]thiazol-2-yl)acetamide (2) and *N*-(6-aryl-benzo[d]thiazol-2-yl)acetamides 3a–3h.

Furthermore, we also optimized other reaction parameters like catalyst loading, solvent, temperature and base used in the reactions producing 3a–3h. Thus, we tried various solvents like toluene, dimethylformamide (DMF) and 1,4-dioxane at different temperatures (80–100 °C). It was noted (in Table 1) that the solvent has an effect on overall reaction yield. 1,4-Dioxane was found to be the best solvent because of its better solvation of the reactants. In our studies, these cross coupling reactions progressed efficiently, even in the presence of known sensitive groups such as CN, to give the desired products in very good yields. Finally after optimization, we investigated the coupling of 2 with different arylboronic pinacol esters/acids in the presence of Pd(PPh$_3$)$_4$ as catalyst.

Table 1: Synthesis of *N*-(6-arylbenzo[d]thiazol-2-yl)acetamides 3a–3h

Entry	Arylboronic Pinacol Ester/Arylboronic Acid	Product	H$_2$O/Solvent (1:4)	Yields %
1	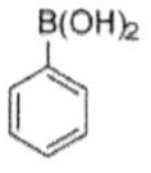 B(OH)$_2$, phenyl	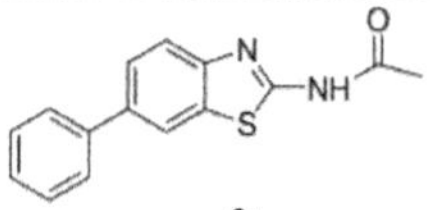 3a	Toluene 1,4-Dioxane	75 80
2	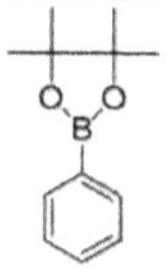 Pinacol ester, phenyl	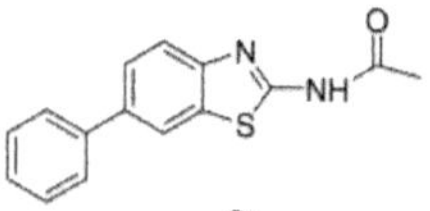 3a	1,4-Dioxane	77
3	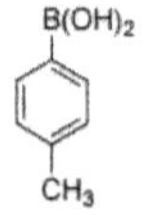 B(OH)$_2$, 4-CH$_3$-phenyl	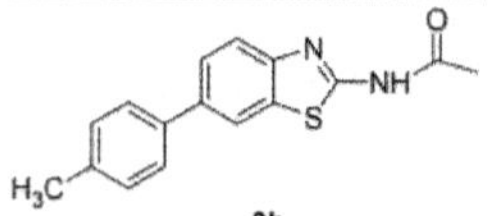 3b	1,4-Dioxane	85
4	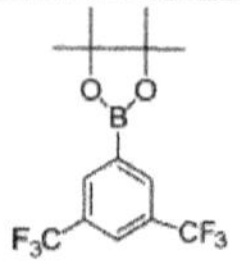 Pinacol ester, 3,5-bis(CF$_3$)-phenyl	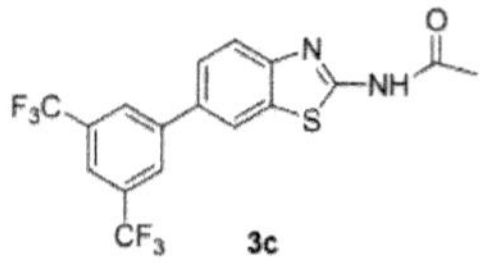 3c	1,4-Dioxane	81
5	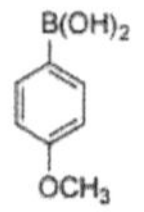 B(OH)$_2$, 4-OCH$_3$-phenyl	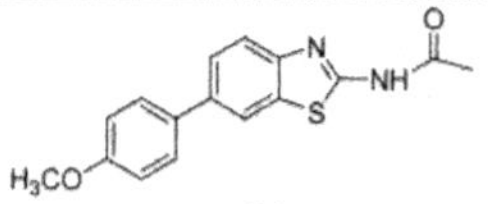 3d	1,4-Dioxane	79
6	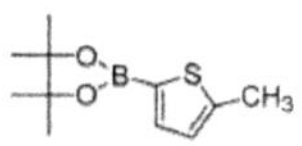 Pinacol ester, 5-CH$_3$-thiophene	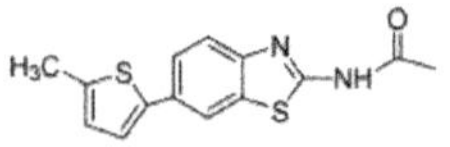 3e	1,4-Dioxane	75
7	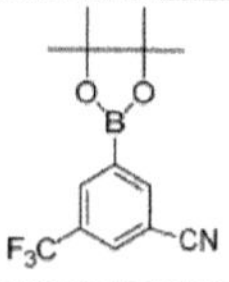 Pinacol ester, 3-CF$_3$-5-CN-phenyl	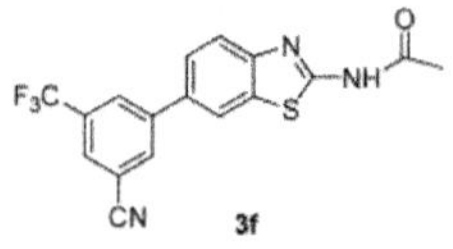 3f	1,4-Dioxane	77
8	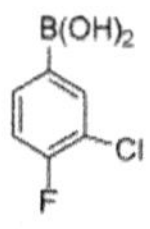 B(OH)$_2$, 3-F-4-Cl-phenyl	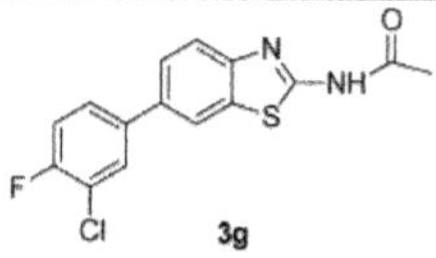 3g	1,4-Dioxane	79
9	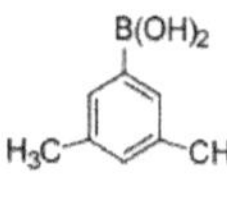 B(OH)$_2$, 3,5-di-CH$_3$-phenyl	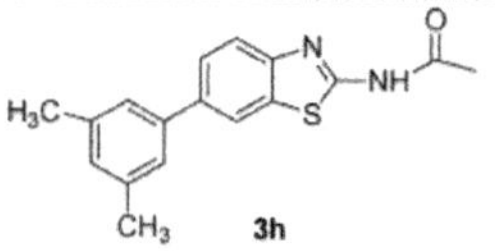 3h	1,4-Dioxane	83

Product 3a was prepared with 80% yield, when 2 was coupled with phenyl boronic acids under the set reaction conditions. The highest yield obtained in this series of reactions (85%) corresponded to product 3b. Product 3h was also obtained in excellent yield (83%) with this cross coupling method. It was seen that the product 3c with an electron withdrawing moiety showed a high yield (81%). Our studies showed that overall the acetylated 2-amino-6-bromobenzothiazole (2) gave good yields in these coupling reactions.

Biological Studies

Urease Inhibitory Activity

Urease is a nickel-containing metalloenzyme that catalyzes the hydrolysis of urea to form ammonia and carbamate, which further decomposes to yield a second equivalent of ammonia and carbon dioxide [20]. Bacterial ureases have been reported as an important virulence factor in the development of many harmful clinical conditions for human and animal health. Urease is directly involved in the formation of infectious stones and contributes to pathogenesis [21]. It is the major cause of pathogenesis induced by *Helicobacter pylori*, which plays an important role in peptic ulcers and may lead to stomach cancer. In recent years, a number of compounds have been proposed as urease inhibitors [22,23], which are considered interesting new targets for anti-ulcer drugs and for the treatment of infections caused by urease producing bacteria [24].

In the current study, urease inhibitory activity assays were performed following a previously reported protocol [25]. Thiourea was used as standard in the assay with an IC_{50} value of 23.1 µg/mL. All of the synthesized benzothiazole derivatives were examined for their urease inhibitory activities at concentration of 15 and 40 µg/mL. All of our synthesized compounds exhibit good to excellent urease inhibitory activities (Table 2, Figure 2 and Figure 3).

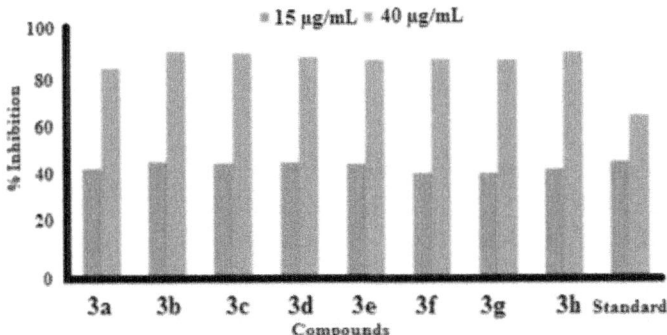

Figure 2: The urease percentage inhibition values at 15 µg/mL and 40 µg/mL.

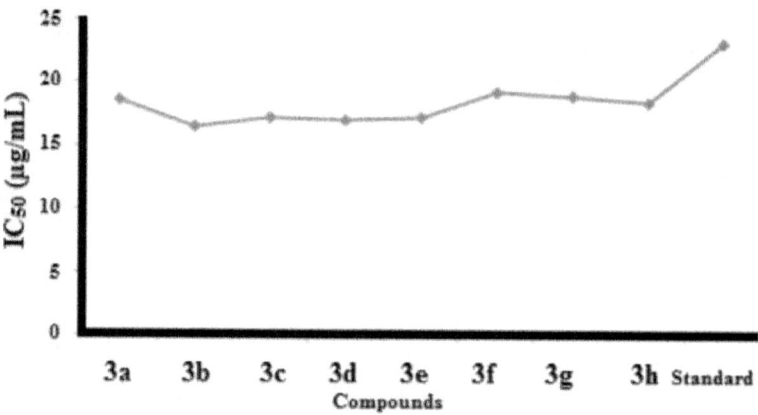

Figure 3: IC_{50} values of anti-urease activity.

Table 2: Antiurease activity of N-(6-arylbenzo[d]thiazole-2-yl)acetamides 3a–3h at 15μg/mL and 40 μg/mL

Compound	% Inhibition at 15 μg/mL	% Inhibition at 40 μg/mL	IC_{50} (μg/mL)
3a	44 ± 0.12	83.9 ± 0.12	18.6
3b	47 ± 0.11	90.51 ± 0.19	16.5
3c	46.09 ± 0.10	90.07 ± 0.20	17.2
3d	46.5 ± 0.15	88.5 ± 0.24	17
3e	46 ± 0.12	87 ± 0.26	17.2
3f	42 ± 0.12	87.5 ± 0.26	19.2
3g	42 ± 0.14	87 ± 0.26	18.9
3h	43.59 ± 0.13	90.33 ± 0.20	18.4
Standard	47 ± 0.31	65 ± 0.01	23.1

Each values is mean ± Standard deviation of three parallel measurements.

Urease enzyme has an active binding site and it was believed that the newly synthesized benzothiazole derivatives have the capability to bind to these active sites of urease enzyme. In this way the hydrolysis of enzyme is stopped and activity of enzyme is inhibited. Compound 3b showed the highest urease inhibition activity (90.51 ± 0.19) with an IC_{50} value of 16.5 μg/mL at 40 μg/mL.

Compounds 3h and 3c exhibited excellent urease inhibitory activities (90.07 ± 0.20 and 90.33 ± 0.20 at 40 μg/mL) with IC_{50} values of 17.2 μg/mL and 18.4 μg/mL, respectively. Molecules 3a, 3d, 3e, 3f and 3g exhibited good urease inhibition with IC_{50} values of 18.6, 17, 17.2, 19.2 and 18.9 μg/mL, respectively. Notably we observed that the presence of electron donating methyl functional groups produced high urease inhibition. We also noted that different functional groups are responsible for variable antiurease activities of the compounds.

In Silico Studies with Urease

To understand the binding of the synthesized compounds with urease *in-silico* studies were performed. Only compounds with IC_{50} values were analyzed for docking studies using the freely available software AutoDock 4.2 and others as described in the Experimental Section. All the compounds were screened at different sites (A & B) of the enzyme. The nickel-containing catalytic site A is the most commonly tested site in the literature [26], while site B is less commonly targeted [20,27]. Our compounds bind more strongly to Site B than to site A. The docking studies results were compared with the experimental results and listed in Table 3 and Table 4. Binding energy, inhibition constant and moldock scores are reported in Table 3, whereas the moldock H-binding energy and number of H-bonding interactions between the test compounds and enzyme are reported in Table 4.

Table 3: Experimental and docking comparative data

Compound	* IC_{50} (μg/mL)	Inhibition Constant	Binding Energy	Moldock Score
3a	18.6	267.93	−4.87	−83.39
3b	16.5	232.29	−4.96	−81.68
3c	17.2	232.56	−4.96	−68.81
3d	17.0	434.76	−4.59	−69.20
3e	17.2	278.10	−4.85	−87.00
3f	19.2	90.570	−5.52	−87.90
3g	18.9	153.46	−5.20	−79.91
3h	18.4	150.67	−5.21	−80.11

* Experimentally measured *in vitro*.

Table 4: H-bonding parameters calculated by Mol-Dock Molegero Molecular Viewer 2.5

Compound	Number of H-bonds	H-Bonding Type (K−H−L) *	H-Bond Distance (K−L) (Å)	IC_{50} (μg/mL)	H-Binding Energy
3a	1	O—H−N	2.957	18.6	−2.50
3b	2	N—H−N N−H—O	2.837 3.544	16.5	−3.45
3c	1	N−H—O	2.782	17.2	−2.98
3d	2	N−H—O N—H−N	3.023 3.23	17.0	−3.30
3e	2	N−H—O N—H−N	2.906 3.388	17.2	−3.01
3f	1	N−H—O	2.739	19.2	−1.67
3g	1	N−H—O	2.696	18.9	−1.99
3h	zero	−	−	18.4	−

* K atom from the compound and L atom from the protein residue.

The number of H-bonding interactions was estimated from the most stable complex formed between the test compound and the enzyme. We found a linear correlation between the experimental IC_{50} values and the calculated binding energy, as shown in Figure 3. This linear correlation of results is in agreement with the data reported in the literature [26,28]. More interestingly a better correlation was found between IC_{50} values and the moldock H-binding energy (Figure 4).

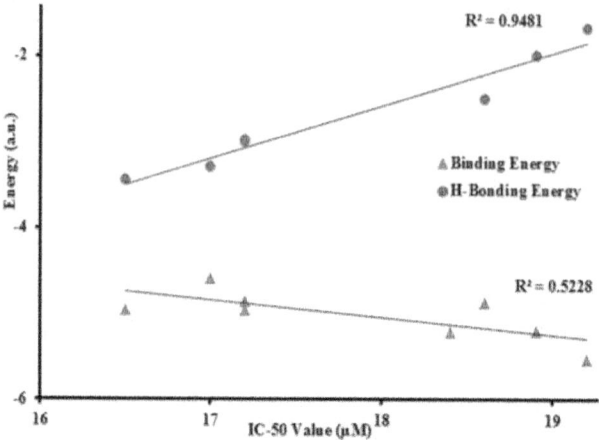

Figure 4: Correlation between docking predicted energies (in arbitrary unit) and *in vitro* IC_{50} values.

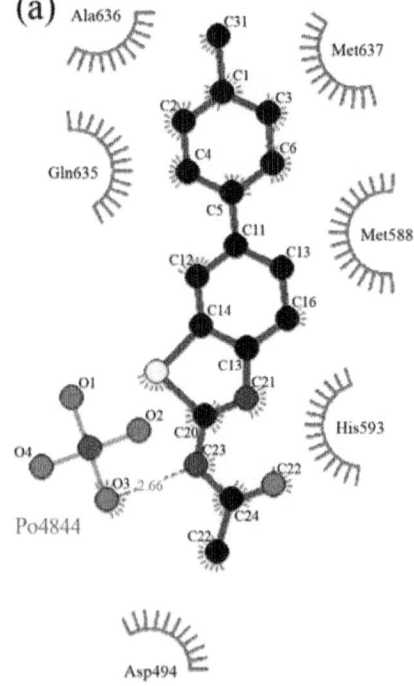

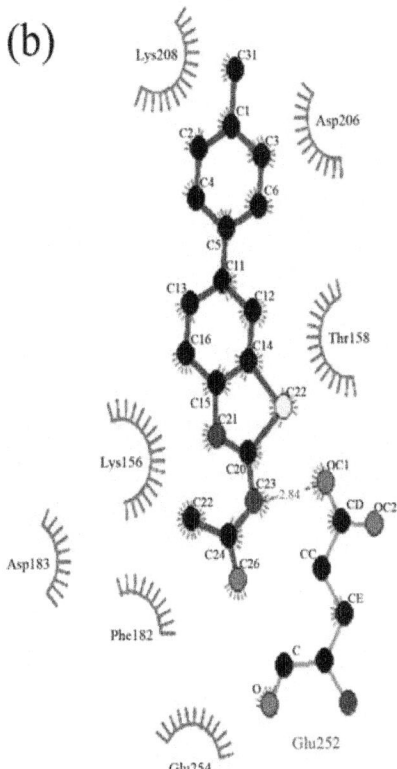

Figure 5: LIGPLOT images of compound 3b with urease enzyme (a) at the catalytic site A; and (b) at the catalytic site B.

These results show that the H-bonding is more important, in urease inhibition mechanism, than other factors involved in this biological reaction. Compound 3b, with the highest *in vitro* activity, is represented here for modeling analysis and its most active conformations are explained in the following paragraphs. As urease is a nickel-dependent enzyme, the active site A of the enzyme shows weak hydrophobic interaction with compound 3b therefore, a drug with hydrophobic substituents would be able to bind strongly as it would project into the hydrophobic grooves of the enzyme and thus effectively inhibit its activity. The LIGPLOT interaction images show that compound 3b has a total of seven interactions with enzyme site A (Figure 5a).

The amino acids His593, Met637, Ala636, Gln635, Met588 and Asp494 form hydrophobic interactions with compound 3b. The hydrophobic interactions favor ligand binding with proteins having metal ions. Furthermore, the study showed hydrogen bonding (2.66 Å, N–H—O type) between compound 3b and

PO$_4$ group of the enzyme nickel catalytic site A. Figure 6a represents the most interacting conformation of 3b in the active pocket (site A) of the enzyme at the electrostatic surface. Figure 6 is generated by Molegro Molecular viewer and analyzed by moldock score. The moldock analysis shows that there are two H-bonding interactions in both cases these are also reported in Table 4.

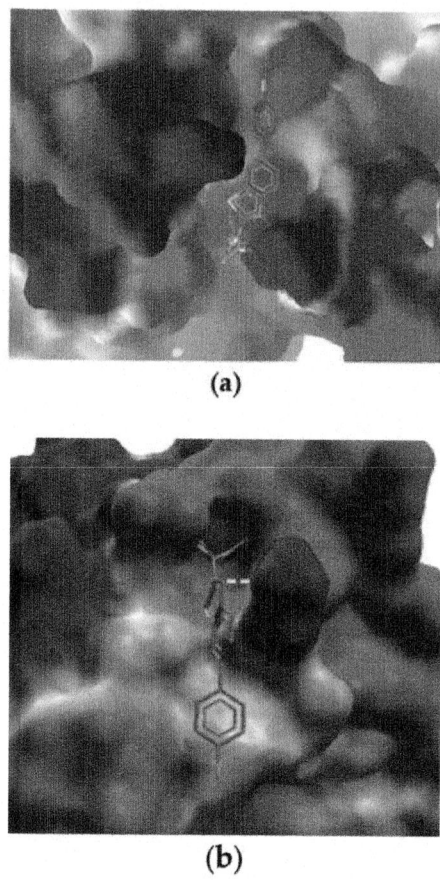

Figure 6. Compound 2c in the Molegro molecular viewer generated electrostatic surface of urease enzyme (a) at the catalytic site A; and (b) at the catalytic site B. The yellow dashed line represents the H-bonding.

All of the synthesized compounds 3a–3h show better interaction at site B. The LIGPLOT interaction diagram of compound 3b illustrates that this inhibitor has better interactions with the protein. The LIGPLOT interaction images show that the compound 3b has a total of eight interactions with enzyme site B (Figure 5b). The amino acids Lys208, Asp206, Thr158, Glu254, Phe182, Lys156 and Asp183 form cationic—π interactions with compound 3b, while Glu252 interacts via hydrogen bonding (2.84Å, N–H—O type) with compound 3b. Figure 5b presents the most interacting conformation of compound 3bin the active pocket of the enzyme (site B). The diagram shows that the enzyme provides enough space for the accommodation of 3b inside the pocket. The backbone dose sterically favors the 3b molecule to interact with catalytic site.

As shown in Figure 4, a strong correlation with experimental results is found between the IC_{50} values and H-bonding data calculated by moldock in the Molegero docking software at site B of enzyme. In the moldock analysis for 3b at site A two H-bonding interactions (N–H—O and N–H—N type) were observed between 3b and $PO_4$844, His593, respectively, while two strong H-bonding interactions between 3b and enzyme residues Glu252 and Lys156 were observed at site B. The H-bonding distances and moldock scores for all compounds, at site B, are listed in Table 4. Figure 7 shows the H-bonding interactions of all of the compounds (except 3h) with the active residue of urease enzyme at site B. Compound 3h does not show any H-bonding interaction with the active site residue. In the linear correlation with *in vitro* IC_{50}, stronger H-bonding is found for compounds with lower IC_{50} value. All the compounds have H-bonding distances in the 2.696 Å–3.544 Å range and H-bonding energies in the −1.674–−3.45 a.u. range.

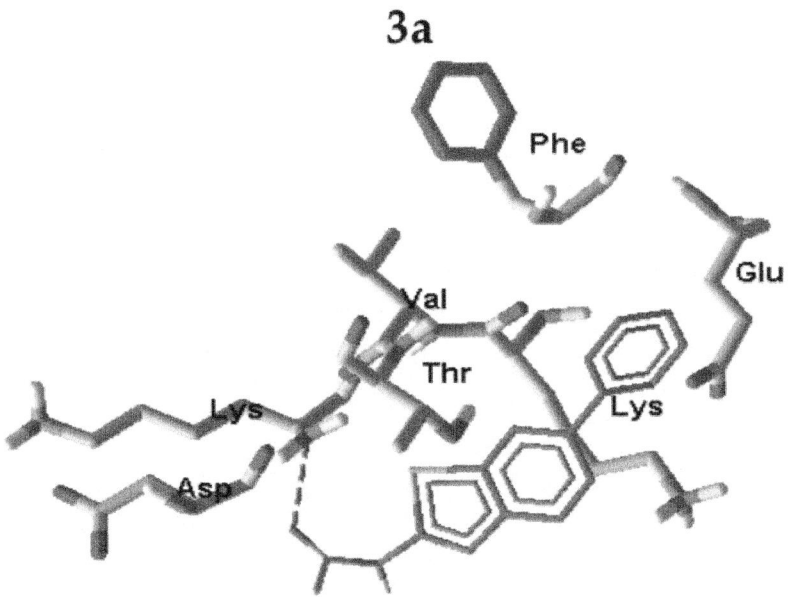

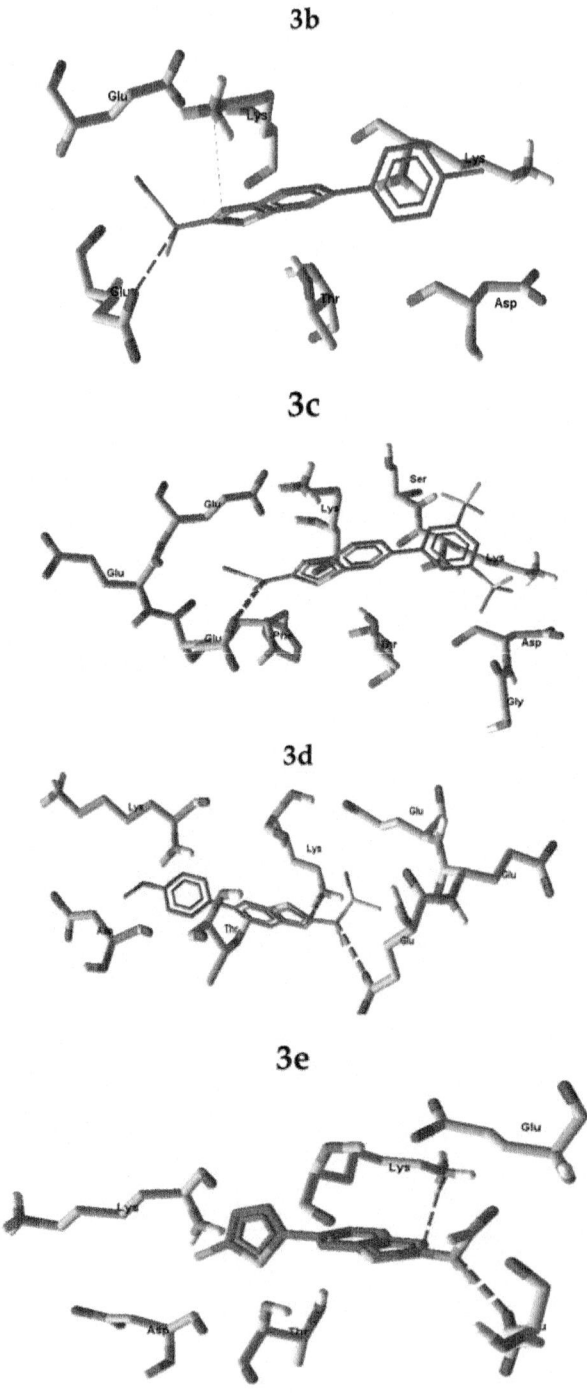

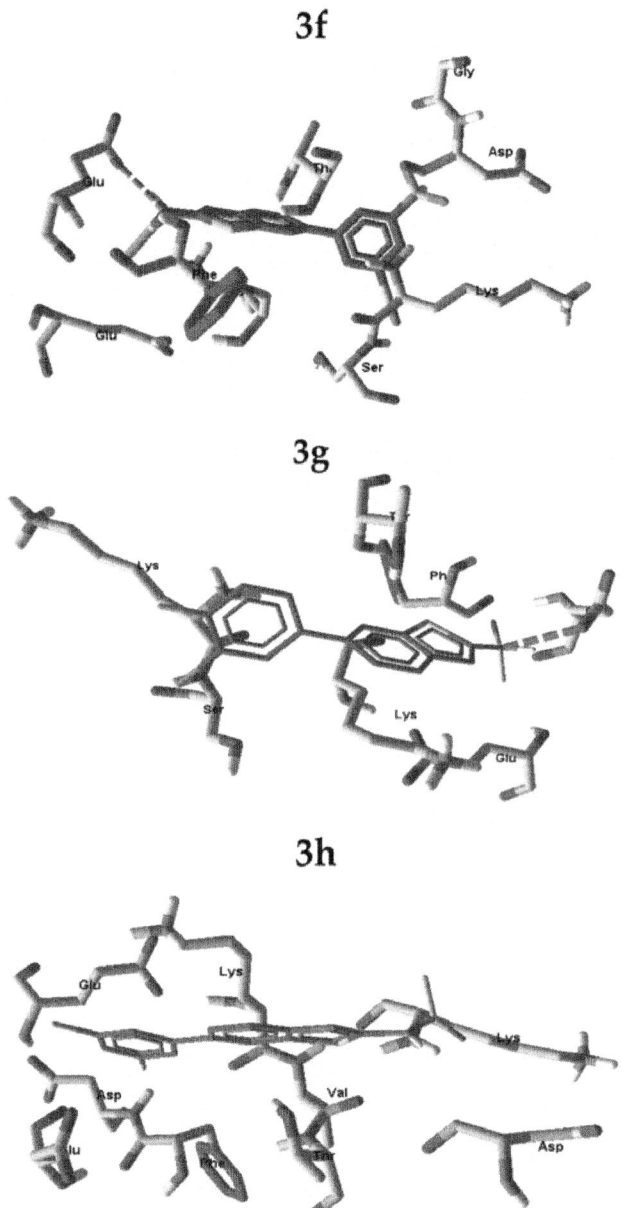

Figure 7: Molegro molecular viewer generated sketches between compounds 3a–3h and the active residue of urease at site B. The blue dashed lines represent the H-bonding.

Nitric Oxide Scavenging Percentage Assay

The literature contains reports on the antioxidant activities of 6-flourobenzothiazole-substituted triazoles using DPPH assays [29]. A survey of the literature showed that benzothiazole molecules along with pyrozoline rings showed the highest antioxidant activities. Having a phenyl ring on the pyrozoline increased the antioxidant activity in the ferric ion reduction and DPPH solution methods [30]. Our newly synthesized N-protected benzothiazole derivatives exhibit nitric oxide scavenging activities. Ascorbic acid was used as a standard in the assay with 38.5 ± 0.16 and 84.1 ± 0.12 percent nitric oxide scavenging at 20 μg/mL and at 50 μg/mL with an IC_{50} value of 50.43 μg/mL. Synthesized compounds with their calculated IC_{50} values are listed in Table 5 (Figure 8 and Figure 9).

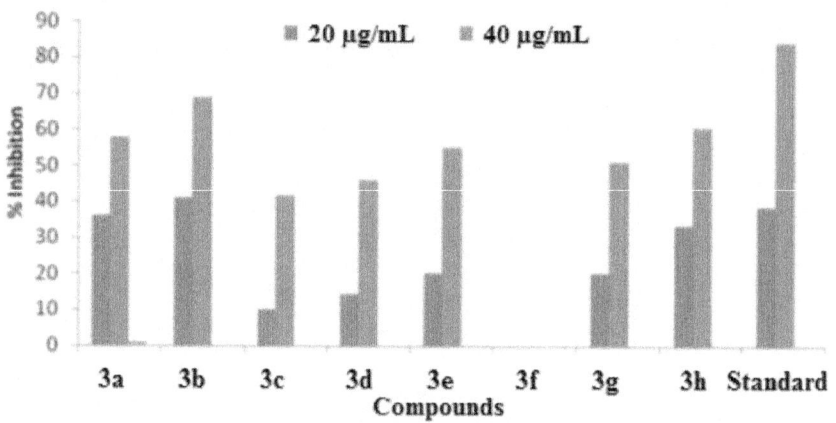

Figure 8: The percentage nitric oxide inhibition at 20 μg/mL and 40 μg/mL.

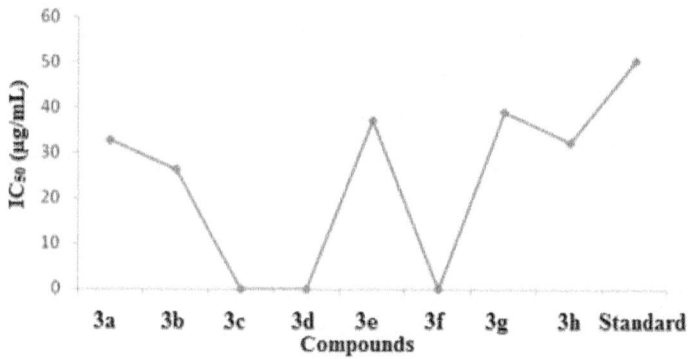

Figure 9. IC_{50} values of nitric oxide scavenging activity.

Table 5: Nitric oxide scavenging activity of *N*-(6-arylbenzo[d]thiazole-2-yl)acetamides 3a–3h at 20 μg/mL and 40 μg/mL

Compound	% Activity at 20 μg/mL	% Activity at 40 μg/mL	IC_{50} (μg/mL)
3a	36.25 ± 0.12	57.75 ± 0.12	32.7
3b	41 ± 0.11	69 ± 0.12	26.4
3c	10 ± 0.18	41.75 ± 0.20	NC
3d	14.25 ± 0.17	46 ± 0.2	NC
3e	20.25 ± 0.15	55 ± 0.31	37.1
3f	0	0	NC
3g	20.25 ± 0.15	51.25 ± 0.15	39.1
3h	33.5 ± 0.13	60.5 ± 0.1	32.3
Standard	38.5 ± 0.16	84.1 ± 0.12	50.43

Each value is mean ± Standard deviation of three parallel measurements. NC stands for not calculated due to less activity.

It was found that acetyl-protected amino group products were more active in the nitric oxide scavenging assay, than the previously reported non-acetylated compounds [19]. Molecules 3a, 2b, 3e, 3g and 3h were found to be the most active for nitric oxide scavenging activity, with percentage inhibitions of 57.75 ± 0.12, 69 ± 0.12, 55 ± 0.31, 51.25 ± 0.15 and 60.5 ± 0.1 at 40μg/mL with IC_{50} values of 32.7, 26.4, 37.1, 39.1 and 32.3, respectively. Compound 3f, however, was found to be inactive in the nitric oxide scavenging assay. We are unable to account for this inactivity.

Haemolytic Activity

The haemolytic activity of benzothiazole derivatives has already been reported. A literature survey reveals that amino-substituted derivatives of benzothiazole have high cytotoxicites. Benzothiazole compounds with halogen substitutions show cytotoxicity towards cancer cell lines [31]. The haemolytic activity of the newly synthesized benzothiazole derivatives were studied against Triton X-100 by a reported method [32].

The newly synthesized benzothiazole derivatives exhibit moderate to high haemolytic activities (see in Table 6, Figure 10). Compound 3c exhibits the highest haemolytic activity (47.089 ± 0.130). Fluorinated analog 2d also displayed the highest toxicity among all the tested compounds. The antitumor activity of a compound might be considered as corresponding to the highest haemolytic activity. It was observed that substitution does not markedly affected the haemolytic activity of these newly synthesized *N*-protected benzothiazole derivatives. Compounds 2, 3b and 3d–3g showed good haemolytic activities. The lowest haemolytic activity was found for compound 3h. It was concluded that halogen substitution on *N*-protected benzothiazole molecules promotes haemolytic activity. These compounds have potential to be used as future anticancer agents.

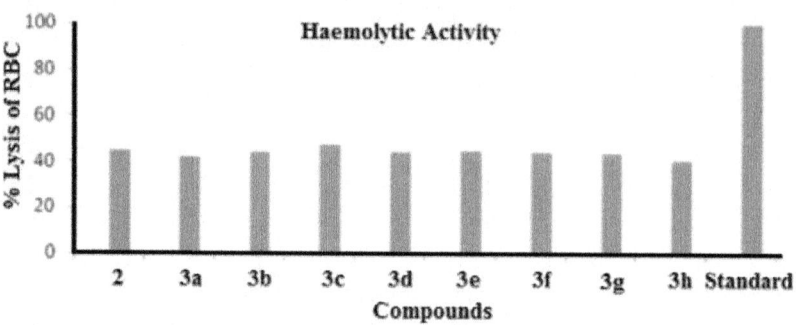

Figure 10: Haemolytic activity of the synthesized compounds.

Table 6: Haemolytic activity of newly synthesized N-(6-bromobenzo[d]thiazol-2-yl) acetamide (**2**) and N-(6-arylbenzo[d]thiazole-2-yl)acetamides (**3a–3h**)

Entry	% lysis of RBC	Entry	% lysis of RBC
2	44.628 ± 0.369	3e	44.425 ± 0.181
3a	42.123 ± 0.479	3f	44.063 ± 0.314
3b	44.179 ± 0.157	3g	43.614 ± 0.157
3c	47.089 ± 0.130	3h	40.661 ± 0.216
3d	44.078 ± 0.279		
Standard		99.78 ± 0.912	

Each value is mean ± Standard deviation of three parallel measurements.

Antibacterial Activity

The synthesized benzothiazole derivatives were examined for their antibacterial activity against two Gram positive-bacterial strains (*Baccilus subtiles, Staphylococcus aureus*) and four Gram-negative strains (*Escherichia coli, Psuedomonas aeruginosa, Shigella dysenteriae, Salmonella typhae*) at concentrations of 40 and 80 μg/mL (Table 7 and Table 8). It was concluded that the potent antibacterial activities of these compounds might be due to electron withdrawing groups present on the aryl moiety in N-protected benzothiazole molecule. Similar observations are also reported by other groups which suggest that the presence of electron releasing and electron withdrawing groups substantially affects the antibacterial activity [33].

Table 7: Antibacterial activity (40 μg/mL) of N-(6-arylbenzo[d]thiazole-2-yl) acetamides 2b–2i.

Entry	% Activity at 40 μg/mL					
	B. subtilis	S. aureus	P. aeruginosa	S. dysenteriae	S. typhae	E. coli
3a	-	-	-	0.94 ± 0.45	-	37.52 ± 0.38
3b	-	-	-	-	-	34.04 ± 0.40
3c	-	-	-	0 ± 0.45	6.0 ± 0.47	49.05 ± 0.32
3d	-	-	-	2.92 ± 0.44	-	42.10 ± 0.36
3e	-	-	-	-	-	37.82 ± 0.38
3f	-	-	7.54 ± 0.6	2.59 ± 0.44	1.2 ± 0.50	33.53 ± 0.41
3g	-	-	-	1.45 ± 0.44	-	45.93 ± 0.3
3h	-	-	-	3.51 ± o.43	-	22.89 ± 0.47
Ampicillin	23 ± 0.1	29 ± 0.61	25 ± 0.12	35 ± 0.32	29 ± 0.61	19 ± 0.31

Each value is mean ± standard deviation of three parallel measurements.

Table 8: Antibacterial activity (80 μg/mL) of N-(6-arylbenzo[d]thiazole-2-yl)acetamides 2b–2i.

Entry	% Activity at 80 μg/mL					
	B. subtilis	S. aureus	P. aeruginosa	S. dysenteriae	S. typhae	E. coli
3a	6.08 ± 0.571	15.25 ± 0.5	-	8.47 ± 0.44	18.5 ± 0.58	57.97 ± 0.25
3b	-	-	-	6.57 ± 0.45	17.7 ± 0.5	56.13 ± 0.32
3c	-	-	-	-	26.2 ± 0.53	50.49 ± 0.30
3d	7.31 ± 0.5635	-	-	5.92 ± 0.46	28.86 ± 0.5	51 ± 0.30
3e	12.66 ± 0.531	3.75 ± 0.65	-	-	19.2 ± 0.5	50.8 ± 0.30
3f	4.27 ± 0.582	8.89 ± 0.5	2.96 ± 0.6	10.93 ± 0.43	25 ± 0.54	55 ± 0.32
3g	-	-	-	11.51 ± 0.42	21.05 ± 0.5	57.84 ± 0.25
3h	5.67 ± 0.5735	-	-	7.48 ± 0.45	17.43 ± 0.5	53.96 ± 0.40
Ampicillin	50.5 ± 0.31	52.9 ± 0.29	52 ± 0.26	56 ± 0.26	42.9 ± 0.29	45.9 ± 0.21

Each value is mean ± standard deviation of three parallel measurements.

The results show that the benzothiazole compounds 3a–3h exhibit higher activities than the standard against some species. Functional group changes in the benzothiazole molecule led to differences in activity. The newly synthesized compounds were found to be inactive against *Baccilus subtilis* and *Staphylococcus aureus*. Only compound 3f showed activity against *Psuedomonas aeruginosa* with a very small value (7.54 ± 0.6). These newly synthesized compounds do not exhibit considerable antibacterial activity and the highest value (3.51 ± 0.43) was observed for compound 3h against *Shigella dysenteriae* at concentration of 40 μg/mL. These new benzothiazole molecules showed weak activities against *Salmonella typhae* at 40 μg/mL. It was found that all newly synthesized benzothiazole compounds gave good to very good activity against *E. coli* at 40 μg/mL. Compounds 3c and 3ge exhibited very good activities against the *E. coli* strain with values of 49.05 ± 0.32 and 45.93 ± 0.3, respectively. These differences in activities may be attributed to the presence of electron loving atoms/groups on the aryl moiety of these N-protected benzothiazole derivatives.

The synthesized compounds were also checked for antibacterial activities at 80 µg/mL and compared against ampicillin. It was shown that these compounds showed moderate activities against *Bacillus subtiles* with the highest value (12.66 ± 0.531) corresponding to compound 3e. The authors concluded that these benzothiazole derivatives showed non-significant activity against *Staphylococcus aureus*. In addition, these compounds were to be found active against *Shigella dysenteriae* and *Salmonella typhae* at 80 µg/mL. These compounds displayed very good activity against *E. coli* at 80 µg/mL. The benzothiazole derivatives were discovered to be the most potent against the *E. coli* strain. Compounds 3a, 3b and 3g proved to be the most potent at the concentration of 80 µg/mL with the highest antibacterial activities with values of 57.97 ± 0.25, 57.84 ± 0.25 and 56.13 ± 0.32, respectively. The results of this study revealed that electron withdrawing group substitution on the aryl moiety on the benzothiazole molecule enhanced the antimicrobial activity of the compounds.

EXPERIMENTAL SECTION

General Information

All reagents and chemicals were brought from Alfa-Aesar Chemical Co. (Ward Hill, MA, USA) and Sigma-Aldrich Chemical Co. (St. Louis, MO, USA). Solvents $CDCl_3$ and CD_3OD were used for ^{13}C-NMR and ^1H-NMR spectra on an Aspect AM-400 instrument at 400/100 MHz (Bruker, Billerica, MA, USA). The coupling constant was determined in Hz and chemical shift in δ in ppm. A JMS-HX-110 spectrometer (JEOL, Peabody, MA, USA) was used for EI-MS spectra. Melting points of benzothiazole compounds were measured on a B-540 melting point apparatus (Büchi, New Castle, DE, USA). Column chromatography with silica gel (mesh size 70 to 230 and 230 to 400) was used for compound purification. TLC (silica gel 60 PF 254 cards, Merck, Kenilworth, NJ, USA) was used for reaction monitoring. Plates were visualized using a UV lamp (254 to 365 nm) (Spectronics Corporation, Westbury, NY, USA).

Procedure for the Preparation of N-(6-arylbenzo[d]thiazole-2-yl) acetamides 3a–3h

The preparation of products 3a–3h was carried out under a nitrogen atmosphere. Compound **2** (synthesized by literature reported method [19]) (2.183 mmol) and 5 mol % $Pd(PPh_3)_4$ in 1,4-dioxane (20 mL) was mixed and stirred for 30 min. After 30 min K_3PO_4 (4.366 mmol), aryl boronic pinacol esters/aryl boronic acids (2.401 mmol), and H_2O (1.5 mL) was added under inert atmospheric conditions. The mixture was stirred for 30 h at 95 °C and cooled down to room temperature. Ethyl acetate was used for work up and the organic layer was separated and dried under vacuum. For purification purposes, column chromatography was done. The desired product was obtained by using ethylacetate and *n*-hexane (20% and 80%respectively) as eluents. The desired products were characterized by various spectroscopic techniques [34].

N-(6-Phenylbenzo[d]thiazol-2-yl)acetamide(3a).m.p. 203–205 °C; ^{1}H-NMR (CDCl$_3$ + CD$_3$OD) δ 9.46 (s, 1H), 7.92 (s, 1H), 7.67–7.50 (m, 4H), 7.46–7.31 (m, 3H), 2.31 (s, 3H); ^{13}C-NMR (CDCl$_3$ + CD$_3$OD) δ 168.3, 159.2, 151.9, 145.6, 132.4, 130.1, 129.8 (2C), 128.4, 128.4 (2C), 124.2, 121.7, 116.8, 23.7; EIMS (*m/z* + ion mode): 269.32 [M + H$^+$] 269.08; Anal Calcd for C$_{15}$H$_{12}$N$_2$OS: C, 67.13; H, 4.53; N, 10.43; found C, 67.23; H, 4.54; N, 10.37.

N-(6-(p-Tolyl)benzo[d]thiazol-2-yl)acetamide(3b).m.p. 205–207 °C; ^{1}H-NMR (CDCl$_3$ + CD$_3$OD) δ 9.33 (s, 1H), 8.11 (d, *J* = 10 Hz, 3H), 7.61 (d, *J* = 10.4 Hz, 1H), 7.31–7.18 (m, 3H), 2.41 (s, 3H), 2.35 (s, 3H); ^{13}C-NMR (CDCl$_3$ + CD$_3$OD) δ 168.2, 159.3, 148.4, 146.8, 133.1, 131.3, 129.6 (2C), 128.4 (2C), 126.1, 124.4, 121.2, 116.8, 22.9, 21.6; EIMS (*m/z* + ion mode): [M + H$^+$] = 283.08; [M − COCH$_3$]$^−$ = 241.01. Anal Calcd for C$_{16}$H$_{14}$N$_2$OS: C, 68.07; H, 5.01; N, 9.93 found C, 68.12; H, 5.04; N, 9.81.

N-(6-(3,5-bis(Trifluoromethyl)phenyl)benzo[d]thiazol-2-yl) acetamide (3c). m.p. 203–205 °C; ^{1}H-NMR (CDCl$_3$ + CD$_3$OD) δ 9.41 (s, 1H), 7.92 (s, 1H), 7.66–7.61 (m, 2H), 7.55–7.43 (m, 3H), 2.27 (s, 3H); ^{13}C-NMR (CDCl$_3$ + CD$_3$OD) δ 168.7, 159.4, 153.7, 146.3, 137.1, 132.1 (2C), 129.3, 128.4 (2C), 124.1 (2C), 123.3, 121.4, 120.12, 118.6, 25.2; EIMS (*m/z* + ion mode): 405.32, [M + H]$^+$; 405.16. Anal Calcd for C$_{17}$H$_{10}$F$_6$N$_2$OS: C, 50.51; H, 2.48; N, 6.92 found C, 50.44; H, 2.52; N, 6.82.

N-(6-(4-Methoxyphenyl)benzo[d]thiazol-2-yl)acetamide (3d). m.p.: 218–220 °C; ^{1}H-NMR (CDCl$_3$ + CD$_3$OD) δ 9.41 (s, 1H), 8.01 (s, 1H), 7.66 (d, *J* = 8, 2H), 7.57–7.47 (m, 2H), 7.08 (d, *J* = 8, 2H), 3.57 (s, 3H), 1.91 (s, 3H). ^{13}C-NMR (CDCl$_3$ + CD$_3$OD) δ 168.2, 158.8, 151.3, 149.5, 145.3, 132.2, 131.8 (2C), 129.5, 123.7, 121.3, 116.6, 115.5 (2C), 49.2, 22.5; EIMS (*m/z* − ion mode): [M − H]$^−$ = 297.25; [M − COCH$_3$]$^−$ = 255.34; [M − OCH$_3$ and COCH$_3$]$^−$ = 227.01; Anal Calcd for C$_{16}$H$_{14}$N$_2$O$_2$S$_2$: C, 64.42; H, 4.73; N, 9.38 found C, 64.51; H, 4.71; N, 9.42.

N-(6-(5-Methylthiophen-2-yl)benzo[d]thiazol-2-yl)acetamide (3e). m.p.: 192–194 °C; ^{1}H-NMR (CDCl$_3$ + CD$_3$OD) δ 9.41 (s, 1H), 7.90 (s, 1H), 7.67–7.35 (m, 4H), 2.34 (s, 3H), 2.11 (s, 3H); ^{13}C-NMR (CDCl$_3$ + CD$_3$OD) δ 166.2, 159.6, 148.7, 145.2, 132.4, 132.3, 129.5, 128.7, 128.3, 123.6, 121.8, 116.8, 22.7, 14.03; EIMS (*m/z* − ion mode): [M − H]$^−$ = 286.93; [M − COCH$_3$]$^−$ = 224.93; [M − CH$_3$]$^−$ = 271.02; Anal Calcd for C$_{14}$H$_{12}$N$_2$OS$_2$: C, 58.32; H, 4.20; N, 9.72 found C, 58.41; H, 4.12; N, 9.79.

N-(6-(3-Cyano-5-(trifluoromethyl)phenyl)benzo[d]thiazol-2-yl) acetamide (3f). m.p.: 212–213 °C; ^{1}H-NMR (CDCl$_3$ + CD$_3$OD) δ 9.48 (s, 1H), 7.92 (d, J = 1.2, 1H), 7.72–7.41 (m, 4H), 7.36 (s, 1H), 2.27 (s, 3H); ^{13}C-NMR (CDCl$_3$ + CD$_3$OD) δ 167.7, 158.8, 148.7, 146.4, 132.3, 132.0, 132.0, 131.8, 129.6, 128.4, 124.2, 123.3, 121.8, 120.3, 117.7, 114.8, 22.5; EIMS (m/z − ion mode): 360.33 [M − H]$^-$ 360.01; Anal Calcd for C$_{17}$H$_{11}$F$_3$N$_3$OS: C, 56.52; H, 2.78; N, 11.62 found C, 56.60; H, 2.77; N, 11.57.

N-(6-(3-Chloro-4-fluorophenyl)benzo[d]thiazol-2-yl)acetamide (3g). m.p.: 166–167 °C; ^{1}H-NMR (CDCl$_3$ + CD$_3$OD) δ 9.32 (s, 1H), 8.11 (d, J = 10 Hz, 1H), 7.90 (s, 1H), 7.80–7.41 (m, 4H), 2.30 (s, 3H); ^{13}C-NMR (CDCl$_3$ + CD$_3$OD) δ 168.5, 156.3, 151.2, 147.8, 137.3, 135.1, 131.0, 129.7, 129.6, 124,6, 124.0, 121.6, 117.5, 116.3, 22.5; EIMS (m/z + ion mode): [M + H]$^+$ = 321.16; [M − Cl and F-benzene and COCH$_3$]$^+$ = 150.07; Anal Calcd for C$_{15}$H$_{12}$ClFN$_2$OS: C, 56.16; H, 3.13; N, 8.72 found C, 56.30; H, 3.15; N, 8.76.

N-(6-(3,5-Dimethylphenyl)benzo[d]thiazol-2-yl)acetamide (3h). m.p. 205–207 °C; ^{1}H-NMR (acetone-d_6) δ 9.41 (s, 1H), 8.14 (d, J =2 Hz, 1H), 7.63 (d, J = 1.5, 1H), 7.72–7.53 (m, 4H), 2.31 (s, 6H), 2.28 (s, 3H); ^{13}C-NMR (acetone-d_6) δ 167.2, 158.0, 147.7, 140.3 (2C), 133.0, 129.9, 129.5, 128.3, 127.4 (2C), 123.7, 121.3, 119.2, 23.1, 22.2 (2C); EIMS (m/z − ion mode): [M − H]$^-$ = 294.91; [M − COCH$_3$ − 2OCH$_3$ and benzene]$^-$ = 149.24; [M − 2OCH$_3$ and benzene]$^-$ = 190.6; Anal Calcd for C$_{17}$H$_{16}$N$_2$OS: C, 68.88; H, 5.43; N, 9.44 found C, 68.83; H, 5.46; N, 9.34.

Procedure for Urease Inhibition Activity

Urease inhibitory assay of newly synthesized compounds 3a–3h were determined as follows: Enzyme (1 unit) in phosphate buffer (200 μL, pH 7) was combined with a particular stock solution (20 μL, a test compound or thiourea) and phosphate buffer (230 μL). The solution was incubated for 5 min at 25 °C. After incubation period 400 μL of urea stock solution (20 mM) was added to the solution. Calibration solution was synthesized without urea solution. For the action of urease, test tubes were incubated for 10 min at 40 °C. The solution of phenol hypochlorite reagent (1150 μL) was added. These tubes were incubated for 25 min at 56 °C. Absorbance of the blue colored compound was noted at 625 nm after 5 min of cooling. Then percentage urease inhibition was determined. While EZ-fit kinetic data base was used to obtain IC$_{50}$ values [25,35].

Molecular Docking Study

The PDB structure of 3LA4 was retrieved for docking purposes as a complex co-crystallized with inhibitor 2-amino-3-(2-(2-hydroxyethyl)disulfanyl) propan-1-ol and (5-amino-6-hydroxyhexyl)carbamic acid at the nickel-containing catalytic site [36]. Then, the amino acid chain was retained and the water molecules and co-crystallized ligands were removed and subsequently the missing atom types were repaired using Modeller 9.11 (University of California San Francisco, San Francisco, CA, USA). Afterwards, the polar hydrogen was added to the receptor and the resulting protein was subjected to minimization using OPLS 2005 force field. The prepared protein was saved in pdbqt format using Autodock Tools 1.5.4 [21,37,38,39]. The ligand coordinates were generated using MarvineSketch 5.8.3, 2012 (ChemAxon LLC, Cambridge, MA, USA) [40], which was converted to 3D structure using Openbabel version (2.3.1). Finally the pdbqt formats (The input format of docking software) of the ligands were prepared with Autodock Tools 1.5.4 using default parameters. AutodockVina ver. 1.1.1 (The Scripps Research Institute, La Jolla CA, USA) was used for docking calculations with default parameters except for exhaustiveness that was set to 80. For all the docking calculations, a grid box size of 40 × 40 × 40, centered at the geometrical center of co-crystallized ligands was used. Co-crystalized ligands were attached at two different sites, one near the nickel catalytic site (A) and the other site where the inhibitor 2-amino-3-(2-(2-hydroxyethyl)disulfanyl)propan-1-ol was attached (Site B) [20]. The coordinates x, y, z for the center of grid box were (Site A) −39.86, −45.06, 72.52 and (Site B) −75.03, 20.84, 81.83 respectively. To validate our docking procedure, the co-crystallized ligands were re-docked into their respective site of the enzyme and the reasonable RMSD value of 1.947 A was obtained. Finally, the conformations with the most favorable free energy of binding were selected for analyzing the interactions between urease and its inhibitor. All of the 3D models are generated using the Molegro Molecular Viewer 2.5 (CLC bio company, Aarhus N, Denmark) [41] and LigPlot + (The European Bioinformatics Institute, Hinxton, Cambridge, UK) [42] software.

Nitric Oxide Scavenging Activity

The activity of newly synthesized benzothiazole derivatives was determined using the Garrat method. Griess reported the Garrat method which is followed by a diazotization reaction [43]. Under acidic conditions, this method utilizes sodium nitroprusside as the source of nitric oxide, sulfanilamide and *N*-1-naphthylethylenediamine dihydrochloride to detect NO_2^-, produced at the expense of nitric oxide. A known amount of tested compounds 3a–3h was dissolved in sodium nitroprusside solution (20 mM, 100 μL) and then the volume was made up to 1000 μL with phosphate buffer (200 mM, pH 7.4). This solution was incubated for 2 h at 37 °C and Griess reagent (100 μL) was added. This solution was stored for 20 min at room temperature. At 528 nm, optical density of this colored solution was observed. For positive control, ascorbic acid was used. Negative control was used to form the standard curve [43].

Haemolytic Activity

Haemolytic activity of newly synthesized benzothiazole derivatives 3a–3h was determined using a reported method [44]. Solutions of compounds were prepared at concentrations of 1 mg/mL in 10% DMSO with 90% water Heparinized human fresh blood (3 mL) was used that was homogeneously mixed and added into a 15 mL sterile Falcon tube. It was centrifuged for 5 min and the supernatant was removed. Chilled sterile isotonic phosphate buffer saline solution (5 mL, 7.4 pH) at 4 °C was used three times. Washed red blood cells were suspended in chilled RBS (20 mL). A haemacytometer was used for counting erythrocytes. For each assay 7.068×10^8 red blood cells per mL count were maintained and then diluted blood cells (180 µL) were added to the test compound (20 µL) and suspended in Eppendorf tubes. It was incubated for 35 min at 37 °C then the tubes were kept in an ice bath for 5 min and centrifuged again for 5 min. After centrifugation, the obtained supernatant was collected carefully and diluted with 900 µL of chilled PBS. All these tubes were kept in ice bath and solution (200 µL) was added into 96 well plates from each Eppendrof tube. For each essay, Triton X-100 (0.1%) was taken as positive control. For negative control, phosphate buffer was used. A microplate reader was used for determining the absorbance at 576 nm [32].

Antibacterial Activity

These newly benzothiazole derivatives were tested for their antibacterial activities against two Gram positive strains (*Baccilus subtilis, Staphylococcus aureus*) and four Gram negative strains (*Escherichia coli, Psuedomonas aeruginosa, Shigella dysenteriae* and *Salmonella typhae*) using a reported protocol [45]. Streptomycin was used as positive control. The 96 well plate method was optimized for measuring the antibacterial activities of these compounds. In each well, sterilized broth (175 µL) was added and inoculated with glycerol stock (5 µL) of a specific bacterial strain. The initial absorbance was observed between 0.12–0.19. The bacteria were allowed to grow in an incubator overnight. After a certain waiting time (12 h), test sample (20 µL) was added to the pre-determined wells. Concentration of test sample was 20 µg/well. Total volume was 200 µL in each well. These plates were incubated for 16–24 h at 37 °C. Absorbance was observed at 630 nm by using an ELISA plate reader. The difference in absorbance values were observed and were used as an index of bacterial growth. The following formula was used to calculate percentage inhibition:

Percentage Inhibition = (O.D of + ve control − O.D of sample × 100)/O.D of +ve control   (1)

CONCLUSIONS

This study reports C-C coupling reactions of 2 with various arylboronic pinacol ester/aryl boronic acids using palladium catalyst. These new products 3a–3h were prepared in moderate to good yields. These Suzuki coupling benzothiazole derivatives were checked for their biological (urease inhibitory, nitric oxide scavenging, haemolytic and antibacterial) activities. The urease inhibition results showed that product 3b was an excellent urease inhibitor. Products with electron releasing groups on the aryl moiety of the benzothiazole molecule showed the highest inhibition of urease activity. Molecular docking studies of the urease inhibitory activity showed that the H-bonding ability present in these *N*-protected benzothiazoles prevents the catalytic activity of the enzyme. Nitric oxide scavenging assays were done for these compounds. Compound 3b also exhibited highest nitric oxide scavenging activity. All newly synthesized compounds showed haemolytic activity. It was found that electron withdrawing substitution on the aryl produced the highest haemolytic activity. The newly synthesized benzothiazole derivatives 3a–3h showed excellent antibacterial activities against *E. coli*.

ACKNOWLEDGMENTS

This study reported herein is part of Ph.D. thesis work of Yasmeen Gull. Financial assistance for this study was provided by the Higher Education Commission (HEC), Pakistan and HEC Scholarship awarded to Ms. Yasmeen Gull.

AUTHOR CONTRIBUTIONS

Conceived and designed the experiments: Y.G., N.R., M.N., A.A.A. Performed the experiments: (Synthesis and bioassay—Y.G., M.N.); (Enzyme inhibition—A.Y.); (Docking and other computational—A.A.A.). Analyzed the data: Y.G., A.A.A., M.N., M.Z., S.G.M. Contributed to manuscript preparation N.R., F.-H.N., M.Z., V.D.F.

REFERENCES

1.   Gajdos, P.; Magdolen, P.; Zahradnik, P.; Foltinova, P. New

conjugated benzothiazole-*N*-oxides: Synthesis and biological activity. *Molecules* 2009, *14*, 5382–5388.

2. Mishra, A.; Behera, R.K.; Behera, P.K.; Mishra, B.K.; Behera, G.B. Cyanines during the 1990s: A review. *Chem. Rev.* 2000, *100*, 1973–2012.

3. Michaelidou, A.S.; Hadjipavlou-Litina, D. Nonsteroidal anti-inflammatory drugs (NSAIDs): A comparative QSAR study. *Chem. Rev.* 2005, *105*, 3235–3271.

4. Yadav, P.S.; Devprakash; Senthikumar, G.P. Benzothiazole: Different method of synthesis and diverse biological activities. *Int. J. Pharm. Sci. Drug Res.* 2011, *3*, 1–7.

5. Aiello, S.; Wells, G.; Stone, E.L.; Kadri, H.; Bazzi, R.; Bell, D.R.; Stevens, M.F.; Matthews, C.S.; Bradshaw, T.D.; Westwell, A.D. Synthesis and biological properties of benzothiazole, benzoxazole, and chromen-4-one analogues of the potent antitumor agent 2-(3,4-dimethoxyphenyl)-5-fluorobenzothiazole (PMX 610, NSC 721648). *J. Med. Chem.* 2008, *51*, 5135–5139.

6. Cho, Y.; Ioerger, T.R.; Sacchettini, J.C. Discovery of novel nitrobenzothiazole inhibitors for mycobacterium tuberculosis ATP phosphoribosyl transferase (HisG) through virtual screening. *J. Med. Chem.* 2008, *51*, 5984–5992.

7. Bondock, S.; Fadaly, W.; Metwally. Recent trends in the chemistry of 2-aminobenzothiazoles. *J. Sulfur Chem.* 2009, *30*, 74–107.

8. Saeed, A.; Rafique, H.; Rasheed, S. Synthesis and antibacterial activity of some new 1-Aroyl-3-(substituated-2-benzothiazolyl)thioureas. *Pharm. Chem. J.* 2008, *48*, 191–195.

9. Bele, D.S.; Singhvi, J. Synthesis of some mannich bases of 6-substituted-2-aminobenzothiazole as analgesic. *Res. J. Pharm. Technol.* 2014, *7*, 316–321.

10. Hutchinson, I.; Chua, M.; Bradshaw, T.D.; Matthews, C.S.; Stevens, F.G.; Westwell, A. Antitumor benzothiazoles, part 20: 3'-Cyano and 3'-alkynyl-substituted 2-(4'aminophenyl) benzothiazoles as new potent and selective analogues.*Bioorg. Med. Chem. Lett.* 2003, *13*, 471–474.

11. Trapani, G.; Franco, M.; Latrofa, A.; Reho, A.; Liso, G. Synthesis *in vitro* and *in vivo* cytotoxicity, and prediction of the intestinal absorption of substituted 2-ethoxycarbonyl-imidazo[2,1-*b*]benzothiazoles. *Eur. J. Pharm. Sci.* 2001, *14*, 209–216.

12. Kamal, M.; Dawood. Microwave-assisted Suzuki-Miyaura and Heck-Mizoroki cross-coupling reactions of aryl chlorides and

bromides in water using stable benzothiazole-based palladium(II) precatalysts. *Tetrahedron* 2007, *63*, 9642–9651.

13. Ali, S.; Rasool, N.; Ullah, A.; Nasim, F.; Yaqoob, A.; Zubair, M.; Rashid, U.; Riaz, M. Design and synthesis of arylthiophene-2-Carbaldehydes via suzuki-miyaura reactions and their biological evaluation. *Molecules* 2013, *18*, 14711–14725.

14. Rizwan, K.; Zubair, M.; Rasool, N.; Ali, S.; Zahoor, A.F.; Rana, U.A.; Khan, S.; Shahid, M.; Zia-Ul-Haq, M.; Jaafar, H.Z. Regioselective synthesis of 2-(bromomethyl)-5-aryl-thiophene derivatives via palladium (0) catalyzed suzuki cross-coupling reactions: As antithrombotic and haemolytically active molecules. *Chem. Cent. J.* 2014, *8*.

15. Noreen, M.; Rasool, N.; Khatib, M.E.; Molander, G.A. Arylation and heteroarylation of thienylsulfonamides with organotrifluoroborates. *J. Org. Chem.* 2014, *79*, 7243–7249.

16. Ikram, H.M.; Rasool, N.; Ahmad, G.; Chotana, G.A.; Musharraf, S.G.; Zubair, M.; Rana, U.A.; Zia-Ul-Haq, M.; Jaafar, H.Z. Selective C-arylation of 2,5-dibromo-3-hexylthiophene via suzuki cross coupling reaction and their pharmacological aspects. *Molecules* 2015, *20*, 5202–5214.

17. Rasheed, T.; Rasool, N.; Noreen, M.; Gull, Y.; Zubair, M.; Ullah, A.; Rana, U.A. Palladium (0) catalyzed Suzuki cross-coupling reactions of 2,4-dibromothiophene: Selectivity, characterization and biological applications. *J. Sulfur Chem.* 2015, *36*, 240–250.

18. Majo, V.J.; Prabhakaran, J.; Mann, J.J.; Kumar, J.S.D. An efficient palladium catalyzed synthesis of 2-arylbenzothiazoles. *Tetrahedron Lett.* 2003, *44*, 8535–8537.

19. Gull, Y.; Rasool, N.; Noreen, M.; Nasim, F.; Yaqoob, A.; Kousar, S.; Rashid, U.; Bukhari, I.H.; Zubair, M.; Islam, M.S. Efficient synthesis of 2-amino-6-Arylbenzothiazoles via Pd(0) suzuki cross coupling reactions: Potent urease enzyme inhibition and nitric oxide scavenging activities of the products. *Molecules* 2013, *18*, 8845–8857.

20. Benini, S.; Rypniewski, W.R.; Wilson, K.S.; Miletti, S.; Ciurli, S.; Mangani, S. A new proposal for urease mechanism based on the crystal structures of the native and inhibited enzyme from Bacillus pasteurii: Why urea hydrolysis costs two nickels. *Structure* 1999, *7*, 205–216.

21. Frishman, D.; Argos, P. Knowledge-based protein secondary structure assignment. *Prot. Struct. Funct. Bioinform.* 1995, *23*, 566–579.

22. Jamil, M.; Zubair, M.; Farid, M.A.; Altaf, A.A.; Rasool, N.; Nasim, F.U.H.; Ashraf, M.; Rashid, M.A.; Ejaz, S.A.; Yaqoob, A.; *et al.* Study

of Antioxidant, Cytotoxic, and Enzyme Inhibition Activities of Some Symmetrical $N^3,N^{3'}$-Bis(disubstituted)isophthalyl-bis(thioureas) and $N^3,N^3,N^{3'},N^{3'}$-Tetrakis(disubstituted)isophthalyl-bis(thioureas) and Their Cu(II) and Ni(II) Complexes. *J. Chem. Soc. Pak.* 2014, *36*, 491–497.

23. Jamil, M.; Zubair, M.; Altaf, A.A.; Farid, M.A.; Hussain, M.T.; Rasool, N.; Bukhari, I.H.; Ahmad, V.U. Synthesis, Characterization and Antibacterial Activity of Some Novel Symmetrical *N*-3,*N*-3'-Bis (disubstituted) isophthalyl-bis (thioureas) and *N*-3,*N*-3,*N*-3',*N*-3'-Tetrakis (disubstituted) isophthalyl-bis (thiourea) and Their Cu (II) and Ni (II) Complexes. *J. Chem. Soc. Pak.* 2014, *35*, 737–743.

24. Onoda, Y.; Iwasaki, H.; Magaribuchi, T.; Tamaki, H. Effects of the new anti-ulcer agent 12-sulfodehydroabietic acid monosodium salt on healing of acetic acid-induced gastric ulcers in rats. *Arzneim. Drug Res.* 1991, *41*, 546–548.

25. Serwar, M.; Akhtar, T.; Hameed, S.; Khan, K.M. Synthesis urease inhibition and antimicrobial activities of some chiral 5-aryle-4-(1-phenylpropyl)-2*H*-1,2,4-trizole-3(4*H*)- thiones. *ARKIVOC* 2009, *7*, 210–221.

26. Mao, W.J.; Lv, P.C.; Shi, L.; Li, H.Q.; Zhu, H.L. Synthesis, molecular docking and biological evaluation of metronidazole derivatives as potent *Helicobacter pylori* urease inhibitors. *Bioorg. Med. Chem.* 2009, *17*, 7531–7536.

27. Aslam, M.A.S.; Mahmood, S.U.; Shahid, M.; Saeed, A.; Iqbal, J. Synthesis, biological assay *in vitro* and molecular docking studies of new schiff base derivatives as potential urease inhibitors. *Eur. J. Med. Chem.* 2011, *46*, 5473–5479.

28. Gul, R.; Rauf, M.; Badshah, A.; Azam, S.S.; Tahir, M.N.; Khan, A. Ferrocene-based guanidine derivatives: *In vitro*antimicrobial, DNA binding and docking supported urease inhibition studies. *Eur. J. Med. Chem.* 2014, *85*, 438–449.

29. Naresh, P.; Pattanaik, P.; Rajeshwar, B. Synthetic characterization and antioxidant screening of some novel 6-fluorobenzothiazole substituted [1,2,4] triazole analogues. *Int. J. Pharm. Sci.* 2013, *3*, 170–174.

30. Hazra, K.; Nargund, L.; Rashmi, P.; Chandra, J.N.S.; Nandha, B. Synthesis and antioxidant activity of some novel fluorobenzothiazolopyrazoline. *Der. Chem. Sin.* 2011, *2*, 149–157.

31. Devmurari, V.P.; Shivan, P.; Goyani, M.B.; Jivani, N.P. Synthesis and anticancer activity of some novel 2-substituared benzothiazole. *Int. J. Chem. Sci.* 2010, *8*, 663–675.

32. Powell, W.A.; Catranis, C.M.; Maynard, C.A. Design of self-processing antimicrobial peptides for plant protection.*Lett. Appl. Microbiol.* 2000, *31*, 163–168.
33. Saeed, S.; Rashid, N.; Jones, P.G.; Ali, M.; Hussain, R. Synthesis, characterization and biological evaluation of some thiourea derivatives bearing benzothiazole moiety as potential antimicrobial and anticancer agents. *Eur. J. Med. Chem.*2010, *45*, 1323–1331.
34. Miyaura, N.; Suzuki. Palladium-catalyzed cross-coupling reactions of organoboron compounds. *Chem. Rev.* 1995, *95*, 2457–2483.
35. Pervez, H.; Rmzan, M.; Yaqub, M.; Nasim, F.H.; Khan, K. Synthesis and biological evaluation of some new *N*-4-aryl substituted 5-chloroisatin-3-thiosemicarbazones. *Med. Chem.* 2012, *8*, 505–514.
36. Balasubramanian, A.; Ponnuraj, K. Crystal structure of the first plant urease from jack bean: 83 Years of journey from its first crystal to molecular structure. *J. Mol.Biol.* 2010, *400*, 274–283.
37. Srinivasan, R.; Rose, G.D. A physical basis for protein secondary structure. *Proc. Nat. Acad. Sci. USA* 1999, *96*, 14258–14263.
38. Baker, N.A.; Sept, D.; Joseph, S.; Holst, M.J.; Mccammon, J.A. Electrostatics of nanosystems: Application to microtubules and the ribosome. *Proc. Nat. Acad. Sci. USA* 2010, *98*, 10037–10041.
39. Harris, R.; Olson, A.J.; Goodsell, D.S. Automated prediction of ligand-binding sites in proteins. *Struct. Funct. Bioinform.* 2003, *70*, 1506–1517.
40. ChemAxon. Available online: http://www.chemaxon.com (accessed on 31 December 2015).
41. CLC Drug Discovery Workbench. Available online: http://www.molegro.com/mmv-product.php (accessed on 31 December 2015).
42. Laskowski, R.A.; Swindells, M.B. LigPlot+: Multiple ligand-protein interaction diagrams for drug discovery. *J. Chem. Inf. Mod.* 2011, *51*, 2778–2786.
43. Garrat, D. *The Quantitative Analysis of Drugs*, 3rd ed.; Chapman and Hall Ltd. Tokyo: Tokyo, Japan, 1996; pp. 456–458.
44. Riaz, M.; Rasool, N.; Bukhari, I.H.; Shahid, M.; Zubair, M.; Rizwan, K.; Rashid, U. *In vitro* antimicrobial, antioxidant, cytotoxicity and GC-MS analysis of Mazus goodenifolius. *Molecules* 2011, *17*, 14275–14287.
45. Rehman, A.U.; Awais, U.R.; Muhammad, A.B.; Hira, K.; Parsa, D. Synthesis and biological screening of *N*-substituted derivatives of *N*-benzyl-4-chlorobenzenesulfonamide. *Asian J. Pharm. Health Sci.* 2012, *2*, 384–389.

Chapter 6

A SURVEY OF CHEMICAL COMPOSITIONS AND BIOLOGICAL ACTIVITIES OF YEMENI AROMATIC MEDICINAL PLANTS

Bhuwan K. Chhetri[1], Nasser A. Awadh Ali[2,3] and William N. Setzer[1]

[1]Department of Chemistry, University of Alabama in Huntsville, Huntsville, AL 35899, USA
[2]Pharmacognosy Department, Faculty of Pharmacy, Sana'a University, Yemen
[3]Pharmacognosy Department, Faculty of Clinical Pharmacy, Albaha University, Saudi Arabia

ABSTRACT

Yemen is a small country located in the southwestern part of the Arabian Peninsula. Yemen's coastal lowlands, eastern plateau, and deserts give it a diverse topography, which along with climatic factors make it opulent in flora. Despite the introduction of Western medicinal system during the middle of the twentieth century, herbal medicine still plays an important role in Yemen. In this review, we present a survey of several aromatic plants used in traditional medicine in Yemen, their traditional uses, their volatile chemical compositions, and their biological activities.

INTRODUCTION

From prehistoric times people have been using medicinal plants for the treatment of a wide variety of ailments. This traditional use of plants is based upon an empirical, timeless trial and error, correlating certain plants to the management and cure of particular diseases. These plants were used as herbal formulations in crude forms like tinctures, teas, powders, and poultices. The traditional way by which these plants were used can still be found in communities, passed down through natural history, and still prevail to be a successful form of medication, despite the advent of modern medicine.

Aromatic medicinal plants are plants with aroma characteristics as well as having medicinal properties and are "chemical goldmines" because of the diverse range of secondary metabolites that they possess and the wide range of pharmacological activities that they show [1]. The use of aromatic medicinal plants has been increasing steadily with notable use in the pharmaceutical, cosmetic, and food industries [2].

The republic of Yemen was formed in 1990 by merging the northern and southern states of Yemen [3]. It is located in the southwest part of Arabian Peninsula, between 12°40′ and 19°00′ N latitude and 42°30′ to 53°05′ E longitude (see Figure 1) [4]. Unlike the Arabian Peninsula, which is known for its vast inhospitable deserts, Yemen has rugged highlands and mountains with several peaks that are more than 3000 m (10,000 ft). The Yemeni highlands experience relatively abundant rainfall and have a temperate climate. Yemen's coastal lowlands, eastern plateau and deserts give it a high topographic diversity, which along with climatic factors, has resulted in a diverse and rich flora. Yemen occupies a very vital position on the southern Arabian Peninsula sharing borders with two countries, Saudi Arabia to the north and Oman towards the northeast. The east and south of Yemen border the Red Sea and the Gulf of Aden with a total coastline measuring 1906 km (1184 miles). Yemen also has more than 100 small islands scattered in the nearby Red Sea and Arabian Sea. The total area of Yemen including these islands is 527,970 sq km (203,850 sq mi) [4].

There are two major forces that determine the climate of Yemen. Winters in Yemen are dominated by dry northerly winds whereas moist monsoons prevail in spring and summer. The other major factor that is responsible for the climatic condition is the elevation, with the highlands experiencing a temperate climate with mild, dry winters and warm summers with abundant rainfall. To take an example, Sana'a, a region in the central highlands, has an average temperature of 14 °C in January and 22 °C in July. The average annual rainfall in Sana'a is 265 mm (10.4 in), whereas the highlands around Ibb and Taiz receive more than 1000 mm (40 in) each year. Sana'a is the capital and the largest city of Yemen. At an altitude of more than 2200 m, the city extends across a very fertile basin near the foot of the mountain called Jabal Nuqum. Taiz is at an altitude of 1400 m in the fertile highlands. It is a productive agricultural region where coffee (*Coffea arabica*) and khat (*Catha edulis*), as mild stimulants, are the main crops. Ibb is also in the southern highlands, north of Taiz at an elevation of 2050 m. The area gets abundant rainfall and has rich volcanic soil, which makes it green and agriculturally productive, the chief crops being grains, coffee, khat, fruits and vegetables [4]. There has been an ever-increasing interest in natural products as alternatives for artificial additives of pharmacologically relevant agents in recent years [5], and essential oils from aromatic plants have gained much interest for their use in various foods, cosmetics, and pharmaceutical products. They have widespread use as flavoring materials and stand as "green" alternatives in pharmaceutical, agricultural, nutritional, and many other fields. They show a wide range of biological activities, which account for the use of aromatic plants in traditional medicine and for their growing interest as alternative therapies for the prevention, cure, and alleviation of certain diseased conditions [1,6]. Essential oils have been widely studied and used for their antimicrobial, antiviral, nematicidal, antifungal, insecticidal and antioxidant properties. In this review, we have focused on the traditional medicinal uses and volatile components (Figure 2, Figure 3 and Figure 4) of several aromatic medicinal plants from Yemen.

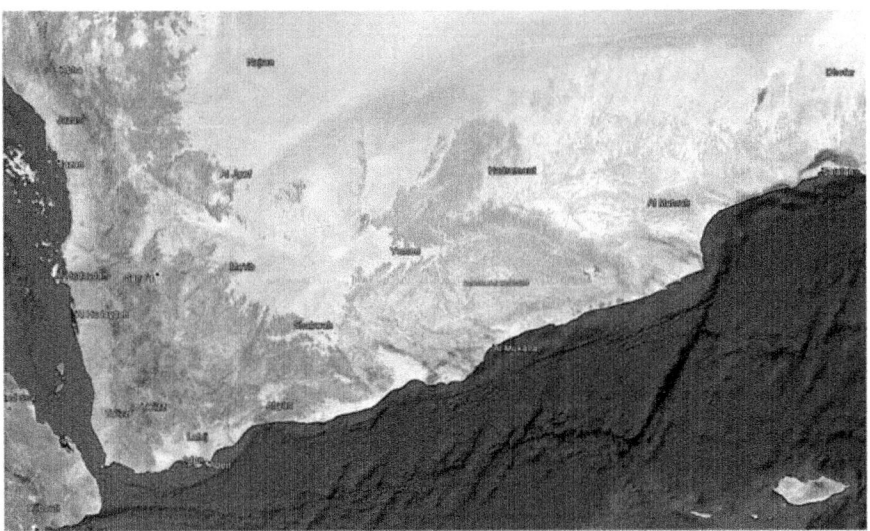

Figure 1: Google Earth© satellite view of Yemen.

AROMATIC MEDICINAL PLANTS FROM YEMEN

Anacardiaceae

Schinus molle L.

There are around 33 species of *Schinus*, all of which are native to the Neotropics [7]. There is only one species, *Schinus molle,* which has been introduced to Yemen. The plant, known as "filfil katheb", is used in Yemeni folk medicine as an expectorant, a diuretic and also for the treatment of stomach upsets [8].

Bioactivity studies of *S. molle* essential oil by various researchers have shown that it possesses antibacterial, antifungal [9], cytotoxic [10], insecticidal, and insect repellent [11] properties. The essential oil of *S. molle* from Yemen, obtained by hydrodistillation contained β-caryophyllene (13.5%), α-pinene (2.8%), carvacrol (2%), germacrene D (16.7%), δ-cadinene (3.2%), spathulenol (3.4%), caryophyllene oxide (6.7%), viridiflorol (3.3%), α-cadinol (2.5%), α-bisabolol (4%), and an unknown compound (18%).

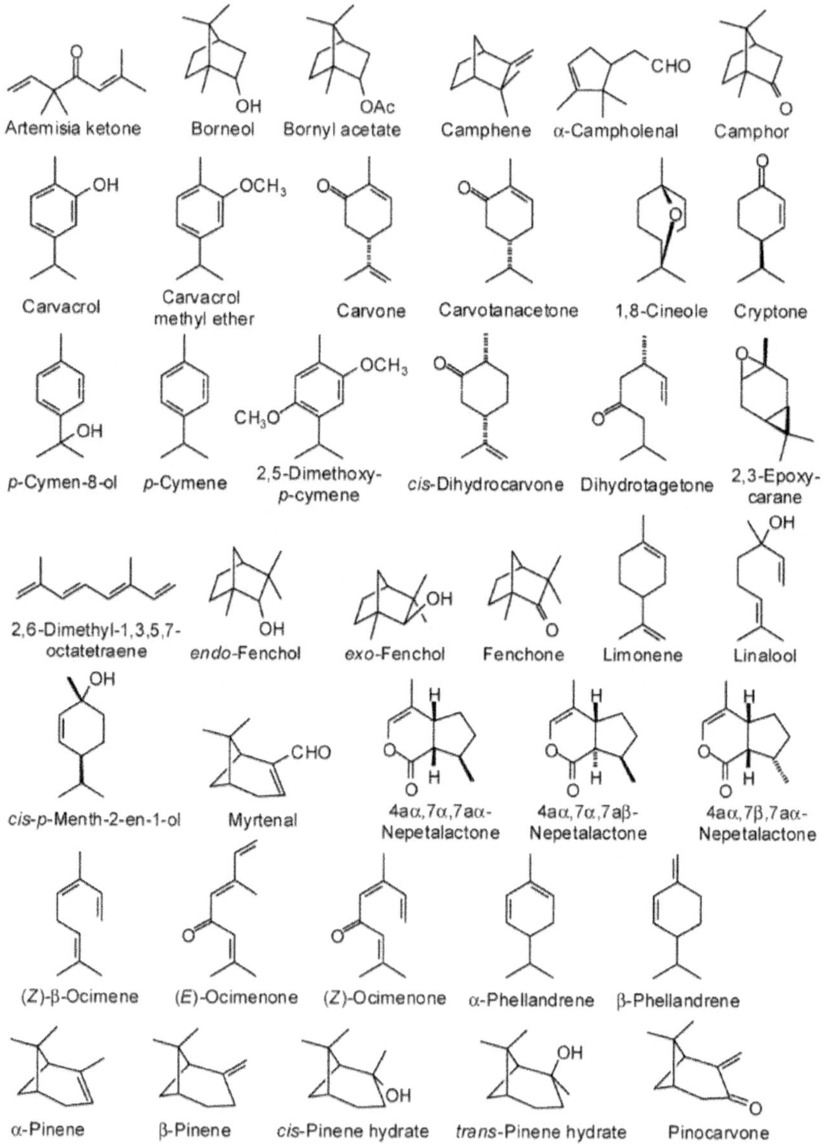

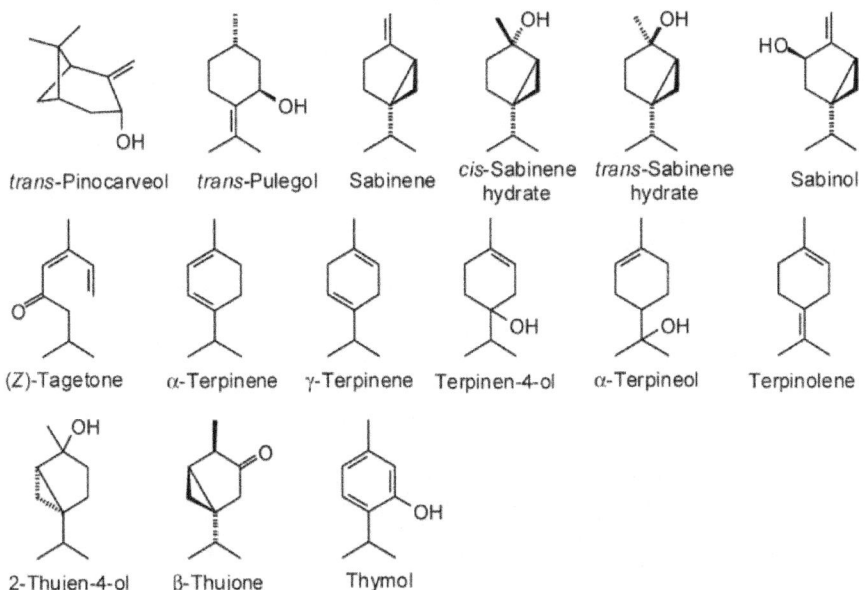

Figure 2: Monoterpenoid essential oil components discussed in this review.

The same plant was also extracted by supercritical CO_2 and analyzed by GC-MS. Here the major components were comparable with only with some minor differences in their concentrations. The notable compounds identified were α-pinene (1.6%), β-caryophyllene (9.1%), germacrene D (13.7%), bicyclogermacrene (2%), spathulenol (2.7%), shybunol (3.3%), abietol (5.0%) and an unknown compound (37%) [12]. The observed bioactivities of *S. molle* essential oil can be attributed to the relatively large concentrations of the sesquiterpenoids β-caryophyllene, germacrene D, and caryophyllene oxide, which have shown antimicrobial and cytotoxic activities [13].

Asteraceae

Artemisia abyssinica Sch. Bip. ex A. Rich.

There are around 400 species of *Artemisia*, distributed typically in Asia, Europe, and North America [7]. *Artemisia abyssinica*, known as "boitheran" in Yemen, is an aromatic, grey, silky-hairy plant with pale yellow flower-heads and is well known as a stimulant and an analgesic. It is short lived perennial plant, with sparingly branched stems that are grooved especially above. Leaves are alternate, grey-green, deeply bipinnatisect with linear segments, and 4–10 cm long. It is widely spread on the high plateau from 2200 to 3600 m and often abundant on roadsides, alluvial plains, and abandoned fields. It is used in Yemen for treating headache and as insect repellent. In Saudi Arabia, a decoction of fresh whole plant is traditionally used to treat diabetes mellitus [14]. The plant has also been used in folk medicine as an anthelmintic, antispasmodic, antirheumatic and antibacterial agent [15]. Antioxidant, antileishmanial and antitrypanosomal activities have also been recorded for *Artemisia abyssinica* essential oil [15].

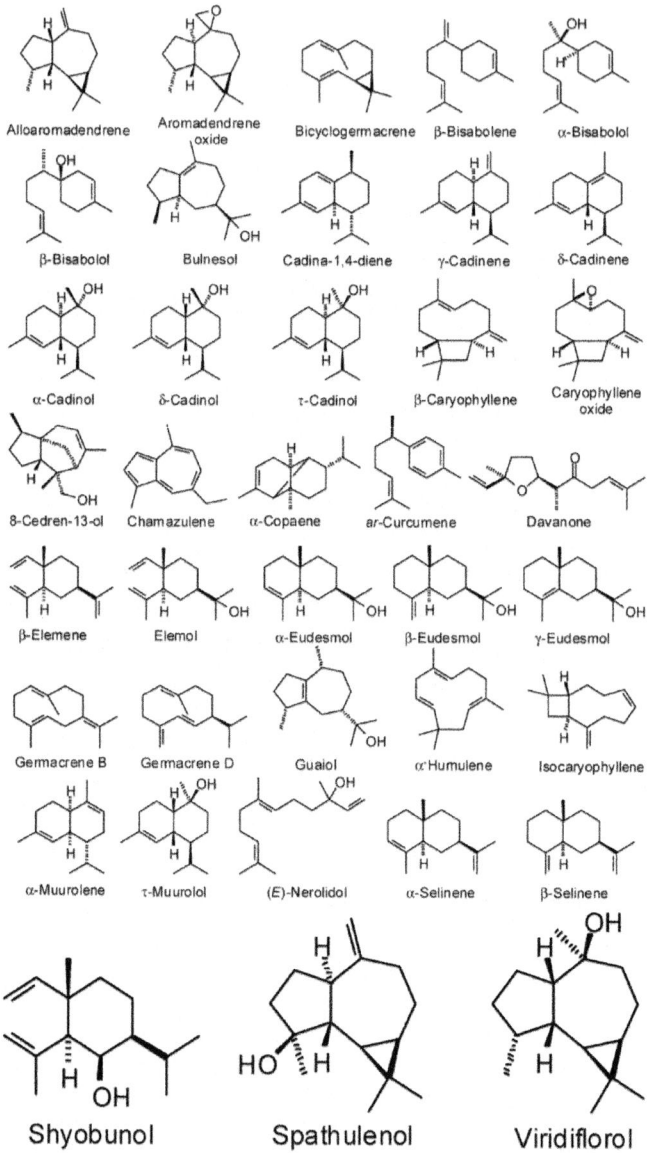

Figure 3: Sesquiterpenoid essential oil components discussed in this review.

Essential oil compositions have been determined for *Artemisia abyssinica* from three different regions of Yemen, namely Taiz (higher than 1500 m), Sana'a (higher than 3000 m) and Alhodiadah (coastal region). The major components of *Artemisia abyssinica* essential oils from Yemen are listed in Table 1 [16,17]. *Artemisia abyssinica* essential oil was rich in camphor and davanone with lesser amounts of (*E*)-nerolidol, *cis*-sabinene hydrate, terpinen-4-ol, linalool, and bornyl acetate. *Artemisia abyssinica* essential oil showed marginal cytotoxic activity against MCF-7 cells (30% cell viability reduction at a concentration of 100 µg/mL) [16]. *Artemisia abyssinica* oil from Yemen is remarkably different in composition from that reported from Ethiopia, which was rich in yomogi alcohol (38.5%), artemisyl acetate (24.9%), and artemisia alcohol (6.7%) [18].

Table 1: Major components (%) in essential oils of *Artemisia abyssinica* from Yemen

Compound	Taiz (above 1500 m)	Sana'a (above 3000 m)	Alhodiadah (Coastal)
α-Pinene	1.1	0.5	1.1
cis-Sabinene hydrate	2.8	3.6	5.8
Linalool	2.1	2.7	1.6
Camphor	29.6	42.1	42.5
Terpinen-4-ol	2.0	3.8	3.9
Bornyl acetate	1.3	2.3	2.3
(*E*)-Nerolidol	5.1	4.0	4.5
Davanone	49.4	34.5	32.3

Artemisia Arborescens L.

Artemisia arborescens has been used traditionally as an anti-inflammatory remedy. The essential oil of *Artemisia arborescens* has been reported to have antibacterial [19] and antifungal activities [15] as well as antiviral activity against HSV-1 and HSV-2 [20]. The essential oil of *Artemisia arborescens* from Yemen has been reported to have α-terpinene (8.7%), artemisia ketone (51.5%), camphor (14.1%), α-bisabolol (12.6%) and palmitic acid (2.4%) as the major constituents [21]. The oil showed strong *in vitro* cytotoxic activity with an IC_{50} of 16.9 µg/mL against HT 29 human colorectal adenocarcinoma cells. *Artemisia arborescens* essential oil also exhibited antifungal activity against *Cladosporium cucumerinum*. The cytotoxic activity of *Artemisia arborescens* can be attributed to the presence of α-bisabolol [22] and palmitic acid [23]. The essential oil composition of *Artemisia arborescens* from Yemen [21] was very different from samples from Sicily [19] or Algeria [24], which were dominated by β-thujone and chamazulene.

Conyza Bonariensis (L.) Cronquist

Conyza bonariensis (L.) Cronquist is known as "sadaf" in Yemen. It is an annual or short lived perennial weed. A variety of phytochemicals have been reported from the genus that includes alkaloids, volatile oils, terpenoids, phenolic acids, flavonoids and hydrolyzable tannins. *Conyza bonariensis* is widely used in folk medicine for the treatment of rheumatism, cystitis, gout, nephritis, dysmenorrhea, tooth pain and headache. Other studied biological activities of *Conyza bonariensis* include molluscicidal activity against *Biomphalaria* snails, anti-inflammatory activity, antipyretic effects, and antimicrobial activity [25]. Essential oil from the aerial plant parts of *Conyza bonariensis* collected from Yemen was dominated by the sesquiterpenoids 8-cedren-13-ol (18.5%) and aromadendrene oxide (18.8%) [26].

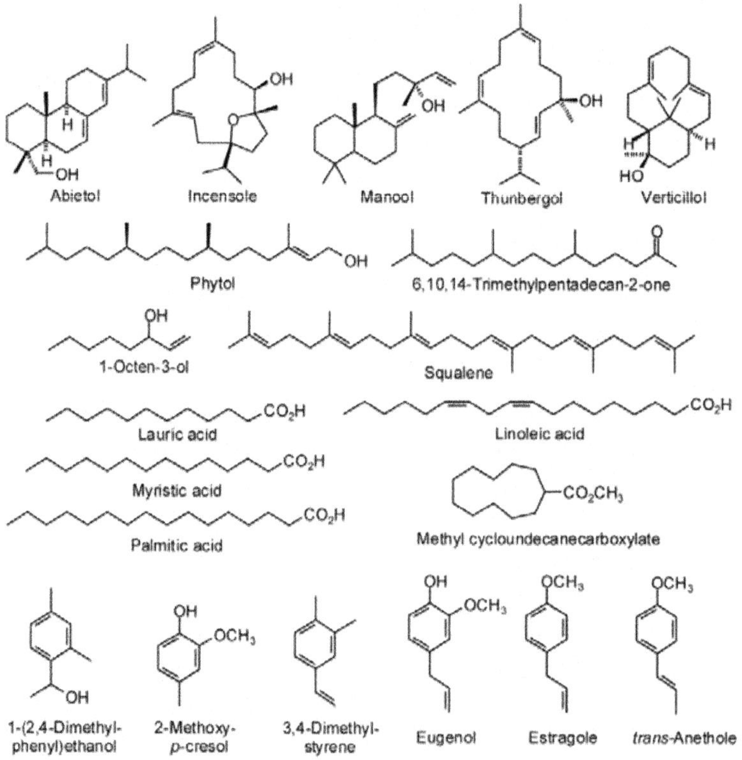

Figure 4: Miscellaneous essential oil components discussed in this review.

Pulicaria Inuloides (Poir.) DC.

There are 77 species of *Pulicaria*, distributed in temperate and warm Eurasia [7], and in folk medicine *Pulicaria* species have been used as insect repellents, galactagogues, antiepileptics, and as tonics [27]. *Pulicaria inuloides* is distributed from Morocco east throughout North Africa, south to Chad, Niger, and Somalia, and east to the Arabian Peninsula [28]. The plant, known as "sekba", is used traditionally in Yemen to treat wounds [29]. The essential oil of *Pulicaria inuloides* from Yemen was composed largely of carvotanacetone (47.3%) and palmitic acid (12.8%) [30]. The oil has demonstrated antimicrobial activity against *Staphylococcus aureus*, *Streptococcus pneumoniae*, *Bacillus subtilis*, *Escherichia coli*, and *Candida albicans*, and antioxidant activity in the 2,2-diphenyl-1-picrylhydrazyl (DPPH) radical-scavenging and β-carotene bleaching assays [31].

Pulicaria Jaubertii Gamal-Eldin

Pulicaria jaubertii is a perennial fragrant herb with erect branches growing up to 50 cm [32]. The plant, known as "anssif" in Yemen, is distributed in the southern Arabian Peninsula, and is used traditionally as a diuretic and antipyretic. Leaves of *Pulicaria jaubertii* are also used flavoring agents for cooking. Similar to *Pulicaria inuloides* (above), the essential oil of *Pulicaria jaubertii* from Yemen was dominated by carvotanacetone (64.0%) [33]. The essential oil of *Pulicaria jaubertii* from southern Saudi Arabia had a higher proportion of carvotanacetone (98.6%), and this essential oil had excellent *in vitro* cytotoxic activity against MCF-7 and Hep-G2 cells (IC_{50} = 3.8 and 5.1 µg/mL, respectively), but was only marginally antibacterial [32].

Pulicaria Stephanocarpa Balf. f.

On Soqotra island of Yemen, *Pulicaria* is represented by seven species, some of which have traditional importance as antiseptic agents and as food additives [34]. *Pulicaria stephanocarpa*, known as "derbeb" in Soqotra, has been traditionally used in a variety of health conditions including headache, abscesses, boils and sores. The essential oil of *Pulicaria stephanocarpa* from Soqotra has a high percentage of oxygenated sesquiterpenoids. One study showed the oil to be rich in α-cadinol (42.5%), spathulenol (22.0%), and β-caryophyllene (10.8%) [35], while a second revealed the major constituents to β-caryophyllene (13.4%), (*E*)-nerolidol (8.5%), caryophyllene oxide (8.5%), α-cadinol (8.2%), spathulenol (6.8%), and τ-cadinol (4.7%) [34]. The essential oil had moderate to high antimicrobial activity against *Staphylococcus aureus* (MIC= 3.12 µL/mL),*Bacillus subtilis* (12.5 µL/L) and *Candida albicans* (6.25 µL/mL). The oil showed free radical scavenging activity with an IC_{50} of 330 µg/mL as assessed by DPPH assay. Furthermore, the oil was tested for acetylcholine esterase (AChE) inhibitory activity where it showed an inhibition of 47% at concentration of 200 µg/mL using Ellman's method [34].

Pulicaria Undulata (L.) C.A. Mey.

An aromatic tea of *Pulicaria undulata*, known as "kho'ah", is used in the central Sahara to treat chills, diabetes, cardiac disorders, skin diseases, and abscesses [36], and in Egypt to treat inflammation, as an insect repellent, and an herbal tea [37].*Pulicaria undulata* oil has shown antibacterial [38], sedative [39], and insecticidal activities [40]. *Pulicaria undulata* essential oil from Yemen has been reported to have an extremely high concentration of carvotanacetone (91.4%) along with 2,5-dimethoxy-*p*-cymene (2.6%) [41]. Although carvotanacetone has been found in *Pulicaria undulata* oil collected from different geographical regions, comparison shows that the highest concentration is found in Yemeni *Pulicaria undulata*. Thus, for example, *Pulicaria undulata* oil from Sudan was composed of 55.9% carvotanacetone [42].

The essential oil from Yemen did show moderate *in vitro* cytotoxic activity on MCF-7 breast tumor cells (IC_{50} = 64.6 µg/mL), which can be attributed to the high concentration of carvotanacetone (see above). The oil also exhibited antibacterial activity against *Staphylococcus aureus* with minimum bactericidal concentration (MBC) of 3.12 µL/mL for*Staphylococcus aureus* and methicillin-resistant *Staphylococcus aureus*. The MBC against *Candida albicans* and *Bacillus subtilis*was 6.25 µL/mL, whereas MBC against *Escherichia coli* was 12.5 µL/mL [41].

Tagetes Minuta L.

Tagetes minuta is native to southern South America, but it has been introduced to Europe, Asia, the Middle East, and Africa [43]. The plant is known as "nirgis" in Yemen and used traditionally as a tea to treat the common cold, upper and lower respiratory inflammations, and for digestive ailments [43]. *Tagetes minuta* essential oil is an article of commerce and is used in baked goods, candy, ice cream, and soft drinks [44]. *Tagetes minuta* is used in the Arabian Peninsula as an antimicrobial, anthelmintic, diuretic, and antispasmodic [45], and the essential oil has shown biocidal, acaricidal, antifungal and antimicrobial activities [46,47,48,49,50,51]. The essential oil of *Tagetes minuta* from Yemen has been reported to have (*Z*)-ocimenone (15.9%), (*E*)-ocimenone (34.8%), (*Z*)-β-ocimene (8.3%), limonene (2.3%), (*Z*)-tagetone (1.8%), and dihydrotagetone (1.4%). The oil had moderate cytotoxic activity against MCF-7 cancer cell line with an IC_{50} of 54.7 µg/mL. However, the oil had a potent DPPH radical-scavenging activity with an IC_{50} of 36 µg/mL. Moreover, the oil was highly active against methicillin-resistant *Staphylococcus aureus* (MRSA) [51]. Similarly, *Tagetes minuta* essential oil showed considerable antioxidant activity in the thiobarbituric acid reactive substances (TBARS) assay [52].

Tarchonanthus Camphoratus L.

Tarchonanthus camphoratus, known locally as "bayad" or "mukar" in Yemen, is a strongly aromatic shrub found growing in the hillsides between 1200–2500 m [53]. Traditionally, its aromatic leaves are used for wounds and urinary tract infections [29]. An infusion of the leaves is used to treat stomach ailments and bronchitis. Burning of the aerial parts of the plant liberates smoke that is inhaled for the treatment of sinus-related complaints and for headaches. The burned seeds and leaves are used traditionally as fumigating agents in funeral rituals. Leaves can also be used as hot poultices for treating chest complaints whereas chewing the leaves is believed to alleviate toothache. There have been numerous studies on the essential oil of *Tarchonanthus camphoratus* and a wide variety of biological properties (e.g., antimicrobial, antifungal, antioxidant, antidiabetic, insecticidal, insect repellant, analgesic, and antipyretic) have been reported [53]. *Tarchonanthus camphoratus* leaf essential oil from Yemen is composed largely of oxygenated monoterpenoids (48.3%) and oxygenated sesquiterpenoids (32.7%). The principle compounds were *endo*-fenchol (21.2%), *trans*-pinene hydrate (8.8%), caryophyllene oxide (7.5%), α-terpineol (6.4%), τ-cadinol (6.4%) and α-cadinol (5.2%). The essential oil was also reported to moderate antimicrobial activity in disc diffusion assay against a number of pathogenic bacteria including methicillin-resistant*Staphylococcus aureus* (MRSA) and *Candida albicans*. The oil also showed moderate *in vitro* cytotoxic activity against HT29 (human colonic adenocarcinoma) tumor cells with an IC_{50} of 84.7 µg/mL [53].

Burseraceae

Boswellia Ameero Balf. f.

The genus *Boswellia* belongs to the Burseraceae and has about 24 species [7]. They are typically small trees and are found to grow mainly in Arabia, on eastern coast of Africa and in India. *Boswellia* trees contain a natural oleogum resin (frankincense) that exudes upon scoring the bark of the tree. On Soqotra Island of Yemen, there are eight endemic *Boswellia*species. Oleogum resin has been used traditionally by the native inhabitants for relieving toothache pain, sweetening the breath, and to sooth a disturbed stomach [54]. *Boswellia ameero* is endemic to Soqotra and is known locally as "ameero". It is dominant in woodlands on granite substrata above 600 m [55]. *Boswellia ameero* essential oil is composed of (3*E*,5*E*)-2,6-dimethyl-1,3,5,7-octatetraene (34.9%), 1-(2,4-dimethylphenyl)-ethanol (20.3%), 3,4-dimethylstyrene (17.3%), α-campholenal (13.4%), and α-terpineol (12.4%) and as the major components [56]. This oil showed DPPH radical-scavenging activity ($IC_{50}=$

175.2 µg/mL) and AChE enzyme inhibitory activity (41.5% inhibition at 200 µg/mL).

Boswellia Elongata Balf. f.

On Soqotra, the endemic *Boswellia elongata* is found on the limestone plateau between 300 and 450 m, on stony soils, and is the *Boswellia* species producing the most valuable frankincense [55]. Both *Boswellia ameero* and *Boswellia elongata* are used on Soqotra to treat the common cold, bronchitis, asthma, and rheumatism [57]. The essential oil of *Boswellia elongata* was found to be dominated by verticillol (52.4%), β-caryophyllene (39.1%), and methyl cycloundecanecarboxylate (7.9%) [56]. *Boswellia elongata* essential oil showed relatively weak DPPH radical-scavenging activity (IC_{50} = 211 µg/mL) and AChE enzyme inhibitory activity (29.6% inhibition at 200 µg/mL).

Boswellia Socotrana Engl.

Another *Boswellia* endemic to Soqotra, *Boswellia socotrana*, known as "samahno", is found primarily on the coastal plain of the island [55]. The essential oil was composed of (*E*)-2,3-epoxycarane (51.8%), 2-thujen-4-ol (31.3%), *p*-cymene (7.1%), 4-terpinenyl acetate (3.9%), and *p*-menth-1(7)-en-2-one (2.6%) [56]. Of the *Boswellia* essential oils tested, *Boswellia socotrana* showed the best DPPH radical-scavenging (IC_{50} = 211 µg/mL) and AChE inhibitory (59.3% inhibition at 200 µg/mL) activities.

Commiphora Habessinica (O. Berg) Engl.

Commiphora is also a member of the Burseraceae and the genus has more than 150 species [7]. The resinous exudates (myrrh) of *Commiphora* species have long been used in the form of perfumes and incense as well as for health problems such as stomachache, colds, wounds, malaria, fever, and as an antiseptic and against skin infections. *Commiphora habessinica* is known in Yemen as "khdash". The major components of the hydrodistilled oil of *Commiphora habessinica* oleogum resin from Yemen were β-elemene (32.1%), α-selinene (18.9%), cadina-1,4-diene (7.5%), germacrene B (3.6%), α-copaene (3.5%), τ-muurolol (3.0%), caryophyllene oxide (2.9%), α-cadinol (2.6%) [58].

Commiphora Kua Vollesen

The Soqotra island of Yemen is represented by five species of *Commiphora*, namely *Commiphora socotrana*, *Commiphora planifrons*, *Commiphora parvifolia*, *Commiphora ornifolia*, and *Commiphora kua* [59]. In Yemeni traditional medicine, the powdered resin of *Commiphora kua* ("bisham") is administered with warm milk or water for stomachache in young children [60]. The leaves of this plant are used to cleanse the mouth and throat, to treat cough and bronchitis, and as an antiseptic on the skin [61]. The major constituents of the hydrodistilled *Commiphora kua* resin from Soqotra were α-cadinol (33.0%), γ-cadinene (22.5%), δ-cadinene (17.0%), isocaryophyllene (3.7%), alloaromadendrene (2.8%), α-muurolene (2.7%), and α-humulene (2.4%) [60]. This essential oil showed moderate antifungal activity against *Cladosporium cucumerinum*.

Commiphora Ornifolia J.B. Gillett

Commiphora ornifolia is endemic to the island of Soqotra and is known as "ikshah". The bark of *Commiphora ornifolia* is used in Yemeni traditional medicine as an antiseptic, as an emmenagogue, as well as to treat diarrhea and dysentery [62]. The major components of the bark essential oil of *Commiphora ornifolia* were determined to be camphor (27.3%), *endo*-fenchol (15.5%), caryophyllene oxide (6.5%), thunbergol (6.4%), fenchone (4.4%), incensole (3.8%), borneol (2.9%) and manool (2.7%) [63]. *Commiphora ornifolia* bark oil had antibacterial against Gram-positive *Staphylococcus aureus* and *Bacillus subtilis* (MIC = 810 and 400 µg/mL, respectively), but demonstrated only weak DPPH radical-scavenging activity.

Commiphora Parvifolia Engl.

Commiphora parvifolia, known as "likham", is also endemic to Soqotra and is traditionally used as an antiseptic, uterine stimulant, emmenagogue, and to treat diarrhea and dysentery [57]. The essential oil from bark of *Commiphora parvifolia* showed weak DPPH radical-scavenging activity and antimicrobial activity, and the major components were palmitic acid (18.4%), caryophyllene oxide (14.2%), camphor (9.1%), β-eudesmol (7.7%), phytol (5.8%), bulnesol (5.7%), *endo*-fenchol (3.9%), and τ-cadinol (3.7%) [63].

Celastraceae

Catha Edulis Forssk.

Catha edulis, a member of the Celastraceae, is a dicotyledonous evergreen shrub known locally known as "khat". The major cultivation areas of khat is Ethiopia, particularly the Harar district, and Yemen. Chewing khat is a primary recreation in Yemen [4,64]. *Catha edulis* leaves contain several chemical groups of compounds including alkaloids, tannins, flavonoids, terpenes and sterols, and essential oils. Many studies have confirmed the presence of a complex set of alkaloids called kathdulinat, which are responsible for its stimulant effects especially phenylalkylamine compounds (cathinone and cathine) [65,66]. The leaves of khat are chewed to increase activity and elevate the mood. The roots and leaves are used traditionally for the treatment of influenza, cough, gonorrhea, chest and stomach problems [67]. The essential oil of *C. edulis* leaves from Yemen has been reported to have a high concentration of carvotanacetone (84.4%) in addition to *trans*-pulegol (2.2%) and 2,5-dimethoxy-*p*-cymene (1.9%) [67].

Lamiaceae

Ajuga Bracteosa Wall. ex Benth.

The Lamiaceae is a family of great importance due to its extensive use in folk medicine, cosmetics, culinary, and for the commercial production of essential oils [68]. *Ajuga bracteosa* is known locally as "hodam" and is a traditionally used medicinal plant that belongs to the Lamiaceae. The genus *Ajuga* is represented by only one species, *Ajuga bracteosa*, in Yemen. It is an erect and aromatic perennial herb used traditionally in Yemeni medicine as an antiseptic agent and to treat toothache [29]. The essential oil of *Ajuga bracteosa* from Yemen has been reported with high concentrations of oxygenated monoterpenoids (34.0%), fatty acids (30.3%), and oxygenated sesquiterpenoids (11.4%). The major components of the oil were 1-octene-3-ol (4.8%), linalool (1.9%), *endo*-fenchol (3.1%), camphor (4.4%), borneol (20.8%), eugenol (1.5%), lauric acid (2.3%), caryophyllene oxide (5.1%), myristic acid (3.3%), 6,10,14-trimethyl-pentadecane-2-one (3.0%), palmitic acid (16.0%), phytol (5.6%) and linoleic acid (7.0%) [69]. The oil showed some antibacterial activity against *Staphylococcus aureus* and *Bacillus subtilis* (MIC = 330 and 670 µg/mL, respectively). Moreover, it showed significant antioxidant activity in the DPPH- radical-scavenging assay (78% inhibition at 1.0 mg/mL).

Lavandula Dentata L.

The genus *Lavandula* consists of about 39 species [7] of which five are found growing naturally in Yemen [69]. In Yemeni folk medicine *Lavandula* species have been extensively used as a diuretic, an antiseptic, and for broncho-pulmonary infections [27]. *Lavandula dentata* is one of the species of *Lavandula* that are found growing in Yemen. It has been traditionally used in Yemeni folk medicine for the treatment of wounds, as a carminative, and to treat rheumatism [29]. *Lavandula dentata* from Yemen has been reported to have high concentrations of oxygenated monoterpenoids (51.8%), sesquiterpene hydrocarbons (13.9%), and oxygenated sesquiterpenoids (22.5%), with the major components being α-pinene (1.7%), camphene (1.8%), β-pinene (3.0%), fenchone (2.0%), linalool (3.7%), *endo*-fenchol (3.0%), *exo*-fenchol (2.4%), camphor (12.4%),*trans*-pinocarveol (7.5%), pinocarvone (3.1%), cryptone (3.0%), myrtenal (3.0%), myrtenol (3.9%), β-selinene (4.5%), caryophyllene oxide (3.1%), α-guaiol (6.1%), γ-eudesmol (2.4%), β-eudesmol (7.1%), and β-bisabolol (2.1%) [69]. The essential oil did not show any significant antimicrobial or antioxidant activity. 1,8-Cineole (41.3%) and sabinene (13.9%), sabinol (6.8%), and myrtenal (5.1%) were the major constituents of *Lavandula dentata* oil from Morocco [70]. Similarly, 1,8-cineole (33.5%), camphor (18.9%), and fenchone (8.4%), were major components found in *Lavandula dentata* oil from Tunisia [71]. The chemical composition of *Lavandula dentata* from Yemen differed qualitatively from samples growing in Morocco, Tunisia, Algeria and Saudi Arabia, but it most likely belongs to the camphor/*trans*-pinocarveol chemotype [69].

Lavandula Pubescens Decne.

Lavandula pubescens is known as "fahita" and has been traditionally used in Yemeni folk medicine as a carminative, insect repellent, and antiseptic [8]. *Lavandula pubescens* from Taiz, Yemen, is made up largely of carvacrol (20.6%), caryophyllene oxide (15.3%), β-bisabolene (12.0%), *p*-cymen-8-ol (11.8%), β-caryophyllene (10.7%), carvacrol methyl ether (7.2%), and terpinolene (6.0%) [15]. *Lavandula pubescens* essential oil has shown notable antibacterial activity against *Staphylococcus aureus* and *Salmonella enterica* (serovar Abony), and antifungal activity against *Aspergillus fumigatus* and *Candida albicans* [72]. The high concentration of carvacrol in this oil probably has some correlation to its use in folk medicine as an antiseptic, owing to the fact that the antimicrobial activity of carvacrol has been well established [73,74].

Leucas Virgata Balf. f.

There are around 100 species of *Leucas* distributed in Africa, Arabia, and Indomalaysia [7]. The Soqotra island of Yemen has 10 different species of *Leucas* of which *Leucas virgata* is endemic and known locally as "sa-lel hon". It is an abundant aromatic shrub and grows up to 1 m in height [75]. The leaves and sprigs are used for tea and it has been traditionally used for the treatment of several stomach problems. The essential oil of *Leucas virgata* is rich in oxygenated monoterpenoids (50.8%), the major ones being camphor (20.5%), *exo*-fenchol (3.4%), fenchone (5.4%), and borneol (3.1%). It also has a high percentage of oxygenated sesquiterpenoids like β-eudesmol (6.1%) and caryophyllene oxide (5.1%). The essential oil exhibits very good antibacterial activity against *Staphylococcus aureus*, *Bacillus subtilis* and *Esherichia coli*. A methanolic extract of *Leucas virgata* has shown good activity against *Trypanosoma brucei* with an IC_{50} of 8.1 µg/mL [76]. *Leucas virgata* extracts have been reported to have moderate antimicrobial activity against various bacterial strains with MIC values ≤ to 250 µg/mL [62].

Mentha Spicata L.

Mentha is a comparatively small genus having 18–19 species [7]. It is a well-known genus for its medicinal properties and is found in the temperate regions of Eurasia, Australia, and South Africa, and is cultivated from tropical to temperate climates of America, Europe, China, Brazil, and India [77]. *Mentha spicata* (spearmint) is a perennial, rhizomatous and glabrous herb that has a strong aromatic odor. The spearmint odor of *Mentha spicata* comes from the high concentration of (*R*)-carvone present in it. There are various ethno-medicinal uses of *Mentha spicata* essential oil and the genus is one of the most researched for its components as well as for the biological activities. It has been extensively used for flatulence, acidity neutralization, gastro stimulation and digestive problems. It has also been used for cough and cold, as a diuretic and spasmolytic [78]. It is considered to have analgesic, antipyretic and anti-inflammatory effects as well [79]. The essential oil of *Mentha spicata* from the Sana'a region of Yemen was composed largely of carvone (63.0%), with lesser amounts of limonene (7.9%), 1,8-cineole (4.8%), terpinen-4-ol (2.9%), *cis*-dihydrocarvone (6.1%), and β-caryophyllene (2.4%) [16]. Carvone is the characteristic volatile component of *Mentha spicata* from which the plant species gets its distinct smell. There are various reports on the presence of carvone as the major component from *Mentha spicata* of various geographical origins.

The Yemeni oil sample was assayed for antimicrobial, antifungal and cytotoxicity against MCF-7 and MDA-MB-231 cells. The oil showed considerable activity against *Bacillus cereus*, *Esherichia coli*, and *Botrytis cinerea* with MIC of 312.5 µg/mL, 156 µg/mL and 78 µg/mL, respectively. However, the oil did not show any cytotoxic activity against either MCF-7 cells or MDA-MB-231 cells. Apart from these, *Mentha spicata* oil has been investigated for other biological activities like antioxidant activity, analgesic, anti-inflammatory and antipyretic effects. In a recent study, Liu and co-workers reported that *Mentha spicata* oil showed considerably strong cytotoxicity on HeLa cells with an IC_{50} of 2.08 µg/mL [80]. The oil was also reported to have high antibacterial effect against *Escherichia coli*, *Saccharomyces cerevisiae* and *Penicillium citrinum*. The major compounds in the essential oil were carvone (65.3%), limonene (18.2%), dihydrocarvone (3.0%) and camphene (2.3%). Agrawal and co-workers have examined the antimicrobial activities of both enantiomers of limonene and carvone [81]. These investigators found that both optical isomers of carvone showed activity against a wide spectrum of human pathogenic fungi and bacteria. Carvone has been found to inhibit the transformation of *Candida albicans* from a coccus to the filamentous form, so making them a potentially good therapeutic agent against infections caused by fungus. Other properties of carvone include its strong insect repellent activity and a promising sprouting inhibitory action on potatoes. (*S*)-carvone is a good potato sprout inhibitor, and is commercialized in the Netherlands under the name "Talent" [82].

Meriandra Bengalensis (Konig ex Roxb.) Benth.

Meriandra bengalensis, known as "dharo", is an aromatic shrub that grows to a height of about 2 m and is highly branched. It has been used traditionally as an antiseptic, astringent, antirheumatic, and carminative [83]. *Meriandra bengalensis* from Yemen showed camphor (43.6%), 1,8-cineole (10.7%), borneol (3.4%), caryophyllene oxide (5.8%) and α-eudesmol (5.8%) as the major constituents [83]. Although the essential oil sample did not show significant antibacterial or antifungal activity, some of its major constituents do have biological activity. Camphor is considered to have good activity against human- and soil-born fungi. At a concentration of 500 µL/L, camphor has been shown to reduce the radial growth of *Sclerotinia sclerotiorum* and *Rhizoctonia solani* by 51.7% and 64.9%, respectively [84]. Camphor has been found to have good fungicidal activity against *Botrytis cinerea* where it showed complete inhibition of mycelia growth at 1.75 g/L [85].

Nepeta Deflersiana Schweinf. ex Hedge

Nepeta deflersiana is known locally as "mokerker alkotat". It is an aromatic perennial herb traditionally used for wounds, as a carminative, for rheumatic disorders and for fever and colic [29]. Mothana examined *Nepeta deflersiana* essential oil from Sana'a and found the principle components to be palmitic acid (8.0%), caryophyllene oxide (6.4%), 2-methoxy-*p*-cresol (5.6%), camphor (4.7%), and eugenol (4.7%), but was devoid of nepetalactones [86]. This oil was weakly antibacterial against*Staphylococcus aureus* and *Bacillus subtilis* (MIC = 400 µg/mL), but was inactive against *Esherichia coli, Pseudomonas aeruginosa*, or *Candida albicans*. In contrast, another study has revealed *Nepeta deflersiana* essential oils from

Taiz and from Sana'a to have very different compositions [16,87]. *Nepeta deflersiana* oil from Taiz was composed largely of germacrene D (40.5%), as well as nepetalactones, 4aα,7α,7aβ-nepetalactone (19.2%), 4aα,7β,7aα-nepetalactone (4.6%), and 4aα,7α,7aα-nepetalactone (3.0%), while an oil from Sana'a was dominated by 4aα,7α,7aα-nepetalactone (77.7%) with a smaller amount of germacrene D (6.0%). The oil from Taiz was somewhat antifungal against *Aspergillus niger* (MIC = 156 µg/mL) while the Sana'a sample was active against *Staphylococcus aureus* (MIC = 156 µg/mL) [16].

Ocimum Basilicum L.

The genus *Ocimum* is made up of around 65 species found throughout tropical and subtropical regions [7]. The genus is represented by seven species in Yemen, namely *Ocimum basilicum*, *Ocimum tenuiflorum*, *Ocimum suave*, *Ocimum spicatum*,*Ocimum gratissimum*, and *Ocimum forskolei*. *Ocimum basilicum* is used in Yemeni traditional medicine to treat various ailments, including abdominal cramps, gastroenteritis, dysentery, and diarrhea. In northern Oman and Saudi Arabia, juice of leaves or crushed leaves is used in the treatment of wounds, acne, and vitiligo. It is used also as a deodorant and is considered to be an aphrodisiac [27,88]. *Ocimum basilicum* from Yemen was dominated by linalool (74.5%) with lower concentrations of 1,8-cineole (7.4%) and estragole (7.2%) [16]. The Yemeni basil oil was screened for *in vitro* cytotoxic activity against MCF-7 (human mammary ductal carcinoma) cells, but was inactive; only 18.5% reduction of cell viability at a concentration of 100 µg/mL.

Origanum Majorana L.

Origanum is a perennial herbaceous genus that is native to North America, Europe, and temperate Asia [89]. The *Origanum* genus consists of over 44 species of which *Origanum majorana* (sweet marjoram), *Origanum vulgare* (oregano) and *Origanum maru* (Egyptian marjoram) are of medicinal importance in Yemen [90]. The genus is important medicinally due to its antimicrobial, antifungal, antioxidant, antibacterial, antithrombin, antimutagenic, angiogenic, antiparasitic, and antihyperglycemic activities. *Origanum majorana* has been used extensively throughout the world for its medicinal value and for its flavor properties. Dried leaves and flowering tips are used in formulation of vermouths and bitters. The essential oil is widely used for flavoring various food products such as sauces, condiments, and other products. It has great importance as a diuretic, anti-asthmatic, and an antipyretic drug in India. It has been used for the management of cancer as well [89]. The essential oil of *Origanum majorana*, known locally as "azab", from the Al-Mahweet region of Yemen had sabinene (4.2%),*p*-cymene (9.8%), limonene (2.8%), γ-terpinene

(7.7%), *cis*-sabinene hydrate (3.2%), *trans*-sabinene hydrate (6.8%), *cis-p*-menth-2-en-1-ol (2.3%), terpinen-4-ol (35.2%), α-terpineol (4.5%) and bornyl acetate (2.9%) as the major constituents [16].

Plectranthus Barbatus Andrews

Plectranthus is a large genus that belongs to the Lamiaceae and contains around 200 species of herbs and shrubs, most of which are aromatic plants tropical and sub-tropical areas of the Old World [7]. There are 12 species of *Plectranthus* found growing in Yemen, and this genus has been widely used for its medicinal properties. They have been used for treating skin, digestive, and respiratory complications [27,91]. *Plectranthus barbatus* is a perennial shrub that grows over the southern and subtropical regions of India, Pakistan, Sri Lanka, tropical east Africa, Asia (South of Arabian Peninsula), China, and Brazil [92].

The main constituents isolated from *Plectranthus barbatus* have been diterpenoids and essential oils. The essential oil composition varies according to the location and the date of harvest and contains mainly mono- and sesquiterpenes [92]. A wide range of biological activity has been reported for *Plectranthus* species. They are very important in ethno-medicinal uses to treat a range of ailments, particularly digestive, skin, infective, and respiratory problems. The plant has also been used widely for food, flavor and fodder.

Plectranthus barbatus is locally known as "baydat" in Yemen. The essential oil of *Plectranthus barbatus* from Taiz region of Yemen has been found to have α-terpinene (2.8%), *p*-cymene (9.3%), γ-terpinene (20.0%), thymol (48.7%), β-caryophyllene (6.4%) and β-selinene (2.1%) as the major constituents [16]. The chemical composition of Yemeni *Plectranthus barbatus* was very different from a *Plectranthus barbatus* sample from Portugal, which was devoid of thymol [93]. The Taiz oil sample was subjected to antimicrobial, antifungal, and cytotoxic assays against MCF-7 and MDA-MB-231 cancer cells. *Plectranthus barbatus* (Taiz) showed considerable activity against *Bacillus cereus* (MIC = 156 μg/mL) and *Esherichia coli* (MIC = 156 μg/mL). It also showed good activity against *Aspergillus niger* (MIC = 156 μg/mL) and *Botrytis cinerea* (MIC = 312.5 μg/mL). On the other hand, *Plectranthus barbatus* from the Ibb region of Yemen showed some activity against *Esherichia coli* (312.5 μg/mL).*Plectranthus barbatus* (Ibb) showed very good cytotoxic activity against MCF-7 cells (IC_{50} = 38.62 μg/mL) and had a 100% growth inhibition of MDA-MB-231 cells at concentration 100μg/mL [16]. The bioactivities of *Plectranthus barbatus* essential oils can be attributed to the high concentrations of thymol [74,94,95].

Plectranthus Cylindraceus Hochst. ex Benth.

Plectranthus cylindraceus is a strong aromatic plant found growing in different parts of United Arab Emirates, Saudi Arabia, Oman, Yemen, and East African countries. GC-MS analysis of the essential oil *Plectranthus cylindraceus* growing in Yemen showed high concentrations of thymol (68.5%), terpinolene (5.3%), β-selinene (4.7%), β-caryophyllene (4.0%), δ-cadinol (2.1%) and *ar*-curcumene (1.7%). The oil showed very good antibacterial and antifungal activity against *Staphylococcus aureus*, *Bacillus subtilis* and *Candida albicans* with MIC of 390, 180 and 180 μL/mL, respectively. It also showed considerable 2,2-diphenyl-1-picrylhydrazyl (DPPH) radical-scavenging activity as an antioxidant with IC_{50} of 34.5 μg/mL [83]. A similar study on *Plectranthus cylindraceus* oil from Oman has shown carvacrol (46.8%) and terpinolene (18.2%) to be the major constituents [96]. Note that thymol and carvacrol have similar retention indices on DB-1 or DB-5 columns and virtually identical mass spectra; these may be the same compound or overlapping mixtures of thymol and carvacrol in the *Plectranthus cylindraceus* samples. The oil also showed antimicrobial [96] and nematicidal activity [97]. The antibacterial and antifungal activity of *Plectranthus cylindraceus* is probably due to the presence of high concentration of thymol/carvacrol (see above).

Stachys Yemenensis Hedge

Stachys species are mostly distributed in the Mediterranean regions and Southwest Asia [7]. *Stachys* is represented by two species in Yemen, namely the endemic *S. yemenensis* [98] and *S. aegyptiaca* [99]. Many *Stachys* species have been used in folk medicine for the treatment of genital tumors, sclerosis of the spleen, inflammatory tumors and cancerous ulcers. A comparative study done on the *Stachys yemenensis* essential oil extracted by supercritical carbon dioxide extraction at 90 bar and by hydrodistillation [100] showed α-pinene (2.4%, 4.6%), α-phellandrene (13.9, 20.7%), *o*-cymene (5.3, 8.5%), β-phellandrene (11.7, 16.8%), bicyclogermacrene (4.3, 3.4%), elemol (12.0, 7.5%), spathulenol (6.7, 4.7%), γ-eudesmol (1.7, 3.2%), β-eudesmol (5.0, 5.1%), α-eudesmol (4.7, 6.4%), shyobunol (6.0, 1.5%) and squalene (4.9, 0%) as the major compounds. The exhausted matrix after the first supercritical extraction was further subjected to 250 bar supercritical CO_2 extraction. The major constituents of the second run were elemol (5.9%), spathulenol (3.3%), γ-eudesmol (1.6%), β-eudesmol (5.1%), α-eudesmol (5.2%) and squalene (49.7%). The oil showed good antimicrobial activity against *E. coli* and *S. aureus*. Studies have shown that only the positive enantiomer of α-pinene is active. Time of kill curves have shown (+)-α-pinene to be highly toxic to *Candida albicans*, killing 100% of the inoculum within 60 min. It has also been reported to have synergistic activity against MRSA with commercial antibiotics like ciprofloxacin [101].

Teucrium Yemense Deflers

Teucrium has about 250 species [7] of which Yemeni flora are represented by three: *T. balfouri*, *T. sokotranum,* and *T. yemense* [102]. The genus plays an important role in Yemeni folk medicine, being used as insect repallant, antispasmodic, and for kidney disease, rheumatism and diabetes [45]. *T. yemense* is known locally as "khodas", and the essential oil from Taiz, Yemen, has been reported to have high concentrations of sesquiterpene hydrocarbons (73.9%), the dominant ones being δ-cadinene (34.9%), β-caryophyllene (22.7%), α-humulene (6.1%), and α-selinene (5.4%) [102]. In another study, *T. yemense* from Taiz was reported to have α-pinene (6.6%), β-caryophyllene (19.1%), α-humulene (6.4%), δ-cadinene (6.5%), caryophyllene oxide (4.3%) and α-cadinol (9.5%) as the major constituents [87].

Thymus Laevigatus Vahl

Thymus laevigatus is an endemic species to Yemen and is the only species that represents this genus in Yemen [103]. The genus *Thymus* has about 250 species that are mostly perennial, aromatic and evergreen plants [7]. *Thymus laevigatus*, known locally as "zatar", is mostly found in the higher mountains in North Yemen in Haggah (2500 m) and in Dhamar (2200 m). In Yemeni folk medicine, *Thymus laevigatus* dried leaves are used as powder in warm milk, sesame oil, or olive oil for the treatment of different stomach diseases, cough, tonsillitis, pharyngitis and renal colic [103]. The essential oils from many species of *Thymus* have been reported to have very good antibacterial and antifungal activities, which are attributed to the high concentrations of thymol and carvacrol is those oils [104,105]. An essential oil study on *Thymus laevigatus* from Yemen has shown a very high concentration of carvacrol (84.3%), along with *p*-cymene (4.1%), γ-terpinene (4.0%) and *trans*-anethole (3.6%) as the major constituents [103]. *Thymus laevigatus* oil was found to have a wide range of antimicrobial activity and a potent fungicidal effect against *Candida albicans* with a MIC of 0.0313% (*v/v*) [104].

CONCLUSIONS

Yemen is a land of diverse landscapes including mountains, desert, gorge-like wadis, and coastal escarpments. The variety of habitats of mainland Yemen, coupled with the endemism of the Soqotra archipelago, give Yemeni flora a great deal of diversity. There are currently around 2930 species of higher plants in Yemen [106,107] of which 699 species are endemic [108]. Despite the introduction of Western medicinal system during the middle of the twentieth century, traditional herbal medicine continues to play an important role in many parts of Yemen. Many of these medicinal plants and their essential oils have shown notable biological activities. In addition to research activities on these traditional medicinal plants, it is hoped that future studies will provide new insights into pharmacological activities of understudied Yemeni flora. Unfortunately, increasing environmental degradation due to human activity, invasive plant species, and climate change threaten the native flora of Yemen. It is hoped that steps be undertaken to safeguard the fragile ecology of Yemen, protect the native flora and fauna, as well as preserve the traditional knowledge of the people.

AUTHOR CONTRIBUTIONS

N.A.A.A. and W.N.S. conceived and organized the review; B.K.C., N.A.A.A. and W.N.S. contributed to the writing and editing of the manuscript. Some of this work was taken from the M.S. Thesis of B.K.C. [16].

REFERENCES

1. Bakkali, F.; Averbeck, S.; Averbeck, D.; Idaomar, M. Biological effects of essential oils—A review. *Food Chem. Toxicol.* 2008, *46*, 446–475.
2. Christaki, E.; Bonos, E.; Giannenas, I.; Paneri, P.F. Aromatic plants as a source of bioactive compounds. *Agriculture* 2012, *2*, 228–243.
3. Dresch, P. *A History of Modern Yemen*; Cambridge University Press: Cambridge, UK, 2000.
4. Hadden, R.L. *The Geology of Yemen: An Annotated Bibliography of Yemen's Geology, Geography and Earth Science*; U.S. Army Corps of Engineers: Alexandria, VA, USA, 2012.
5. Turek, C.; Stintzing, F.C. Stability of essential oils: A review. *Compr. Rev. Food Sci. Food Saf.* 2013, *12*, 40–53.
6. Shaaban, H.A.E.; El-Ghorab, A.H.; Shibamoto, T. Bioactivity of essential oils and their volatile aroma components: Review. *J. Essent. Oil Res.* 2012, *24*, 203–212.
7. Mabberley, D.J. *Mabberley's Plant-Book*; Cambridge University Press: Cambridge, UK, 2008.
8. Dubai, A.; Alkhulaidi, A. *Medicinal and Aromatic Plants in Yemen*; Obadi Centre for Publishing: Sana'a, Yemen, 1997.

9. Gundidza, M. Antimicrobial activity of essential oil from *Schinus molle* Linn. *Cent. Afr. J. Med.* 1993, *39*, 231–234.
10. Díaz, C.; Quesada, S.; Brenes, O.; Aguilar, G.; Cicció, J.F. Chemical composition of *Schinus molle* essential oil and its cytotoxic activity on tumour cell lines. *Nat. Prod. Res.* 2008, *22*, 1521–1534.
11. Abdel-Sattar, E.; Zaitoun, A.A.; Farag, M.A.; El Gayed, S.H.; Harraz, F.M.H. Chemical composition, insecticidal and insect repellent activity of *Schinus molle* L. leaf and fruit essential oils against *Trogoderma granarium* and *Tribolium castaneum*. *Nat. Prod. Res.* 2010, *24*, 226–235.
12. Ali, N.A.A.; Marongiu, B.; Piras, A.; Porcedda, S.; Falconieri, D.; Al-Othman, A.M.R. Comparative analysis of the oil and supercritical CO_2 extract of *Schinus molle* L. growing in Yemen. *Nat. Prod. Res.* 2011, *14*, 1366–1369.
13. Schmidt, J.M.; Noletto, J.A.; Vogler, B.; Setzer, W.N. Abaco bush medicine: Chemical composition of the essential oils of four aromatic medicinal plants from Abaco Island, Bahamas. *J. Herbs Spices Med. Plants* 2007, *12*, 43–65.
14. Mossa, J.S. Phytochemical and biological studies on *Artemisia abyssinica* and anti diabetic herb used in Arabian folk medicine. *Fitoterapia* 1985, *56*, 311–314.
15. Abad, J.M.; Bedoya, L.M.; Apaza, L.; Bermejo, P. The *Artemisia* L. genus: A review of bioactive essential oils. *Molecules* 2012, *17*, 2542–2566.
16. Chhetri, B.K. A Gas Chromatographic/Mass Spectral Analysis of Aromatic Medicinal Plants from Yemen. M.S. Thesis, University of Alabama in Huntsville, Huntsville, AL, USA, 2015.
17. Chhetri, B.K.; Al-Sokari, S.S.; Setzer, W.N.; Ali, N.A.A. Essential oil composition of *Artemisia abyssinica* from three habitats in Yemen. *Am. J. Essent. Oils Nat. Prod.* 2015, *2*, 28–30.
18. Tariku, Y.; Hymete, A.; Hailu, A.; Rohloff, J. Essential-oil composition, antileishmanial, and toxicity study of *Artemisia abyssinica* and *Satureja punctata* ssp. *punctata* from Ethiopia. *Chem. Biodivers.* 2010, *7*, 1009–1018.
19. Militello, M.; Settanni, L.; Aleo, A.; Mammina, C.; Moschetti, G.; Giammanco, G.M.; Amparo Blàzquez, M.; Carrubba, A. Chemical composition and antibacterial potential of *Artemisia arborescens* L. essential oil. *Curr. Microbiol.* 2011, *62*, 1274–1281.
20. Saddi, M.; Sanna, A.; Cottiglia, F.; Chisu, L.; Casu, L.; Bonsignore, L.;

de Logu, A. Antiherpesvirus activity of *Artemisia orborescens* essential oil and inhibition of lateral diffusion in Vero cells. *Ann. Clin. Microbiol. Antimicrob.* 2007, *6*, 10.

21. Ali, N.A.A.; Wurster, M.; Denkert, A.; Al-Sokari, S.S.; Lindequist, U.; Wessjohann, L. Cytotoxicity and antiphytofungal activity of the essential oils from two *Artemisia* species. *World J. Pharm. Res.* 2014, *3*, 1350–1354.

22. Magnelli, L.; Caldini, R.; Schiavone, N.; Suzuki, H.; Chevanne, M. Differentiating and apoptotic dose-dependent effects in (-)-α-bisabolol-treated human endothelial cells. *J. Nat. Prod.* 2010, *73*, 523–526.

23. Siegel, I.; Liu, T.L.; Yaghoubzadeh, E.; Keskey, T.S.; Gleicher, N. Cytotoxic effects of free fatty acids on ascites tumor cells. *J. Nat. Cancer Inst.* 1987, *78*, 271–277.

24. Abderrahim, A.; Belhamel, K.; Cahlchat, J.C.; Figuérédo, G. Chemical composition of the essential oil from *Artemisia arborescens* L. growing wild in Algeria. *Rec. Nat. Prod.* 2010, *4*, 87–90.

25. Thabit, R.A.S.; Cheng, X.; Al-Hajj, N.; Rahman, M.R.T.; Lei, G. Antioxidant and *Conyza bonariensis*: A review. *Eur. Acad. Res.* 2014, *2*, 8454–8474.

26. Cheng, X.R.; Thabit, R.A.; Wang, W.; Shi, H.W.; Shi, Y.H.; Le, G.W. Analysis and comparison of the essential oil in*Conyza bonariensis* grown in Yemen and China. *Prog. Mod. Biomed.* 2013, *36*, 011.

27. Ghazanfar, S.A. *Handbook of Arabian Medicinal Plants*; CRC Press: Boca Raton, FL, USA, 1994.

28. Flann, C. GCC: Global Compositae Checklist (version 5 (Beta), June 2014). In *Species 2000 & ITIS Catalogue of Life, 15th February 2015*; Roskov, Y., Kunze, T., Paglinawan, L., Orrell, T., Nicolson, D., Culham, A., Bailly, N., Kirk, P., Bourgoin, T., Baillargeon, G., *et al*, Eds.; Species 2000: Reading, UK; Available online: www.catalogueoflife.org/col/ (accessed on 4 April 2015).

29. Mothana, R.A.A.; Gruenert, R.; Bednarski, P.J.; Lindequist, U. Evaluation of the *in vitro* anticancer, antimicrobial and antioxidant activities of some Yemeni plants used in folk medicine. *Pharmazie* 2009, *64*, 260–268.

30. Al-Hajj, N.Q.M.; Ma, C.; Thabit, R.; Al-alfarga, A.; Gasmalla, M.A.A.; Musa, A.; Aboshora, W.; Wang, H. Chemical composition of essential oil and mineral contents of *Pulicaria inuloides*. *J. Acad. Indust. Res.* 2014, *2*, 675–678.

31. Al-Hajj, N.Q.M.; Wang, H.X.; Ma, C.; Lou, Z.; Bashari, M.; Thabit, R.

Antimicrobial and antioxidant activities of the essential oils of some aromatic medicinal plants (*Pulicaria inuloides*–Asteraceae and *Ocimum forskolei*–Lamiaceae. *Trop. J. Pharm. Res.* 2014, *13*, 1287–1293.

32. Fawzy, G.A.; Al Ati, H.Y.; El Gamal, A.A. Chemical composition and biological evaluation of essential oils of *Pulicaria jaubertii*. *Pharmacogn. Mag.* 2013, *9*, 28–32.

33. Algabr, M.N.; Ameddah, S.; Menad, A.; Mekkiou, R.; Chalchat, J.C.; Benayache, S.; Benayache, F. Essential oil composition of *Pulicaria jaubertii* from Yemen. *Int. J. Med. Aromat. Plants* 2012, *2*, 688–690.

34. Ali, N.A.A.; Crouch, R.A.; Al-Fatimi, M.A.; Arnold, N.; Teichert, A.; Setzer, W.N.; Wessjohann, L. Chemical composition, antimicrobial, antiradical and anticholinesterase activity of the essential oil of *Pulicaria stephanocarpa* from Soqotra. *Nat. Prod. Commun.* 2012, *7*, 113–116.

35. Ali, N.A.A.; Al-Haj, M.A.; Wurster, M.; Lindequist, U. Chemical composition and antifungal activity of essential oil of Soqotran *Pulicaria stephanocarpa* Balf. f. *Univ. Aden J. Nat. Appl. Sci.* 2009, *13*, 429–434.

36. Hammiche, V.; Maiza, K. Traditional medicine in Central Sahara: Pharmacopoeia of Tassili N'ajjer. *J. Ethnopharmacol.* 2006, *105*, 358–367.

37. Hegazy, M.E.F.; Matsuda, H.; Nakamura, S.; Yabe, M.; Matsumoto, T.; Yoshikawa, M. Sesquiterpenes from an Egyptian herbal medicine, *Pulicaria undulata*, with inhibitory effects on nitric oxide production in RAW264.7 macrophage cells. *Chem. Pharm. Bull.* 2012, *60*, 363–370.

38. El-Kamali, H.S.; Ahmed, A.H.; Mohammed, A.S.; Yahia, A.A.M.; El-Tayeb, I.H.; Ali, A.A. Antibacterial properties of essential oils from *Nigella sativa* seeds, *Cymbopogon citratus* leaves and *Pulicaria undulata* aerial parts. *Fitoterapia* 1998, *69*, 77–78.

39. Ali, N.A.A.; Makboul, M.A.; Assaf, M.H.; Anton, R. Essential oil of *Pulicaria undulata* L. growing in Egypt and its effect on animal behaviour. *Bull. Pharm. Sci.* 1987, *10*, 37–49.

40. Elegami, A.A.B.; Ishag, K.E.; Mahmoud, E.N.; Abu Alfutuh, I.M.; Karim, E.I.A. Insecticidal activity of *Pulicaria undulata* oil. *Fitoterapia* 1994, *65*, 82–83.

41. Ali, N.A.A.; Sharopov, F.S.; Alhaj, M.; Hill, G.M.; Porzel, A.; Arnold, N.; Setzer, W.N.; Schmidt, J.; Wessjohann, L. Chemical composition and biological activity of essential oil from *Pulicaria undulata* from Yemen. *Nat. Prod. Commun.* 2012, *7*, 257–260.

42. El-Kamali, H.H.; Yousif, M.O.; Ahmed, O.I.; Sabir, S.S. Phytochemical

analysis of the essential oil from aerial parts of *Pulicaria undulata* (L.) Kostel from Sudan. *Ethnobot. Leaf.* 2009, *13*, 467–471.

43. Soule, J.A. *Tagetes minuta*: A potential new herb from South America. In *New Crops*; Janick, J., Simon, J.E., Eds.; Wiley: New York, NY, USA, 1993; pp. 649–654.

44. Duke, J.A.; Bogenschutz-Godwin, M.J.; Ottesen, A.R. *Duke's Handbook of Medicinal Plants of Latin America*; CRC Press: Boca Raton, FL, USA, 2009; pp. 691–692.

45. Al-Musayeib, N.M.; Mothana, R.A.; Matheeussen, A.; Cos, P.; Maes, L. In vitro antiplasmodial, antileishmanial and antitrypanosomal activities of selected medicinal plants used in the traditional Arabian Peninsular region. *BMC Complement. Altern. Med.* 2012, *12*, 49.

46. Bii, C.C.; Siboe, G.M.; Mibey, R.K. Plant essential oils with promising antifungal activity. *East Afr. Med. J.* 2000, *77*, 319–322.

47. Senatore, F.; Napolitano, F.; Mohamed, M.A.; Harris, H.P.J.C.; Mnkeni, P.N.S.; Henderson, J. Antibacterial activity of *Tagetes minuta* L. (Asteraceae) essential oil with different chemical composition. *Flavour Fragr. J.* 2004, *19*, 574–578.

48. Chamorro, E.R.; Sequeira, A.F.; Velasco, G.A.; Zalazar, M.F.; Ballerini, J. Evaluation of *Tagetes minuta* L. essential oils to control *Varroa destructor* (Acari: Varroidae). *J. Argent. Chem. Soc.* 2011, *98*, 39–47.

49. López, S.B.; López, M.L.; Aragón, L.M.; Tereschuck, M.L.; Slanis, A.C.; Feresin, G.E.; Zygadlo, J.A.; Tapia, A.A. Composition and anti-insect activity of essential oils from *Tagetes* L. species (Asteraceae, Helenieae) on *Ceratitis capitata* Wiedemann and *Triatoma infestans* Klug. *J. Agric. Food Chem.* 2011, *25*, 5286–5292.

50. Garcia, M.V.; Matias, J.; Barros, J.C.; de Lima, D.P.; Lopes, Rda.S.; Andreotti, R. Chemical identification of *Tagetes minuta* Linneus (Asteraceae) essential oil and its acaricidal effect on ticks. *Rev. Bras. Parasitol. Vet.* 2012, *21*, 405–411.

51. Ali, N.A.A.; Sharopov, F.S.; Al-kaf, A.G.; Hill, G.M.; Arnold, N.; Al-Sokari, S.S.; Setzer, W.N.; Wessjohann, L. Composition of essential oil from *Tagetes minuta* and its cytotoxic, antioxidant and antimicrobial activities. *Nat. Prod. Commun.* 2014, *9*, 265–268.

52. Al-Mamary, M.; Abdelwahab, S.I.; Al-Ghalibi, S.; Al-Ghasani, E. The antioxidant and tyrosinase inhibitory activities of some essential oils obtained from aromatic plants grown and used in Yemen. *Sci. Res. Essays* 2011, *6*, 6840–6845.

53. Ali, N.A.A.; Al-Fatimi, M.A.; Crouch, R.A.; Denkert, A.; Setzer, W.N. Antimicrobial, antioxidant, and cytotoxic activities of the essential oil of *Tarchonanthus camphoratus*. *Nat. Prod. Commun.* 2013, *8*, 683–686.
54. Miller, G.A.; Morris, M. *Ethnoflora of the Soqotra Archipelago*; Charlesworth Group: Huddersfield, UK, 2004; pp. 457–464.
55. De Sanctis, M.; Adeeb, A.; Farcomeni, A.; Patriarca, C.; Saed, A.; Attorre, F. Classification and distribution patterns of plant communities on Socotra Island, Yemen. *Appl. Veget. Sci.* 2012, *16*, 148–165.
56. Ali, N.A.A.; Wurster, M.; Arnold, N.; Teichert, A.; Schmidt, J.; Lindequist, U.; Wessjohann, L. Chemical composition and biological activities of essential oils from the oleogum resins of three endemic Soqotraen *Boswellia* species. *Rec. Nat. Prod.* 2008, *2*, 6–12.
57. Mothana, R.A.A.; Lindequist, U. Antimicrobial activity of some medicinal plants of the island of Soqotra. *J. Ethnopharmacol.* 2005, *96*, 177–181.
58. Ali, N.A.A.; Wurster, M.; Lindequist, U. Chemical composition of essential oil from the oleogum resin of *Commiphora habessinica* (Berg.) Engl. from Yemen. *J. Essent. Oil Bear. Plants* 2009, *12*, 244–249.
59. Banfield, L.M.; Van Damme, K.; Miller, A.G. Evolution and biogeography of the flora of the Socotra archipelago (Yemen). In *The Biology of Island Floras*; Bramwell, D., Caujapé-Castells, J., Eds.; Cambridge University Press: Cambridge, UK, 2011; pp. 197–225.
60. Ali, N.A.A.; Wurster, M.; Arnold, N.; Lindequist, U.; Wessjohann, L. Essential oil composition from oleogum resin of Soqotraen *Commiphora kua*. *Rec. Nat. Prod.* 2008, *2*, 70–75.
61. Al-Fatimi, M.; Wurster, M.; Schröder, G.; Lindequist, U. Antioxidant, antimicrobial and cytotoxic activities of selected medicinal plants from Yemen. *J. Ethnopharmacol.* 2007, *111*, 657–666.
62. Mothana, R.A.; Lindequist, U.; Gruenert, R.; Bednarski, P.J. Studies of the *in vitro* anticancer, antimicrobial and antioxidant potentials of selected Yemeni medicinal plants from the island Soqotra. *BMC Complement. Altern. Med.* 2009, *9*, 7.
63. Mothana, R.A.; Al-Rehaily, A.J.; Schultze, W. Chemical analysis and biological activity of the essential oils of two endemic Soqotri *Commiphora* species. *Molecules* 2010, *15*, 689–698.
64. Zahran, M.A.; Khedr, A.; Dahmash, A.; El-Ameir, Y.A. Qat forms in Yemen: Ecology, dangerous impacts and future promise. *Egypt J. Basic Appl. Sci.* 2014, *1*, 1–8.

65. Kalix, P. Cathinone, a natural amphetamine. *Pharmacol. Toxicol.* 1992, *70*, 77–86.
66. Baxter, R.L.; Crombie, L.; Simmonds, D.J.; Whiting, D.A.; Braenden, O.J.; Szendrei, K. Alkaloids of *Catha edulis* (khat). Part 1. Isolation and characterization of eleven new alkaloids with sesquiterpene cores (cathedulins); identification of the quinone-methide root pigments. *J. Chem. Soc., Perkin Trans. 1* 1979, 2965–2971.
67. Algabr, M.N.; Al-Wadhaf, H.A.; Ameddah, S.; Menad, A.; Mekkiou, R.; Chalchat, J.C.; Benayache, S.; Benayache, F. Analysis of the essential oil of *Catha edulis* leaves from Yemen. *Int. J. Appl. Res. Nat. Prod.* 2014, *7*, 21–24.
68. Adorjan, B.; Buchbauer, G. Biological properties of essential oils: An updated review. *Flavour Fragr. J.* 2010, *25*, 407–426.
69. Mothana, R.A.; Alsaid, M.S.; Hasoon, S.S.; Al-Mosaiyb, N.M.; Al-Rehaily, A.J.; Al-Yahya, M.A. Antimicrobial and antioxidant activities and gas chromatography mass spectrometry (GC/MS) analysis of the essential oils of *Ajuga brancteosa* Wall. ex Benth. and *Lavandula dentata* L. growing wild in Yemen. *J. Med. Plants Res.* 2012, *6*, 3066–3071.
70. Imelouane, B.; Elbachiri, A.; Ankit, M.; Benzeid, H.; Khedid, K. Physico-chemical compositions and antimicrobial activity of essential oil of eastern Moroccan *Lavandula dentata*. *Int. J. Agric. Biol.* 2009, *11*, 113–118.
71. Touati, R.; Chograni, H.; Hassen, I.; Boussaïd, M.; Toumi, L.; Brahim, N.B. Chemical composition of the leaf and flower essential oils of Tunisian *Lavandula dentata* L. (Lamiaceae). *Chem. Biodivers.* 2011, *8*, 1560–1569.
72. Alkhyat, S.H.; Maqtari, M.A.A.; Alhamzy, E.H.; Saeed, M.A.; Ali, N.A.A. Antimicrobial activity of *Lavandula pubescens* essential oil from two places In Yemen. *J. Adv. Biol.* 2014, *4*, 446–454.
73. Cantore, P.L.; Shanmugaiah, V.; Iacobellis, N.S. Antibacterial activity of essential oil components and their potential use in seed disinfection. *J. Agric. Food Chem.* 2009, *57*, 9454–9461.
74. Marei, G.I.K.; Rasoul, M.A.A.; Abdelgaleil, S.A.M. Comparative antifungal activities and biochemical effects of monoterpenes on plant pathogenic fungi. *Pestic. Biochem. Physiol.* 2012, *103*, 56–61.
75. Mothana, R.A.; Al-Said, M.; Al-Yahya, M.A.; Al-Rehaily, A.J.; Khaled, J.M. GC and GC/MS analysis of essential oil composition of the endemic Soqotraen *Leucas virgata* Balf.f. and its antimicrobial and antioxidant

activities. *Int. J. Mol. Sci.* 2013, *14*, 23129–23139.
76. Mothana, R.A.; Al-Musayeib, N.M.; Al-Ajmi, M.F.; Cos, P.; Maes, L. Evaluation of the *In vitro* antiplasmodial, antileishmanial, and antitrypanosomal activity of medicinal plants used in Saudi and Yemeni traditional medicine.*Evid.-Based Complement. Altern. Med.* 2014, *2014*.
77. Chauhan, R.S.; Kaul, M.K.; Shahi, A.K.; Kumar, A.; Ram, G.; Tawa, A. Chemical composition of essential oils in*Mentha spicata* L. accession [IIIM(J)26] from North-West Himalayan region, India. *Ind. Crops Prod.* 2009, *29*, 654–656.
78. Karousou, R.; Balta, M.; Hanlidou, E.; Kokkini, S. "Mints", smells and traditional uses in Thessaloniki (Greece) and other Mediterranean countries. *J. Ethnopharmacol.* 2007, *109*, 248–257.
79. Yousuf, P.M.H.; Noba, N.Y.; Shohel, M.; Bhattacherjee, R.; Das, B.K. Analgesic, anti-Inflammatory and antipyretic effect of *Mentha spicata* (spearmint). *Br. J. Pharm. Res.* 2013, *3*, 854–864.
80. Liu, K.H.; Zhu, Q.; Zhang, J.J.; Xu, J.F.; Wang, X.C. Chemical composition and biological activities of the essential oil of *Mentha spicata* Lamiaceae. *Adv. Mater. Res.* 2012, *524-527*, 2269–2272.
81. Aggarwal, K.K.; Khanuja, S.P.S.; Ahmad, A.; Kumar, T.R.S.; Gupta, V.K.; Kumar, S. Antimicrobial activity profiles of the two enantiomers of limonene and carvone isolated from the oils of *Mentha spicata* and *Anethum sowa*. *Flavour Fragr. J.* 2002, *17*, 59–63.
82. De Carvalho, C.; da Fonseca, M.M.R. Carvone: Why and how should one bother to produce this terpene. *Food Chem.*2006, *95*, 413–422.
83. Ali, N.A.A.; Wurster, M.; Denkert, A.; Arnold, N.; Fadail, I.; Al-Didamony, G.; Lindequist, U.; Wessjohann, L.; Setzer, W.N. Chemical composition, antimicrobial, antioxidant and cytotoxic activity of essential oils of *Plectranthus cylindraceus* and *Meriandra benghalensis* from Yemen. *Nat. Prod. Commun.* 2012, *7*, 1099–1102.
84. Pitarokili, D.; Tzakou, O.; Loukis, A.; Harvala, C. Volatile metabolites from *Salvia fruticosa* as antifungal agents in soilborne pathogens. *J. Agric. Food Chem.* 2003, *51*, 3294–3301.
85. Moretti, M.D.L.; Peana, A.T. Activity of the oil of *Salvia officinalis* L. against *Botrytis cinerea*. *J. Essent. Oil Res.* 1996, *8*, 399–404.
86. Mothana, R.A. Chemical composition, antimicrobial and antioxidant activities of the essential oil of *Nepeta deflersiana*growing in Yemen. *Rec. Nat. Prod.* 2012, *6*, 189–193.
87. Maqtari, M.A.A.; Alhamzy, E.H.; Saeed M, A.; Ali, N.A.A.; Setzer, W.N.

Chemical composition and antioxidant activity of the essential oils from different aromatic plants grown in Yemen. *J. Glob. Biosci.* 2014, *3*, 390–398.

88. Schopen, A. *Traditionelle Heilmittel in Jemen*; Steiner: Wiesbaden, Germany, 1983; pp. 66–68.
89. Chishti, S.; Kaloo, Z.A.; Sultan, P. Medicinal importance of genus *Origanum*: A review. *J. Pharmacogn. Phytother.* 2013,*5*, 170–177.
90. Lev, E. Eastern Mediterranean pharmacology and India trade as a background for Yemeni medieval medicinal plants. In *Herbal Medicine in Yemen: Traditional Knowledge and Practice, and Their Value for Today's World*; Hehmeyer, I., Schönig, H., Eds.; Brill: Leiden, The Netherlands, 2012; p. 34.
91. Lukhoba, C.W.; Simmonds, M.S.J.; Paton, A.J. *Plectranthus*: A review of ethnobotanical uses. *J. Ethnopharmacol.* 2006,*103*, 1–24.
92. Alasbahi, R.H.; Melzig, M.F. *Plectranthus barbatus*: A review of phytochemistry, ethnobotanical uses and pharmacology—Part 1. *Planta Med.* 2010, *76*, 653–661.
93. Mota, L.; Figueiredo, A.C.; Pedro, L.G.; Barroso, J.G.; Miguel, M.G.; Falerio, M.L.; Ascensão, L. Volatile-oils composition, and bioactivity of the essential oils of *Plecranthus barbatus*, *P. neochilus* and *P. ornatus* grown in Portugal.*Chem. Biodivers.* 2014, *11*, 719–732.
94. Gallucci, M.N.; Oliva, M.; Casero, C.; Dambolena, J.; Luna, A.; Zygadlo, J.; Demo, M. Antimicrobial combined action of terpenes against the food-borne microorganisms *Escherichia coli*, *Staphylococcus aureus* and *Bacillus cereus*. *Flavour Fragr. J.* 2009, *24*, 348–354.
95. Sobral, M.V.; Xavier, A.L.; Lima, T.C.; de Sousa, D.P. Antitumor activity of monoterpenes found in essential oils. *Sci. World J.* 2014, *2014*. ID 953451.
96. Marwah, R.G.; Fatope, M.O.; Deadman, M.L.; Ochei, J.E.; Al-Saidi, S.H. Antimicrobial activity and the major components of the essential oil of *Plectranthus cylindraceus*. *J. Appl. Microbiol.* 2007, *103*, 1220–1226.
97. Onifade, A.K.; Fatope, M.O.; Deadman, M.L.; Al Kindy, S.M. Nematicidal activity of *Haplophyllum tuberculatum* and*Plectranthus cylindraceus* oils against *Meloidogyne javanica*. *Biochem. Syst. Ecol.* 2008, *36*, 679–683.
98. Wood, J.R.I. *A Handbook of the Yemen Flora*; Royal Botanic Gardens: Kew, UK, 1997; pp. 197–198.
99. Trease, G.E.; Evans, W.C. *Trease and Evans' Pharmacognosy*, 13th ed.; Bailliere Tindall: London, UK, 1989.

100. Ali, N.A.A.; Marongiu, B.; Piras, A.; Porcedda, S.; Falconieri, D.; Molicotti, P.; Zanetti, S. Essential oil composition of leaves of *Stachys yemenensis* obtained by supercritical CO_2. *Nat. Prod. Res.* 2010, *24*, 1823–1829.
101. Da Silva, A.C.R.; Lopes, P.M.; de Azevedo, M.M.B.; Costa, D.C.M.; Alviano, C.S.; Alviano, D.S. Biological activities of α-pinene and β-pinene enantiomers. *Molecules* 2012, *17*, 6305–6316.
102. Ali, N.A.A.; Wurster, M.; Arnold, N.; Lindequist, U.; Wessjohan, L. Chemical composition of the essential oil of*Teucrium yemense* Deflers. *Rec. Nat. Prod.* 2008, *2*, 25–32.
103. Al-Fatimi, M.; Wurster, M.; Schröder, G.; Lindequist, U. In vitro antimicrobial, cytotoxic and radical scavenging activities and chemical constituents of the endemic *Thymus laevigatus* (Vahl). *Rec. Nat. Prod.* 2010, *4*, 49–63.
104. Othman, A.M.; Awadh, N.A.; Al-Fadhli, E.; Salama, M. Topical herbal antimicrobial formulation containing *Thymus laevigatus* essential oil. *World J. Pharm. Res.* 2014, *3*, 3693–3703.
105. Loziene, K. Selection of fecund and chemically valuable clones of thyme (*Thymus*) species growing wild in Lithuania.*Ind. Crops Prod.* 2009, *29*, 502–508.
106. Kilian, N.; Hein, P.; Hubaishan, M.A. New and noteworthy records for the flora of Yemen, chiefly of Hadhramout and Al-Mahra. *Willdenowia* 2002, *32*, 239–269.
107. Kilian, N.; Hein, P.; Hubaishan, M.A. Further notes on the flora of the southern coastal mountains of Yemen.*Willdenowia* 2004, *34*, 159–182.
108. Hall, M.; Miller, A.G. Strategic requirements for plant conservation in the Arabian Peninsula. *Zool. Middle East* 2011,*54* (Suppl. 3), 169–182.

Chapter 7

THE CHEMISTRY AND TOXICOLOGY OF DEPLETED URANIUM

Sidney A. Katz

Department of Chemistry, Rutgers University, Camden, NJ 08102-1411, USA

ABSTRACT

Natural uranium is comprised of three radioactive isotopes: ^{238}U, ^{235}U, and ^{234}U. Depleted uranium (DU) is a byproduct of the processes for the enrichment of the naturally occurring ^{235}U isotope. The world wide stock pile contains some 1½ million tons of depleted uranium. Some of it has been used to dilute weapons grade uranium (~90% ^{235}U) down to reactor grade uranium (~5% ^{235}U), and some of it has been used for heavy tank armor and for the fabrication of armor-piercing bullets and missiles. Such weapons were used by the military in the Persian Gulf, the Balkans and elsewhere. The testing of depleted uranium weapons and their use in combat has resulted in environmental contamination and human exposure. Although the chemical and the toxicological behaviors of depleted uranium are essentially the same as those of natural uranium, the respective chemical forms and isotopic compositions in which they usually occur are different. The chemical and radiological toxicity of depleted uranium can injure biological systems. Normal functioning of the kidney, liver, lung, and heart can be adversely affected by depleted uranium intoxication. The focus of this review is on the chemical and toxicological properties of depleted and natural uranium and some of the possible consequences from long term, low dose exposure to depleted uranium in the environment.

INTRODUCTION

Natural uranium is comprised of three radioactive isotopes: ^{238}U, ^{235}U, and ^{234}U. The current generation of nuclear power reactors is based on the controlled fission of the ^{235}U in the fuel at concentrations enriched to some five fold greater than that occurring in nature. The residue from the enrichment process is the depleted uranium. The world wide stock pile of depleted uranium contains more than 1½ million tons. Some of it has been used to dilute weapons grade uranium (~90% ^{235}U) down to reactor grade uranium (~5% ^{235}U) [1]. Among the other uses found for depleted uranium is the fabrication of munitions. Such weapons were used in the Persian Gulf, in the Balkans and elsewhere. The use of depleted uranium in armor-piercing bullets and missiles during wartime has resulted in environmental contamination and human exposure. The chemical properties and the toxicological behavior of depleted uranium are very similar to those of the natural uranium. The chemical and radiological toxicity of depleted uranium can injure biological systems. Normal functioning of the kidney, liver, and lung can be adversely affected by depleted uranium intoxication. This review describes depleted uranium munitions and focuses on the chemistry of depleted and natural uranium, their toxicological effects on several systems in the mammalian body and some consequences of long term, low dose environmental exposure. This review does not resolve the apparent divergence of opinions expressed a decade ago in the reviews prepared by Legget and Pellmar [2] and by Bleise, *et al.* [3] on the biological fate of depleted uranium shrapnel embedded in the soft tissue of wounded military personnel. It does, however, update some of the information in the subsequent reviews prepared by Craft *et al.* [4] and by Briner [5].

OCCURRENCE OF URANIUM

Of the two and a half dozen known uranium isotopes, only three occur in nature. They are ^{234}U, ^{235}U, and ^{238}U. The radiological properties of these three as well as those of the other uranium isotopes have been compiled at Karlsruhe Kernforchungszentrum [6] and at Brookhaven National Laboratory [7]. These radiological properties are listed in Table 1.

Table 1: Uranium isotopes [a,b]

Naturally occurring isotopes			
Isotope	Abundance *	Half life	Principle decay
^{234}U	0.00054%	2.455×10^5 years	α: 4776 MeV
^{235}U	0.07204%	4.468×10^8 years	α: 4.398 MeV
^{238}U	99.2742%	4.468×10^9 years	α: 4.197 MeV

Other known isotopes		
Isotope	Half life	Principle decay
^{217}U	16 ms	α
^{218}U	1.5 ms [a]	α: 8.27 MeV
^{218}U	0.51 ms [b]	α
^{218m}U	0.56 ms	α
^{219}U	~42 μs	α: 9.68 MeV
^{220}U		
^{221}U	700 ns [b]	
^{223}U	18 μs	α: 8.78 MeV
^{224}U	0.7 ms	α: 8.47 MeV
^{224}U	0.9 ms [b]	α
^{225}U	95 ms	α: 7.88, 7.82 MeV
^{226}U	0.2 s	α: 7.57, 7.42 MeV
^{226}U	0.35 s [b]	α
^{227}U	1.1 min	α: 6.86, 7.06, 6.74 MeV
^{228}U	9.1 min	α: 6.68, 6.59 MeV
^{229}U	58 min	α: 6.36, 6.33, 6.30 MeV
^{230}U	20.8 days	α: 5.89, 5.82 MeV
^{231}U	4.2 days	α: 5.46, 5.47, 5.40 MeV
^{232}U	68.9 years	α: 5.32, 5.26 MeV
^{233}U	1.59×10^5 years	α: 4.82, 4.78 MeV
^{234}U	2.46×10^5 years	α: 4.77, 4.72 MeV
^{235}U	7.04×10^8 years	α: 4.40 MeV
^{236}U	2.34×10^7 years	α: 4.49, 4.45 MeV
^{237}U	6.75 days	β⁻: 0.2 MeV
^{238}U	4.47×10^9 years	α: 4.20 MeV
^{239}U	23.5 min	β⁻: 1.2, 1.3 MeV
^{240}U	14.1 h	β⁻: 0.4 MeV
^{242}U	16.8 min	β⁻: 1.2 MeV

* Isotopic abundance; a G. Phennig, H. Klewe—Nebenius, W. Seelmann—Eggebert, Karlsruher Nuklidkarte, 6 Auflage 1995, korrigierten 1998, Institut für Instrumentelle Analytic, 1998, Karlsruhe; b Chart of the Nuclides, National Nuclear Data Center, Brookhaven National Laboratory, Upton, NY. α = radioactive decay is by alpha emission, β⁻ = radioactive decay is by negatron emission

Some of the primary and secondary uranium minerals are listed in Table 2. Major deposits of uranium ores are found in Canada and the U.S.A., in Brazil, in the Russian Federation, in Kazakhstan and Uzbekistan, in Namibia and South Africa and in Australia. The global distribution of uranium in the crust of the earth is approximately 2.3 mg/kg making it as common as tin, 2.1 mg/kg [8]. The mining, milling, refining and enriching of uranium as well as the security issues and the waste management strategies are beyond the scope of this review.

Table 2: Uranium ores

Some primary uranium minerals	
Branneritem	UTi_2O_6
Coffinite	$U(SiO4)_{1-4}(OH)_{4-1}$
Davidite	$(REE)(Y,U)(Ti,Fe)_{20}O_{38}$
Pitchblende	U_3O_8
Urainnite	UO_2
Some secondary uranium minerals	
Autunite	$Ca(UO_2)_2(PO_4)_4 \cdot (8 \text{ to } 12)H_2O$
Camolite	$K_2(UO_2)_2(VO_4)_2 \cdot (1 \text{ to } 3)H_2O$
Seleeite	$Mg(UO_2)_2(PO_4)_2 \cdot 10H_2O$
Torbernite	$Cu(UO_2)_2(PO_4)_2 \cdot 12H_2O$
Tyuyamunite	$Ca(UO_2)_2(VO_4)_2 \cdot (5 \text{ to } 8)H_2O$
Uranocircite	$Ba(UO_2)_2(PO_4)_2 \cdot (8 \text{ to } 10)H_2O$
Uranophane	$Ca(UO_2)_2(HSiO_4)_2 \cdot 5H_2O$
Zeunerite	$Cu(UO_2)_2(AsO_4)_2 \cdot (8 \text{ to } 10)H_2O$

PHYSICAL PROPERTIES OF URANIUM

Elemental uranium is a dense, malleable and ductile, silvery-white metal. Typically depleted uranium contains as much as 70% less ^{235}U and as much as 80% less ^{234}U than does naturally occurring uranium. The enrichment process reduces the radioactivity of depleted uranium to approximately half of that of natural uranium. The isotopic distributions and their respective contributions to the radioactivity are summarized in Table 3 [9]. These data show the radioactivity of natural uranium is 25,280 Bq g^{-1}, and that of 3.5% enriched uranium is 81,508 Bq g^{-1} while that of depleted uranium from the 3.5% enrichment is only 14,656 Bq g^{-1}. This corresponds to a reduction of 42% in total radioactivity. The data presented by Bleise *et al.* [3] also show a reduction of 42% in the total radioactivity of depleted uranium compared to the radioactivity of natural uranium.

The physical properties of uranium are summarized in Table 4.

Table 3: Isotopic distributions in natural uranium, enriched uranium and depleted uranium [9]

Natural uranium	^{234}U	^{235}U	^{238}U
Mass %	0.0053	0.711	99.284
Radioactivity %	48.9	2.2	48.9
Activity, Bq g U^{-1}	12356	568	12356
Enriched (3.5%) uranium	^{234}U	^{235}U	^{238}U
Mass %	0.02884	3.5	96.471
Radioactivity %	81.8	3.4	14.7
Activity, Bq g U^{-1}	66703	2800	12500
Depleted uranium	^{234}U	^{235}U	^{238}U
Mass %	0.0008976	0.2	99.799
Radioactivity %	14.2	1.1	84.7
Activity, Bq g U^{-1}	2076	160	12420

Table 4: Physical properties of uranium metal

Density (highly purified)	19.05 ± 0.02 gm cm^{-3}
Density (industrial grade)	18.85 ± 0.20 gm cm^{-3}
Melting Point	1132 ± 1 °C
Boiling Point	3811 ± 3 °C
Heat of Fusion	19.7 J $mole^{-1}$
Vapor Pressure at 1600 °C	10^{-4} mm
Thermal Conductivity at 70 °C	0.297 J (cm s °C)$^{-1}$
Electrical Resistivity at 25 °C	35×10^6 ohm cm^{-3}
Enthalpy at 25 °C	6364 J $mole^{-1}$
Entropy at 25 °C	58.2 ± 0.2 J (mole °C)$^{-1}$

CHEMICAL PROPERTIES OF URANIUM

Gindler's [10] monograph provides much information on the chemistry of uranium and its compounds, and Roberts et al.[11] subsequently reported detailed analytical methodologies for the determination of uranium. Grenthe et al. [12] contributed a comprehensive chapter on the chemical and physical properties of uranium to a larger work on the chemistry of the actinides. This chapter includes descriptions of the processing and refining of uranium ores as well as material on the chemistry of uranium in solution. In compounds, the oxidation number of uranium can range from 2 to 6. The oxidation-reduction chemistry of uranium is reflected by the standard reduction potentials listed in Table 5 [10,12].

Table 5: Standard reduction potentials

Acidic solution	Gindler [10]	Grenthe et al. [12]
$UO_2^{2+} + e \rightarrow UO_2^{1+}$	0.05 V	0.0878 ± 0.0013 V
$UO_2^{1+} + 4 H^{1+} + e \rightarrow U^{4+} + 2 H_2O$	0.62 V	
$UO_2^{2+} + 4 H^{1+} + 2 e \rightarrow U^{4+} + 2 H_2O$	0.334 V	0.2673 ± 0.0012 V
$U^{4+} + e \rightarrow U^{3+}$	−0.61 V	-0.553 ± 0.004 V
$U^{3+} + 3 e \rightarrow U$	−1.80 V	
$U^{4+} + 4 e \rightarrow U$	−1.38 V	
Alkaline solution		
$UO_2(OH)_2 + 2 e \rightarrow UO_2$	−0.3 V	
$UO_2 + e \rightarrow U(OH)3$	−2.6 V	
$U(OH)_3 + 3 e \rightarrow U$	−2.17 V	

In aqueous solutions of low pH, hexavalent uranium exists primarily as the yellow uranyl or dioxouranium (VI) ion, UO_2^{2+}. Some of its typical salts are: the nitrate, $UO_2(NO_3)_2$, the acetate, $UO_2(CH_3COO)_2$, and the sulfate, UO_2SO_4. Complexes of the uranyl ion with inorganic ligands include $[UO_2(NH_3)_2]^{2+}$, $[UO_2(CN)_4]^{2-}$, $[UO_2(CO_3)_2]^{2-}$, and $[UO_2F_3]^{1-}$[13].

Complexes with bioligands are more relevant to the distribution of uranium in biological systems. The transport of uranium species in the blood most likely takes place as complexes with plasma proteins, erythrocytes and/or low molecular mass species. Gutowski et al. [14] suggested the histidine residues in plasma proteins were responsible for binding of the uranyl ion. On the basis of infrared spectroscopy, Raman spectroscopy, single crystal, X-ray crystallography and computational methods, they determined the binding of the uranyl ion with 1-methylimidazole (meimid) to form a complex of the type $UO_2(meimid)_2(CH_3COO)_2$. Coordination of the uranyl ion is at the nitrogen atoms with bond lengths of 2.528 Å. The lengths of the uranium-oxygen bonds are 1.775 Å.

Vanengelen et al. [15] reported the coordination of the uranyl ion with pyrroloquinoline quinone (PQQ) cofactor and its potential as an inhibitor of flavoproteins. Using ultraviolet-visible spectroscopy and electrospray ionization mass spectroscopy as well as density functional theory computations for geometric structural optimizations, a complex having the general formula of $UO_2(CO_3)(H_2O)_x(PQQ)$ with bonding of the uranyl ion to the carbonyl oxygen, the pyridine nitrogen and the quinone oxygen of the pyrroloquinoline quinone cofactor was proposed. They suggested, "…UO_2^{2+} may also coordinate with enzymes or enzyme cofactors responsible for Mn(II) oxidation." Previously,

Chinni et al. [16] reported the oxidation of UO_2 by biogenic MnO_2, and they suggested the possibility of a catalytic enhancement of UO_2 oxidation-MnO_2 reduction. Such a cycling would impact the environmental fate of tetravalent uranium compounds by formation of more mobile hexavalent uranium compounds.

Pible et al. [17] employed computational tools to identify calcium-dependent interactions between proteins and small molecules likely to be inhibited by complexation of the uranyl ion. Four proteins were selected for experimental evaluation: C-reactive protein (P02741), fructose-binding lectin PA-ILL (Q9HYN5), 3,4-dihydroxy-2-butanone 4-phosphate synthatase (Q60364) and Mannose-binding protein C (P08661). Biochemical experiments confirmed the predicted binding site for UO_2^{2+}, and surface Plasmon resonance assays demonstrated the binding of UO_2^{2+} prevented the calcium mediated binding of phosphorylcholine. Experiments such as these partially elucidate the toxicological responses to uranium and the understanding of uranium toxicity.

The basic uranyl unit is thought to be retained in the uranates, Na_2UO_4, $CaUO_4$, etc. X-ray diffraction studies indicate each uranium atom is bonded to two uranyl oxygen atoms by shorter (1.9 Å) bonds and oxide oxygen atoms by longer (2.3 Å) bonds [18]. Diuranates such as $(NH_4)_2U_2O_7$ are analogous to the dichromates. Uranyl salts are the usual product of uranate hydrolysis; i.e.,

$$UO_4^{2-} + 4\, H^{1+} \leftrightarrows UO_2^{2+} + 2\, H_2O \text{ and}$$

$$UO_2^{2+} + 4\, OH^{1-} \leftrightarrows UO_4^{2-} + 2\, H_2O$$

Pentavalent uranium is oxidized by atmospheric oxygen. In the absence of air, pentavalent uranium undergoes disproportionation to hexavalent and tetravalent compounds; i.e.,

$$U_2O_5 \rightarrow UO_2 + UO_3.$$

Tetravalent uranium compounds include the oxide, UO_2, the binary halides as well as the acetate, the sulfate and the perchlorate. Tetravalent uranium also exists as basic salts such as $UOCl_2$, $UO(NO_3)_2$ and $UO(CH_3COO)_2$. Aqueous solutions of tetravalent uranium compounds are green in color, and the tetravalent uranium compounds are stable in the absence of air. Carbonato- and oxalato-complexes are more stable.

Trivalent uranium compounds are red in color. In aqueous media, trivalent uranium compounds are readily oxidized with the liberation of hydrogen. Among the trivalent compounds are UCl_3 and $UH(SO_4)_2$. Typical examples of compounds containing U(III) are the so called double chlorides such as

RbUCl$_4$ and complexes with 1-phenyl-2,3-dimethyl-5-pyrazolone [19]. Some of the divalent uranium compounds described by Grenthe *et al.* [12] are UO and US.

Elemental uranium is an active metal. It dissolves readily in hydrochloric acid and nitric acid but slowly in sulfuric acid. Uranium metal is unreactive with sodium hydroxide solutions. Finely divided uranium metal is pyrophoric. The metal can spontaneously ignite in air. A mixture of uranium oxides is produced in this combustion.

DEPLETED URANIUM PENETRATORS

The pyrophoricity of uranium is among the properties considered in the selection of depleted uranium for the fabrication of high energy penetrators. A variety of depleted uranium munitions have been developed and deployed. These depleted uranium penetrators are hardened by reducing the carbon content and by alloying with 0.75 percent by mass titanium during fabrication [3]. The high temperature generated by the impact with steel ignites the surface of a depleted uranium penetrator, and the projectile sharpens as it melts making it better able to pierce heavy armor. Depleted uranium projectile impacts are often characterized by round entry holes.

The 30-mm depleted uranium rounds were among the depleted uranium munitions used by the United States Air Force when Desert Shield became Desert Storm. They were able to pierce steel armor up to a thickness of 9 cm. This depleted uranium 30-mm ammunition consists of a conical penetrator 95 mm in length, 16 mm in diameter at the base and approximately 280 g in mass. The penetrator shown in Figure 1a is fixed in aluminum casing or jacket having a diameter of 30 mm and a length of 60 mm. The A-10 aircraft shown in Figure 1b is equipped with one gun capable of firing 3900 rounds per minute. A typical burst of fire from this gun usually has a duration of 2–3 s and involves between 130 and 190 rounds. Normally the depleted uranium ammunition is present in about 75% of the rounds. The remainder is traditional ammunition. The shots hit the ground in a straight line. Depending on the angle of approach, the shots hit the ground 1–3 m apart and cover an area of about 500 m^2. The number of penetrators striking the ground depends upon the type of target. Frequently, not more than 10% of the penetrators hit the target. When the penetrator hits a hard object such as an armored vehicle, the penetrator pierces the metal armor, generally leaving the jacket behind [20].

Figure 1: (a) Depleted uranium round (30 mm); (b) A-10 Warthog.

Among the munitions fired by the M1A1 Abrams tanks were 120 mm depleted uranium penetrators. The depleted uranium content of these munitions was 4.7 kg. The 120 mm ammunition consisted of a family of kinetic energy rounds and a family of high explosive anti-tank rounds. The kinetic energy rounds used a high length over diameter ratio sub-caliber projectile with a depleted uranium fin-stabilized rod as the penetrator element. Traveling at supersonic speed, this penetrator concentrated an extremely high level of kinetic energy over a relatively small surface area of the target. The high specific energy on target enabled the kinetic energy round to penetrate even the most resistive armor plates. All 120 mm rounds used a common combustible case which structurally combined the ammunition's components prior to firing and is completely consumed during firing. The combustible case is the primary reason for the superior interior ballistics performance of the 120 mm ammunition [20]. The fires and explosions resulting from hard-target penetrations of depleted uranium munitions produce UO_2 and U_3O_8. As the uranium oxides weather in the environment, some UO_3 may be formed. Mitchel and Sunder [21] reported an X-ray diffraction analysis of the dust obtained from live firing of depleted uranium munitions showing the uranium was present as 47% U_3O_7, 44% U_3O_8 and 9% UO_2. The composition of the oxides is most likely variable, and it will change over time as weathering takes place.

Individual particles of depleted uranium collected from different sites in Kuwait were examined by scanning electron microscopy with X-ray fluorescence spectrometry for microanalysis and with synchrotron radiation based X-ray absorption near edge spectroscopy. The particles collected at the holes made in armored vehicles by depleted uranium penetrators had a median size of 13 μm. The median size of those collected at the site of a fire at a storage facility for depleted uranium munitions was 44 μm. The compositions of the smaller particles corresponded to UO_2 and U_3O_8 while the mu-XANES (micro X-ray absorption near edge structure) spectra of the larger particles indicated the presence of uranyl compounds [22].

The products from live fire tests with depleted uranium munitions against hard and soft targets and from unfired uranium munitions buried in soils for corrosion studies were collected and examined. On the basis of electron microscopic and X-ray spectrometric examinations, three classes of particles were identified: depleted uranium aerosol particulates with diameters ranging form 1 to 20 μm composed mainly of UO_2 and U_3O_8, fused uranium particles containing iron (most likely from the targets) having diameters between 200 and 500 μm and deposits on particles of sand up to 500 μm in diameter. Particles in the first two classifications were thought to be derived from live fire impacts while those in the last classification originated from corrosion of buried, unfired munitions [23].

A dust of uranium oxides is often formed during impact. This dust can be dispersed and contaminate the environment. An estimated from 10% to 35% of the depleted uranium penetrator can become aerosolized on impact or when the depleted uranium ignites [3].

Depleted uranium dust is black. Many sites impacted by depleted uranium ammunition frequently show this black dust on and around the target. After an attack with depleted uranium ammunition, this black dust this can be deposited on the ground and other surfaces as partially oxidized depleted uranium fragments of different size, and as uranium oxide dust. Most of the depleted uranium dust is deposited within 100 m of the hit target.

The majority of the penetrators impacting sand or clay usually remain intact after penetrating the ground to a depth of more than 50 cm. Those striking soft targets such as unarmored or lightly armored vehicles do not generate significant dust contamination. Weathering of intact and fragmented depleted uranium penetrators and the black dust produced from their detonation is variable depending on the chemical properties of surrounding soils and rocks. In quartz sand, granite, or acidic volcanic rock, solubilization rates may be high enough to lead to local contamination of groundwater. The actions of wind and water may redistribute the fine depleted uranium dust. However, following a sand storm in south western Iraq, Yousefi and Najafi [24] collected and analyzed air and soil samples to determine the transport of depleted uranium dust from Iraq to Iran. None was found. Adsorption onto soil particles, mainly clay particles and organic matter, can reduce mobility and the danger of re-aerosolization. However, concern remains for the potential contamination of ground water after weathering of intact penetrators or large penetrator fragments.

Depleted uranium ordnance has been employed in at least three recent conflicts: the 1991 Iraq-Kuwait conflict, the 1995 Bosnia-Herzegovina conflict and the 1999 Kosovo-Serbia conflict. A summary of the depleted uranium weapons fired in these conflicts is presented in Table 6 [25]. This summary indicates that some 325 tons of depleted uranium have been introduced into the environment.

Table 6: Depleted uranium munitions deployed in recent military actions

Action	Munitions	Total mass, tons
1991 Iraq-Kuwait	US Air Force 30 mm rounds	259
1991 Iraq-Kuwait	US Army 120 mm tank rounds	50
1991 Iraq-Kuwait	US Marine aviation rounds	11
1991 Iraq-Kuwait	UK 120 mm tank rounds	1
1995 Bosnia-Herzegovina	NATO 30 mm rounds	3
1999 Kosovo-Serbia	NATO 30 mm rounds	10

EXPOSURE TO DEPLETED URANIUM

The isotopic composition of depleted uranium, as shown in Table 3 is dominated by ^{238}U at 99.977%. The percentages of ^{235}U and ^{234}U are 0.2% and 0.0008976%, respectively, which are below the natural values. All are long-lived alpha emitters as shown in Table 1. Like all radioactive nuclides, ^{238}U undergoes radioactive decay, and, within about six months, very small amounts of ^{234}Th, ^{234m}Pa, ^{234}Pa and ^{234}U are formed. They exist in a complicated equilibrium system with the ^{238}U. The decay scheme for ^{238}U and its progeny is shown in Figure 2. The progeny undergo sequential radioactive decays involving α and β emissions with some concurrent γ emissions to eventually become a stable isotope of lead, ^{206}Pb. The quantities of the intermediates are small, and their radiations like the α radiation from the parent ^{238}U present little, if any, external exposure hazard.

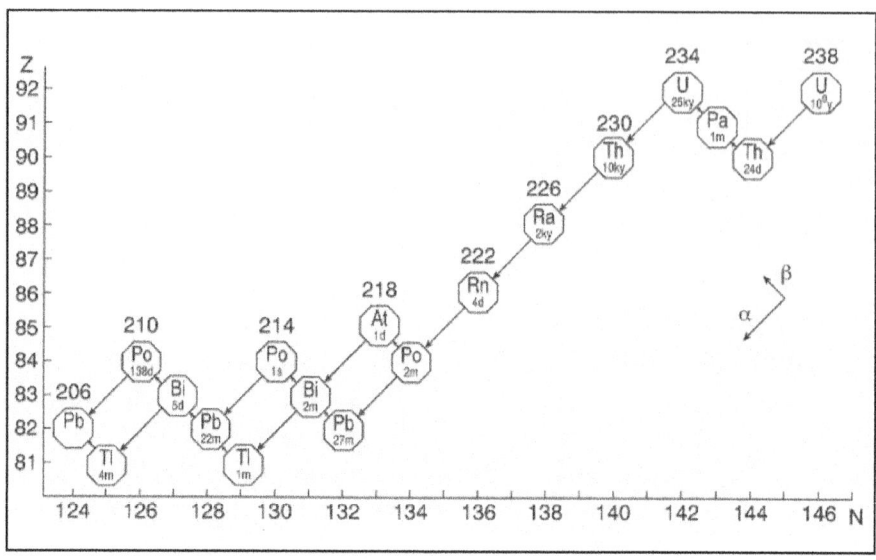

Figure 2: Decay chain for uranium-238.

Depleted uranium can become an internal exposure hazard by inhalation, by ingestion, by percutaneous absorption, and by dermal penetration of shrapnel or other explosion fragments. Tasat and her collaborators [26] have conveniently classified these as the inhalation route, the oral route, the dermal route, and the subcutaneous entry route. The US EPA [27] considers inhalation to be the most likely route for the intake of depleted uranium. Likewise, the World Health Organization [28] considers inhalation to be "…the most likely route of intake during or following the use of depleted uranium munitions in conflict or when depleted uranium in the environment is resuspended in the atmosphere by wind or other disturbances Accidental inhalation may also occur as a consequence of fire in a depleted uranium storage facility, an aircraft crash or the decontamination of vehicles from within or near areas of conflict." The World Health Organization [28] has pointed out that ingestion can occur in a large section of a community or population if drinking water or food supplies become contaminated with depleted uranium. Using thermal ionization mass spectrometry, Sahoo *et al.* [29] determined the uranium contamination of surface and ground water in South Serbia as of both natural and anthropogenic origins. The annual effective dose due to ^{238}U was estimated to be in the range of 9.2 × 10^{-5} to 2.1 × 10^{-3} mSv. The hand to mouth ingestion by children of soil contaminated with the black dust could become an additional potential pathway for the ingestion of uranium. The World Health Organization [28] considered dermal contact as a "... relatively unimportant type of exposure since little of the depleted uranium will pass across the skin into the blood". A similar statement, "Dermal contact is considered unimportant since little of the DU will pass across the skin into the blood" was made by the US EPA [27]. The Toxicological Profile for Uranium [30] contains the tabulated results of numerous animal experiments on the dermal toxicity of many uranium compounds. A large majority of these animal experiments were conducted with water soluble uranium compounds. Evidence for the percutaneous absorption of uranium in the human has been reported [31]. In the course of patch testing veterans exposed to depleted uranium during the Gulf Wars for dermal sensitivity using aqueous uranyl acetate solutions, the detection of elevated uranium concentrations in 24 h urine samples was interpreted as evidence for percutaneous absorption. The uranium in the black dust resulting from the detonation of depleted uranium munitions is of low solubility in water. However, elevated concentrations of uranium in the urine of 32 Gulf war veterans who were known to have been injured by depleted uranium munitions some 12 years earlier were attributed to fragments of depleted uranium that remained imbedded in the tissues [32]. The environmental weathering of unexploded depleted munitions or the *in vivo* corrosion of depleted uranium shrapnel could result in the formation of soluble and physiologically active forms of uranium.

The Inhalation Route

In his review on the toxicity of depleted uranium, Briner [5] has written about a cloud of particles, ranging in size from 0.2 to 15 μm in median aerodynamic diameter and containing a variety of oxides, being produced when a depleted uranium penetrator is detonated. He wrote, "When these particles are inhaled, they are either trapped in the oropharynx, where they are eventually swallowed, or they reach the lower airways where they are subject to alveolar absorption. Alveolar absorption appears to occur in two phases. There appears to be an early rapid phase which results in peak plasma levels and then a decline followed by a prolonged period of steady absorption. It is unclear what accounts for this biphasic pattern. It could be due to the heterogeneous chemistry of depleted uranium particles in the black dust. Some components may be more soluble than others. It could be due to the various sizes of inhaled particulates with those of a greater surface area to volume ratio dissolving quickly leaving behind those that dissolve more slowly. It may be due to an inflammatory response of the lung tissue that begins to retard absorption after a few days. Whatever the cause, the inhaled depleted uranium appears to have a pulmonary half life of about 4 years". Earlier Durakovic *et al.* [33] determined the minimum value for the biological half life of ceramic depleted uranium oxide in the lungs as 3.85 years. Valdéz [34] developed a linear model for estimating the lung burden of depleted uranium from measurement of depleted uranium in the urine. This model considered the intercellular dissolution of depleted uranium particles as well as the precipitation of a significant fraction of the dissolved depleted uranium as uranyl phosphate. Once desorbed from the lung, uranium becomes widely distributed throughout the body.

The Ingestion Route

Ingestion of depleted uranium is an additional exposure pathway. Uptake of uranium with drinking water is one of the major pathways for exposure to natural uranium, and possible for exposure to depleted uranium. Drinking water can become contaminated with depleted uranium from the black dust, munitions' fragments or penetrators buried in soil. Such contamination is strongly dependent on the chemical form of the depleted uranium, and the acidity and oxidation-reduction properties of the contaminated soil and ground water. Correlations between the ^{234}U to ^{238}U ratios in drinking water and the ^{234}U to ^{238}U ratios in the hair, nails, and urine have been reported for a group of 45 subjects who consumed from 0.2 to 2775 µg of uranium per day [35]. However, consuming up to 3000 µg of uranium per day in drinking water was not found to have cytotoxic effects on the kidney [36]. In addition to exposure to uranium in drinking water, direct ingestion of contaminated soil by children must be taken into consideration. Samples of soil from Kuwait and Kosovo known to be contaminated with depleted uranium were subjected to simulated gastric digestion with 0.16 M HCl. Between 73% and 96% of the depleted uranium particles in these soils were dissolved within one week indicating the potential for bioavailability by way of the ingestion route [37].

The Dermal Route

Percutaneous absorption is possibly one of the routes whereby uranium can intoxicate the body systemically. However, the amount of uranium absorbed depends on factors such as the solubility of the uranium compound, the length of time of exposure, the size of the area that is exposed, and other physical and physiological conditions. As described above, percutaneous absorption is not thought to be a major route to systemic intoxication.

The Subcutaneous Route

Wound contamination can occur during combat activities or later in the case of accidentally abrading of skin on contaminated surfaces. In the latter case, wound cleaning will be effective decontamination, and the resulting exposure to depleted uranium can be expected to be negligible. However, embedded fragments not removed by surgical means can result in chronic, internal exposure. Sixty-two American soldiers, wounded with depleted uranium shrapnel when their tanks or armored vehicles were hit by friendly fire during the Gulf War were studied for the effects of embedded fragments of depleted uranium shrapnel in their bodies [38]. It was shown that the depleted uranium metal slowly solubilized in the body fluids, and that several years after the war, blood and urine levels of uranium were elevated by up to two orders of magnitude [39].

TOXICOKINETICS OF DEPLETED URANIUM

Leggett [40] reported on the development of the, International Commission on Radiological Protection (ICRP) age specific biokinetic model for uranium. The compartments for this model are shown in Figure 3 [41]. This model describes the deposition of uranium from the blood into various, compartments, the return of uranium to the blood, and the eventual excretion of uranium. In keeping with ICRP's move towards physiological realism in its models, the uranium model includes recycling, *i.e.*, the possibility for material to pass from compartment to compartment via the blood stream [42]. The model is based on a number of sources which include data from both animal experiments and studies on humans. Clearly human data is preferred, and for uranium, ICRP can draw on a large database. In particular, there are data from the so-called Boston Subjects, a group of terminally ill patients who were injected with uranium in the 1950s. A brief overview of the human data that support the ICRP model is given in ICRP-69 [35]. Other reviews are provided by Leggett and Harrison [43] and Leggett [44]. The principal sites of uranium deposition in the body are the kidneys, the liver, and the bones. In addition, some material is deposited in various other tissues generally at lower concentrations than the main sites of deposition. These are usually referred to as "soft tissues". Of the amount absorbed into the blood stream, the ICRP model assigns 30% to soft tissues (rapid turnover, ST0 in Figure 3). This represents a pool of activity distributed throughout the body which exchanges rapidly with the blood stream. The remaining activity is apportioned as follows; kidneys 12%, liver 2%, bone 15%, red blood cells 1%, soft tissue (intermediate turnover, ST1 in Figure 3) 6.7%, soft tissue (slow turnover, ST2 in Figure 3) 0.3%, with 63% being promptly excreted in urine via the kidneys.

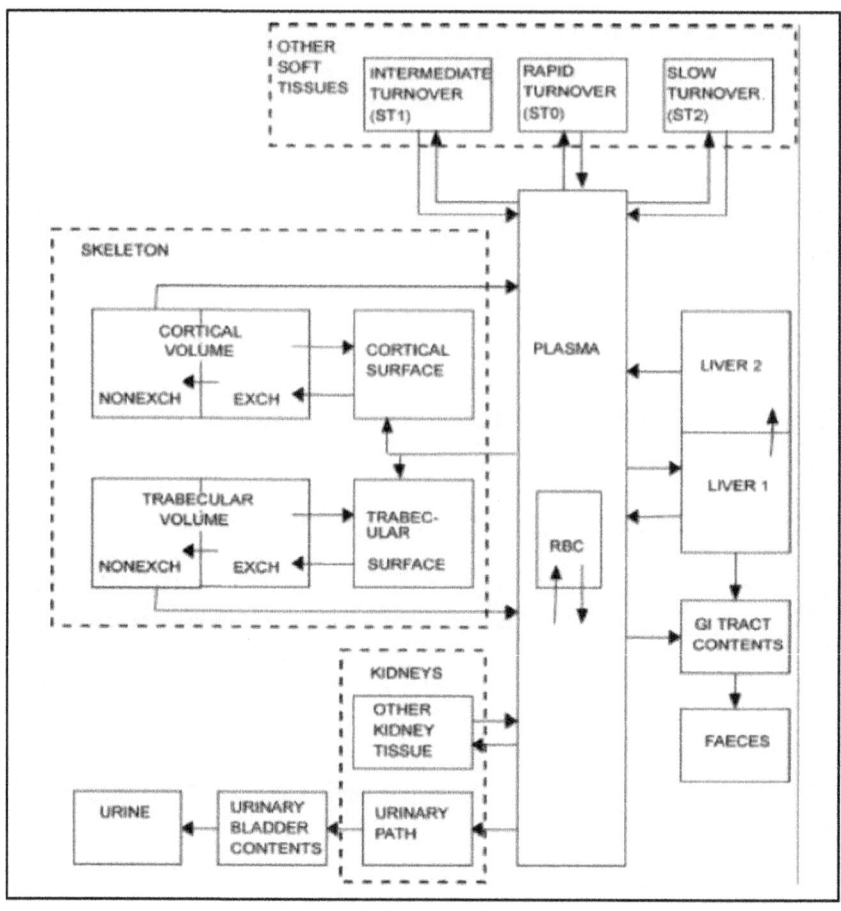

Figure 3: Biokinetic model for uranium [41].

Some of the uranium initially deposited in these organs and tissues can be returned to the blood stream, and some can be transferred to other organs or tissues (Figure 3). For example, uranium in the soft tissue compartments can be returned only to the blood while uranium in the liver can be exchanged with blood or transferred to other regions of the liver (Liver 2 inFigure 3). The bone warrants additional comment. Uranium is initially deposited on the bone surface (either trabecular or cortical), from where it can be transferred to bone volume (exchangeable) or returned to the blood stream. Uranium which does reach the exchangeable bone volume can be buried deeper in the bone volume (non-exchangeable) or returned to the surface. Uranium in the non-exchangeable volume can be transferred slowly to the blood. All the pathways used in the model are illustrated in Figure 3. In time, most of the systemic uranium is excreted in urine via the kidneys. A small fraction is also excreted in the feces. The length of time material remains in these compartments is partly governed by a removal half time, *i.e.*, the time that it takes to remove half of the material present. This time varies from organ to organ. For example, the removal half time for ST0 is as little as two hours, while for ST2 it is one hundred years. The net or apparent time that it takes to halve the amount of material in an organ, however, can be very different from the removal half-time, since uranium is continually being re-deposited by the recycling nature of the model. The net half time thus results from a combination of removing existing uranium and depositing new uranium from the blood stream.

TOXICOLOGY OF DEPLETED URANIUM

The potential for the toxicity of uranium lies in its properties as a heavy metal and as a radioactive substance. It is difficult to distinguish between the chemical toxicity of uranium and its radiotoxicity. The very small amounts of ^{234}Th, ^{234m}Pa, ^{234}Pa and ^{234}U formed during the first few months by the radioactive decay of ^{238}U are most likely of little toxicological consequence. However, radiation-specific damage from depleted uranium has been demonstrated *in vitro* using human osteoblast cells. Significant elevations in dicentric frequency were observed after 24 h exposures to depleted uranium as 50 μM $UO_2(NO_3)_2$ compared to parallel exposures using 50 μM solutions of nickel and tungsten as the nitrates. Exposures to 50 μM ^{238}U-$UO_2(NO_3)_2$ (0.33 μCi g^{-1}), 50 μM DU-$UO_2(NO_3)_2$ (0.44 μCi g^{-1}) and 50 μM ^{235}U-$UO_2(NO_3)_2$ (2.2 μCi g^{-1}), showed that at equivalent micro-molar concentrations (50 μM) of $UO_2(NO_3)_2$, there was a specific activity (μCi g^{-1}) dependent increase in neoplastic transformation frequency suggesting that α radiation can play a role in depleted uranium induced biological effects [45]. This observation, however, does not negate the chemical toxicity of uranium. The ATSDR Toxicological Profile for Uranium [30] states, "The health effects associated with oral or dermal exposure to natural and depleted uranium appear to be primarily chemical in nature and not radiological, while those from inhalation exposure may also include a slight radiological component, especially if the exposure involves prolonged exposure to insoluble uranium compounds. This profile is primarily concerned with the effects of exposure to natural and depleted uranium, but does include limited discussion regarding enriched uranium, which is considered to be more of a radiological than a chemical hazard. Also, whenever the term "radiation" is used, it applies to ionizing radiation and not to non-ionizing radiation". Some examples of the impact on enzyme systems of uranium, as uranyl salts, are the reversible inhibition of glucose-hexosediphosphate fermentation [46], the inhibition of S-adenosylmethionine synthetase [47], the inhibition of lactate dehydrogenase, pyruvate carboxykinase and glucose-6-phosphatase [48] and the inhibition of xenobiotic-metabolizing enzymes, CYP3A in particular [49].

Toxicity to the Lung

Inhalation studies with rats using 38-nm CMD nanoparticles in a nose-only exposure system showed that uranium was deposited in the respiratory tract and distributed to the blood, brain, skeleton, and kidneys. Of the uranium deposited in the lungs, some 20% was cleared with a half time of 2½ h, but the majority of the deposited uranium were cleared more slowly, $t_{1/2} = 141\tfrac{1}{2}$ days [50]. The occurrence of lung cancer among uranium miners has been attributed to radon gas rather than uranium dust. There is a causal relationship between lung cancer and radon exposure among uranium miners even though there is a cancer mortality deficit among all workers exposed to uranium [51]. In the natural state, radon gas is a part of the ^{238}U decay scheme. This is shown in Figure 2. Uranium has been shown to induce oxidative stress in rat lung epithelial cells, decrease the antioxidant potential of these cells, and reduce their proliferation [52]. Human lung epithelial cells lost contact inhibition and anchorage independent growth when exposed to depleted uranium. Cytogenetic analysis showed that exposure to depleted uranium induced neoplastic transformations in more than half of the cells [53]. Orona and Tasat [54] reported generation of the superoxide anion when rat alveolar macrophages were exposed to uranyl nitrate.

Renal Toxicity

High levels of uranium have been reported to accumulate in renal tissue [55]. At higher doses of uranium, kidney damage is the primary concern and the most immediate threat to patient health and survival. A human population of nearly 200 subjects continuously exposed to uranium in drinking water at concentrations as high as 1500 µg L^{-1} showed no evidence of renal damage. No statistically significant differences between the exposed population and age- and gender-matched controls were found for indicators of nephortoxicity in the urinary the levels of calcium, phosphate, glucose, creatinine, N-acetyl-gamma-D-glucosamididase, alkaline phosphatase, gamma-glutamyltransfertase, glutathione-S-transferase and lactate dehydrogenase as well in the serum levels of calcium, phosphate, glucose and creatinine [36]. A subsequent review of the literature, however, reported some evidence of adverse renal effects from exposure to uranium in drinking water as assessed by biomarkers of proximal tubular damage. In addition, indicators of proximal tubular effects, as evidenced by increased urinary β(2)-microglobulin and retinol binding protein levels, were reported in Desert Storm veterans exposed to depleted uranium [56]. It was previously reported that urinary uranium concentrations of veterans with embedded depleted uranium shrapnel ranged from 10 to over 500 times normal levels and that high urinary uranium concentrations were consistent with biomarkers of renal proximal tubular cell function and cytotoxicity associated with elevated urinary protein excretion, and it was proposed that these results supported basing the health protection guidelines for depleted uranium on chemical rather than radiological toxicity [57]. Chronic low dose exposures of up to 12 months to surgically-implanted depleted uranium in the rat has been reported to produce subtle pathologic changes in the kidneys along with blood chemistry changes suggesting renal dysfunction with accompanying anemia due to aberration of the kidneys' erythropoitic function [58]. The uranium concentrations in the bone, 21.64 ± 3.68 µg g^{-1}, and in the kidney, 17.79 ± 2.87 µg g^{-1}, at 360 days post implant were significantly greater than those in other tissues [59]. Using mitochondria isolated from the rat kidney, the nephrotoxicity of uranyl acetate, at concentrations of 50, 100 and 200 µM, was shown to disrupt the electron transfer chain at complex II and III leading to the generation of reactive oxygen species and subsequent oxidative stress [60]. It was further shown by Shaki and Pourahmad that beta-glucan offered some degree of protection against mitochondrial damage by uranyl acetate *in vitro* [61].

Roszell *et al.* [62] have concisely and succinctly described the renal toxicity of uranium. They wrote, "Whether the route of exposure is through inhalation or ingestion, after absorption the kidney is considered to be the target organ for uranium chemical toxicity". Uranium is transported in the blood as complexes with carbonate or bicarbonate, or with transferrin, or with other ligands. The uranium is either deposited in other compartments or excreted by the kidney. The carbonato complexes are filtered in the kidney at the glomerulus. These complexes dissociate as the glomerulate filtrate passes along the proximal tubule and becomes more acidic. The uranyl ion released from the complex can react with components of the filtrate or with components of the tubular membrane. The former may remain dissolved in the filtrate, and the latter may become bound at the ionic sites of the brush border membrane of the proximal tubule. The soluble forms of uranium can pass into the bladder and be eliminated, but the bound forms can alter or destroy renal tubular cells. Roszell *et al.* [62] cited one suggested mechanism for renal damage by which the binding of uranium altered cellular permeability to sodium which, in turn, interfered with the transport of glucose, amino acids, and phosphates and resulted in increased excretion of these compounds. Other possible cellular events could include changes to the plasma membrane, which could in turn affect membrane transport and permeability, lysosomal damage, mitochondrial dysfunction, and DNA damage leading to apoptosis. In preparing their assessment, they recognized the formation of oxides when depleted uranium ammunition impacts upon a hard target, and they recognized their limited solubility in aqueous media. They proposed inhalation as the major route of uranium intake.

Arzuaga, et al [56] have also reviewed the renal effects of uranium, and they too have described mechanisms for renal toxicity. One of the studies they described used proximal and distal tubular cell lines (MD CK and LLC PK1) and suggested the dependence of renal toxicity on the formation of complexes such as $UO_2(PO_4)^{-1}$ and $UO_2(HPO_4)$ with their subsequent uptake by the sodium dependent co-transporter NaPi-II and absorptive endocytosis. They described another study in which isolated human and murine kidney cortex tubules exposed to uranium showed inhabitation of cellular ATP and glucogenesis due to inhabitation of lactate dehydrogenase, pyruvate carboxylase, glucose 6-phosphatase and phoshonolpyruvate carboxykinase. Additional studies suggested uranium exposure of human renal HEK293 cells altered expression of genes associated with calcium-dependent cell signaling such as IP3 cascade kinases PI4K11 and PIK3R1, the intercellular calcium receptor calmodulin and calmodulin-dependent proteins and cell trafficking pathways such as the potassium channel ABC subunits ATP6V1A1 and ABCCC8. Also described are some *in vitro* and *in vivo* studies showing uranium exposure depleted glutathione and glutathione reductase activity, increased reactive oxygen species production and DNA damage and promoted apoptosis. The final description was of studies reporting chronic exposure to uranium enhances the expression of genes associates with oxidative stress responses such as superoxide dismutase 1 and ion transports including NaPi-II and S1c34A1. The involvement of the oxidative stress responses is a factor Lestaevel *et al.* [63] considered in their work on neurotoxicological disturbances attributed to depleted uranium exposure (See section 8.3). Neither Arzuaga *et al.* [56] nor Roszell and her collaborators [62] make mention of radiotoxicity in their overviews on the mechanisms of renal toxicity.

Neurological Toxicity

Houpert *et al.* [64] reported a study comparing neurophysiological and behavioral changes in rats exposed to enriched uranium and depleted uranium. Exposure was with mineral water containing 40 mg U (from uranyl nitrate) L^{-1} as either enriched (95.74% ^{238}U, 4.24% ^{235}U, 0.02 ^{234}U) uranium having a specific activity of 66.3 kBq g^{-1} or depleted (99.74% ^{238}U, 0.026% ^{235}U, 0.001 ^{234}U) uranium having a specific activity of 14.7 kBq g^{-1}. The control rats received the same mineral water. After 45 days of exposure, electroencephalographic (EEG) activity was recorded, and spatial working memory was assessed. The rats were sacrificed at the end of the exposure, and the uranium contents of kidneys, adrenals bones and brains were determined by ICP-MS. The EEG activity of the rats exposed to the enriched uranium showed increased paradoxical sleep episodes relative to the control rats while the rats exposed to the depleted uranium showed no such changes in sleep architecture. In assessing spatial working memory by spontaneous alternation examination, it was observed that the rats exposed to the enriched uranium demonstrated decreased activity while the activity of those exposed to the depleted did not differ from the control rats. As might be expected, the uranium concentrations in the kidneys of the rats exposed to the enriched uranium and the depleted uranium did not differ and both were approximately ten fold greater that those of the control rats. Surprisingly, the mean concentrations of uranium in the hippocampus and hypothalamus of the brain and in the adrenals of the rats exposed to the enriched uranium were significantly greater than those of the rats exposed to the depleted uranium. While the accumulation of uranium in the adrenals and in some brain structures was correlated with decreased neurobiological functions, the differential between the enriched and the depleted uranium remain intriguing.

Jaing and Aschner [65] have described several neurobiological consequences of exposure to depleted uranium in their review on its neurotoxicity. Among these were: differences in electrophysiological studies in hippocampal slices from rats implanted with depleted uranium, increased acetylcholinesterase activity in the cortex of rats receiving intramuscular injections of uranyl acetate at doses of 1 mg kg^{-1} and differences in brain lipid oxidation in rats treated with depleted uranium.

Jaing et al. [66] recognized the gap in the understandings of the specific effects of uranium on cells of the central nervous system and the potential molecular changes resulting from exposure to depleted uranium. Using primary cortical neuron cultures as targets for exposure to depleted uranium (as uranyl acetate), they observed: little if any effect on cell viability and morphology; little if any effect on thiol metabolite levels, redox potential or high energy phosphates; and little if any effect on lipid peroxidation. Neuro degradation was not observed when *Caenorhabditis elegans* was the target for the depleted uranium exposure. One of the conclusions drawn from these observations was the neurotoxic potential of depleted uranium as uranyl acetate was low. No speculations were made regarding the chemical form of the depleted uranium encountered under military combat, fuel processing or other conditions.

Lestaevel et al. [63] reported further on the double toxicity, chemical and radiological, of uranium. Rats were to uranyl nitrate as cited above for Houpert et al. [64]. The cerebral cortex of those exposed to the enriched uranium showed enhanced lipid peroxidation. Increases in the activities of superdioxide dismutase, catalyse and glutathione peroxidase were observed in the cerebral cortex of the rats exposed to the depleted uranium. These results were interpreted as a demonstration that depleted uranium induced increases in several antioxidants while exposure to enriched uranium was associated with oxidative stress, and the reactive oxygen species associated with oxidative stress could induce neurotoxicological disturbances attributed to uranium exposure. Lestaevel *et al*. [67] also reported decreases of the cholesterol and the acetylcholine in the entorhinal cortex of nice exposed to depleted uranium, and suggested exposure to depleted uranium could modify the pathology of the apolipoprotein E associated with Alzheimer's disease.

Reproductive/Developmental Toxicity

A literature review published a decade after the deployment of depleted uranium munitions in Operation Desert Storm reported no data was available on the reproductive effects of embedded depleted uranium shrapnel [68]. Subsequent studies using the offspring of female and male rats surgically implanted with depleted uranium pellets revealed no gross physical abnormalities attributable to prenatal uranium exposure. Elevations of the urinary uranium concentrations, confirming *in vivo* solubilization of the implanted pellets, were observed in the parent generation. Neurodevelopment and immune function assessments of the first generation offspring were normal [69]. In a follow up study of the second generation offspring, development was normal, and no gross abnormalities were observed. As with the first generation offspring, no instances of rib cage malformation were observed at necropsy. Histopathology of kidneys, spleen, thymus, bone marrow and ovaries and testes did not differ from control rats for both the first and second generation offspring. In the necropsy of the parent rats, marked inflammatory responses were observed in the soft tissues surrounding the implanted depleted uranium pellets. The mean heart masses of the first and second generation offspring of the rats receiving the highest doses of depleted uranium were greater than those of the corresponding control animals. In general, it appears that imbedded depleted uranium is not a reproductive or developmental hazard. However, the elevated heart masses of the first and second generation offspring suggest conservatism in totally discounting the possibility of teratogenic effects [70].

As was the observation of the increased masses of hearts in offspring from parents receiving the highest doses of depleted uranium pellets, the mutation frequency in vector recovered from the DNA in bone marrow cells of the offspring of transgenic male mice exposed to depleted uranium also showed a dose dependent increase. In addition, a dose dependent (in terms of specific activity) mutation increase was observed in the offspring of male mice exposed to either enriched or depleted uranium in drinking water as uranyl nitrate at a total uranium concentration of 50 µg L^{-1} [71].

Carcinogenicity

Increased cancer incidence is one of the many concerns about exposure to depleted uranium. A British study involving surveillance for the ten years following the conflict of 52,721 Gulf War veterans and 50,755 age- and gender-matched military personnel who had not been deployed to the Gulf revealed no excesses in site specific cancers in the former cohort. There were 270 incidences of cancer among those who had served in Iraq and 269 incidences of cancer in the control cohort [72]. Four bone cancers among the 20,012 Danish Balkan War veterans exceeded expectations [73]. A review of twelve epidemiological studies found no evidence of excess cancer risk among veterans of military operations in Iraq and Kosovo except for the apparent increased risk of bone cancer in the Danish cohort [74]. In a subsequent study comparing 18,175 male Dutch soldiers who had been deployed to the Balkans with 1,365,355 male members of the Dutch military who had not served in this region, the cancer incidence in the former was 17% lower than that of the later [75].

Twenty-Year Surveillance

In the heat of combat, some three dozen U.S. servicemen were killed by friendly fire, and almost twice that number were wounded. Of the wounded, more than half were left with fragments of depleted uranium embedded in their bodies. They have been under periodic medical surveillance during the two decades following the cessation of hostilities in early 1991.

Eight years after hostilities ended, urinary uranium levels of Gulf War veterans with retained depleted uranium shrapnel were between 0.018 and 39.1 µg per g creatinine while uranium in the urine from other veterans without imbedded shrapnel ranged from 0.002 to 0.231 µg per g creatinine. The persistently elevated urinary uranium excretion was interpreted as an indication of continued mobilization and chronic low dose exposure. This was correlated with an increase in sister chromatid exchange in peripheral lymphocytes [76]. Continued surveillance confirmed the elevated urinary uranium concentrations [77]. Application of the ICRP biokinetic model described above suggested the possibility of renal damage from both the chemical and radiological toxicity of uranium. However, markers for changes in renal glomerular and tubular function did not differ significantly between those with embedded depleted uranium shrapnel and their controls. The absence of firm evidence for renal effects, despite the kidney's reputation as the critical organ, may be attributed to a relatively low uranium burden compared to uranium exposures in the occupational sector. However, measures of renal tubular function and structural integrity have yielded results suggestive of, though not statistically significant for, an early effect from exposure to depleted uranium. Genotoxicity endpoints continued to yield mixed results [32]. On follow up, the concentrations of retinol binding protein, a marker for renal proximal tubular function, in the exposed group showed no significant differences from those of the controls, but evidence of a weak genotoxic effect was reported [78]. Sixteen years after being wounded by friendly fire, military personnel continue to excrete uranium at concentrations proportional to the depleted uranium shrapnel burden in the body. Although subtle trends emerge in renal proximal tubular function and bone formation, the wounded veterans exhibited few of the clinically significant health effects associated to uranium intoxication [79]. Two years later, these parameters showed little change. The elevated urinary excretion of uranium by those with the embedded shrapnel continued. No significant evidence of clinically important changes was observed in kidney and bone, the two principle uranium target organs [80]. Compared with uninjured Gulf War veterans, those with embedded depleted uranium showed no increased frequency of micronuclei formation in the peripheral blood lymphocytes [81]. Likewise, chronic exposure from the embedded depleted uranium did not induce chromosomal aberrations in the peripheral blood lymphocytes [82]. Pulmonary health assessments of veterans with high body burdens of depleted uranium, who most likely sustained inhalation exposure during the friendly fire incidents, were within normal clinical expectations. No significant respiratory symptoms, abnormal pulmonary function values or prevalence of chest CT scan abnormalities were observed [83]. McDiarmid *et al.*, [84] observed few clinically significant health effects related to long-term, low-dose depleted uranium exposure from embedded shrapnel. Renal biomarkers showed minimal effects on proximal tubular function and cytotoxicity, and pulmonary functions remained within the normal clinical ranges.

Collateral Injuries

While American soldiers wounded with shrapnel from depleted uranium penetrators showed few, if any, symptoms or signs of injury due to uranium intoxication during twenty years of observation, increased incidences of congenital birth abnormalities and of cancers have been reported in Iraq. Of 6049 births recorded at the Fallujah General Hospital in 2009, there were 291 instances of congenital abnormality, which corresponds to a rate of 48 per thousand. This frequency of congenital abnormality is higher than the 12½ per thousand reported in neighboring Kuwait, the 8 per thousand reported in the United Arab Emirates and the 32 per thousand reported in Egypt. The higher rates of congenital abnormality in Iraq were attributed to prenatal environmental exposures of the parents to genotoxic agents such as uranium [85]. Environmental exposures to depleted uranium of the Iraqi parents of children with congenital abnormality were assessed by monitoring soil and drinking water and by analyzing scalp hair. (Hair is considered by many to be a bio-indicator of occupational or environmental exposure to toxic elements.) The average uranium concentration for two dozen scalp hair samples from parents of children born with major congenital abnormalities at the Fallujah General hospital were between two to three times higher than that obtained with samples from a hundred residents of southern Israel. The average values were 0.16 *versus* 0.062 mg kg^{-1}, respectively. The uranium concentration of six soil samples collected in different districts of Fallujah ranged from 0.1 to 1.5 mg kg^{-1}, and the concentrations of uranium in tap water, well water and water from the Euphrates River were 2.28, 2.72 and 2.24 µg L^{-1}, respectively. The high concentrations of uranium in the scalp hair from the Fallujah subjects could not be attributed to either the soil or the water concentrations of uranium. "… these results support the belief that the effects in Fallujah follow the development of a uranium-based weapon or weapons of some unknown type" [86]. However, a subsequent review of the literature on congenital abnormalities in Iraq concluded, "As no [*sic*] enough data on pre 1991 Gulf War prevalence of birth defects in Iraq are not available, the ranges of birth defects reported in the reviewed studies from Iraq most probably do not provide a clear indication of a possible environmental exposure including DU or other teratogenic agents although the country has faced several environmental challenges since 1980" [87]. In addition, the frequency of congenital abnormalities among infants born to U.S. service personnel who had served in Iraq between 1990 and 1991 did not differ from that among infants born to members of the U.S. military who were not deployed to Iraq. A slight increase in the prevalence of birth defects was observed among infants born to male war fighters who were deployed for between 153 and 200 days compared to those deployed to Iraq for between 1 and 92 days [88].

A report on a questionnaire designed to substantiate media coverage of increased birth defects and incidences of cancer attributed to the use of depleted uranium weapons in Fallujah did little to clarify the issues. In the words of the authors, "… the results reported here do not throw any light upon the identity of the agent(s) causing the increased levels of illness, and although we have drawn attention to the use of depleted uranium as one potential relevant exposure, there may be other possibilities and we see the current study as investigating the anecdotal evidence of increases in cancer and infant mortality in Fallujah" [89]. Elevated blood uranium concentrations among leukemia victims have been reported in the Basrah-Muthanna-Dhi Qar region south east of Fallujah [90]. The blood uranium concentrations were determined by a track-etch technique based on the neutron induced fission of ^{235}U. The mean blood uranium concentrations with their standard errors for 30 leukemia victims and 30 control subjects matched for gender, age and domicile were 2.87 ± 0.11 and 1.43 ± 0.07 µg L^{-1}, respectively. The higher uranium concentrations in the blood samples from the leukemia victims were attributed to military activities during the Gulf Wars. No environmental data are presented to confirm their exposure, nor is any explanation offered for the lower values in the control subjects who shared this environment. Studies on frequencies of childhood leukemia over time in the Basrah region disagree on the trends. Hagopian *et al.* [91] maintain the rates have doubled over the fifteen year period from 1993 to 2007 while others [92] observed no temporal increases during the six year period from 2004 to 2009. The former cited an annual rate of 12.2 per 100,000 population for 2006 while the 2006 annual rate reported by the latter was 4.49 per 100,000 population. The latter rate is several orders of magnitude less that that cited above for 2009 in Fallujah 2009 [85]. Greiser and Hoffmann [93] challenged the former report citing the work of a WHO mission that found no increase in childhood leukemia for the governorate of Basrah. The authors [94] response was based on uncertainties in the accuracy of the population statistics and did little to resolve the discrepancy. In addition, instances of further increased cancer rates due to depleted uranium have been reported in the northern Iraq city of Mosul [95].

Polarized views from different interest groups maintain a somewhat sustained controversy about the collateral effects of depleted uranium weapons. The issues are sometimes more in the realm of the public media than they are in the scientific community. Regardless of the orientation, the growing body of evidence should encourage putting the problem and its solution high on the list of priorities.

SOME CONCLUDING COMMENTS

Some 325 tons of depleted uranium munitions were detonated during the conflicts in the Persian Gulf and in the Balkans, and additional quantities of depleted uranium were released into the environment during incidents such as the explosion and fire on 11 July 1991 at the ammunition storage depot known as Camp Doha. Uranium is an α emitting radioactive element having both radiological and chemical toxicity. The release of uranium and its decay products into the environment (air, soil, and water) presents a threat to human health and environmental quality. Uranium can enter the body by inhalation, ingestion, transdermal absorption, or injection from injuries. The primary route of entry into the body is inhalation. Research on inhaled, ingested, and/or dermally absorbed industrial uranium compounds has shown that solubility influences the target organ, the toxic response, and the mode of uranium excretion. The overall clearance rate of uranium compounds from the lung reflects both mechanical and dissolution processes depending on the morphological and chemical characteristics of uranium particles. Three kinds of uranium can be considered: natural uranium, enriched uranium, and depleted uranium. While the chemical and radiological toxicities of natural uranium and enriched uranium have been the subject of extensive research for more than a half century, depleted uranium is a relatively new arrival on the scene. The radiological and chemical properties of natural uranium and depleted uranium are similar. In fact, natural uranium has essentially the same chemical toxicity as depleted uranium, but the radiological toxicity is some 60% higher for the former. Depleted uranium has been used in military conflicts, and it has been claimed to contribute to health problems.

The United Nations Environment Programme (UNEP) [96] post conflict environmental assessment of depleted uranium was conducted at 11 locations in southwestern Kosovo along the border with Albania. These sites were selected from among the 112 targets where NATO air strikes fired 30 mm depleted uranium munitions. This assessment concluded, "There was no detectable widespread contamination of the ground surface by depleted uranium. .... The corresponding radiological and toxicological risks are insignificant or even non-existent. Detectable ground surface contamination by DU is limited to areas within a few meters of penetrators and localized points of concentrated contamination caused by penetrator impacts. ... There is no significant risk related to these contamination points in terms of possible contamination of air, water of plants. ...." None the less, Nafezi, *et al.* [97] have reported radon levels ranging from 82 to 432 Bq m^{-3} in 15 dwellings in the village of Planej, one of the 11 sites selected for the UNEP assessment.

During the conflict, Serbs held positions in and around this village. In November 2000, there was heavy fighting in the area, and the village was largely destroyed. According to NATO, the village was attacked on 31 May 1999 by A-10 aircraft which fired 970 rounds. The size of the targeted area is not known. Many penetrators may remain hidden in the ground. Eventually these could dissolve, with the depleted uranium entering the ground water. There is, consequently, a possibility that the drinking water in some nearby wells could become contaminated. The drinking water in nearby wells should be kept under surveillance by taking samples at appropriate intervals for uranium testing.

The UNEP has conducted additional post conflict environmental assessments of depleted uranium in Serbia, Montenegro and Bosnia and Herzegovina [98,99]. The large majority of the NATO air strike targets were in Kosovo. These are considered in the preceding paragraphs. Investigations were made at five of the 11 sites targeted in Serbia and Montenegro. As was the case in Kosovo, the primary concern in Serbia was for the potential contamination of ground water with depleted uranium from the sub-surface corrosion of buried penetrators. In Bosnia and Herzegovina, 10 sites near the cities of Sarajevo and Gorazde were investigated. It is important to point out this investigation was conducted seven years after hostilities ended. Surface β/γ radiation was detected at some 300 sites of depleted uranium penetrator air strikes. Analysis of air samples taken at some of these sites indicated no significant risk was expected to arise from these sites by inhalation of soil particulate matter. Recovery of buried depleted uranium penetrators revealed they were corroded, and the soil in contact with the buried depleted uranium penetrators was contaminated with as much as 45 g of uranium per kilogram of soil. Uranium concentration in the soil 10 cm below the penetrator was some hundred times less, and it was reduced by another factor of ten within the next 30 cm. The low mobility of corroded uranium in the soil does not preclude the contamination of ground water. Traces of depleted uranium were found in two samples of drinking water by mass spectrometry. Although the mobility of depleted uranium corrosion products in soil is low, the potential for drinking water contamination should undergo further monitoring.

The corrosion of depleted uranium and the subsequent transport of the corrosion products were studied in column experiments simulating a sand rich environment with field moist conditions. The depleted uranium was oxidized to metaschoepite, $(UO_2)_8O_2(OH)_{12}$, at a rate of 100 ± 12 mg cm^{-2} years^{-1} [100]. The rate at which buried depleted uranium munitions corroded at the Kirkcudbright and Eskmeals live fire test ranges was reported to be from 130–1900 mg cm^{-2} years^{-1}, and the time for complete corrosion was estimated as between 2½ and 48 years [101]. Experimental remediation of contaminated soil at the latter site showed extraction with citric acid removed between 30 and 42 percent of the depleted uranium while extraction with ammonium bicarbonate solution removed between 42 and 50 percent of the depleted uranium. Sequential extraction with ammonium bicarbonate and then with citric acid improved uranium removal from soil at the Eskmeals site to between 68 and 87 percent [102]. Both of these extractants are environmentally friendly, but after use, they would require management as hazardous wastes.

The results from *in vivo* and *in vitro* investigations on both natural and depleted uranium and the renewed efforts to understand the chemistry and the toxicology of depleted uranium appear to have been compromised somewhat by the political agendas of special interest groups at the national and international levels. The presence of the depleted uranium in the environment, the routes of its entry to the body and its impact on human health and environmental quality will occupy the scientific community and the political arena for decades to come.

REFERENCES

1. Uranium and Depleted Uranium. Available online: http://www.world-nuclear.org/info./Nuclear-Fuel-Cycle/Uranium-Resources/Uranium-and-Depleted-Uranium/ (accessed on 31 December 2013).
2. Legget, R.W.; Pellmar, T.C. The biokinetics of uranium migrating from embedded DU fragments. *J. Environ. Radioact.* 2003, *64*, 205–225.
3. Bleise, A.; Danesi, P.R.; Burkart, W. Properties, use and health effects of depleted uranium (DU): A general overview. *J. Environ. Radioact.* 2003, *64*, 93–112.
4. Craft, E.S.; Abu-Qare, A.W.; Flaherty, M.M.; Garofolo, M.C.; Rincavage, H.L.; Abou-Donia, M.B. Depleted and natural uranium: Chemistry and toxicological effects. *J. Toxicol. Environ. Health B* 2004, *7*, 297–317.
5. Briner, W. The toxicity of depleted uranium. *Int. J. Environ. Res. Public Health* 2010, *7*, 303–313.
6. Phennig, G.; Klewe-Nebenius, H.; Seelmann-Eggebert, W. *Karlsruher Nuklidkarte, 6 Auflage 1995, korrigereten 1998*; Institut für Instrumentelle Analytic: Karlsruhe, Germany, 1998.
7. Sonzogi, A. Chart of the Nuclides. In Proceedings of the International Conference on Nuclear Data for Science and Technology, 22–27 April 2007; Bersillon, O., Gunsing, F., Bauge, E., Jacqmin, R., Leray, S., Eds.; EDP Science: Upton, NY, 2008; pp. 105–106.
8. Emsley, J. *The Elements*; Clarendon Press: Oxford, UK, 1989; p. 197.
9. Neghabian, R.A.; Becker, H.J.; Baran, A.; Binzel, H.-W. *Verwendung von wiederaufgearbeitetem Uran und von abgreichertem Uran*; NUKEM: Alzenau, Germany, 1991.
10. Gindler, J.E. *The Radiochemistry of Uranium*; National Academy of Sciences, National Research Council: Washington, DC, USA, 1962.
11. Roberts, R.A.; Choppin, G.R.; Wild, J.F. *The Radiochemistry of Uranium, Neptunium and Plutonium*; National Academy of Sciences, National Research Council: Washington, DC, USA, 1986.
12. Grenthe, I.; Drożdżyński, J.; Fujino, T.; Buck, E.C.; Albrecht-Schmitt,

T.E.; Wolf, S.F. *Uranium, in The Chemistry of the Actinide and Transactinide Elements*, 4th ed.; Morss, L.R., Edelstein, N.M., Fuger, J., Katz, J.J., Eds.; Springer: Dordrect, the Netherlands, 2011.

13. Emsley, J. *The Elements*; Oxford Press: Oxford, UK, 1989; p. 202.
14. Gutowski, K.E.; Cocalia, V.A.; Griffin, S.T.; Bridges, N.J.; Dixon, D.A.; Rodgers, R.D. Interactions of 1-methylimidazole with $UO_2(CH_3CO_2)_2$ and $UO_2(NO_3)_2$: Structural, spectroscopic and theoretical evidence of imidazole binding to the uranyl ion. *J. Am. Chem. Soc.* 2007, *129*, 526–536.
15. Vanengelen, M.R.; Szilagyi, R.Z.; Gerlach, R.; Lee, B.D.; Apel, W.A.; Peyton, B.M. Uranium exerts acute toxicity by binding to pyrroloquinole quinine cofactor. *Environ. Sci. Techol.* 2011, *45*, 937–942.
16. Chinni, S.; Anderson, C.B.; Ulrich, K.-U.; Giammar, D.E.; Tebo, B.M. Indirect UO_2 oxidation by Mn(II)-oxidizing spores of *Bacillus* sp. Strain SG-1 and the effect of U and Mn concentrations. *Environ. Sci. Technol.* 2008, *42*, 8709–8714.
17. Pible, O.; Vidaud, C.; Plantevin, S.; Pellequer, J.-L.; Quéméneur, E. Predicting the disruption by UO_2^{2+} of a protein-ligand interaction. *Protein Sci.* 2010, *19*, 2219–2230.
18. Roof, I.P.; Smith, M.D.; zur Loye, H.-C. Crystal growth of uranium-containing complex oxides: $Ba_2Na_{0.83}U_{1.17}O_6$, $BaK_4U_3O_{12}$ and $Na_3Ca_{1.5}UO_6$. *Solid State Sci.* 2010, *12*, 1941–1947.
19. Barnard, R.; Bullock, J.I.; Gellatly, B.J.; Larkworthy, L.F. The chemistry of the trivalent actinides. Part II, uranium (III) double chlorides and some complexes with organic ligands. *J. Chem. Soc. Dalton Trans* 1972, 1932–1938.
20. Office of the Special Assistant for Gulf War Illness, Department of Defense. Environmental Exposure Report: Depleted Uranium in the Gulf. 1998. Available online: http://www.gulflink.osd.mil.du (accessed on 31 December 2013).
21. Mitchel, R.E.; Sunder, S. Depleted uranium dust from fired munitions: Physical, chemical and biological properties.*Health Phys.* 2004, *87*, 57–67.
22. Salbu, B.; Janssans, K.; Lind, O.C.; Proost, K.; Gijsels, L.; Danesi, P.R. Oxidation state of uranium in depleted uranium particles from Kuwait. *J. Environ. Radioact.* 2005, *78*, 125–135.
23. Sajih, M.; Livens, F.R.; Alvarez, R.; Morgan, M. Physiochemical characterization of depleted uranium (DU) particles at a UK firing test

range. *Sci. Total Environ.* 2010, *408*, 5990–5996.

24. Yousefi, H.; Najafi, A. Assessment of depleted uranium in South-Western Iran. *J. Environ. Radioact.* 2013, *124*, 160–162.

25. Parliamentary Office of Science and Technology. Depleted Uranium. 2001. Available online: www.parliament.uk/post/home.htm (accessed on 31 December 2013).

26. Tasat, D.R.; Orona, N.S.; Bozal, C.; Ubios, A.M.; Cabrini, R.L. Intercellular Metabolism of Uranium and the Effects of Bisphosphonates on Its Toxicity. In *Cell Metabolism-Cell Homeostasis and Stress Response*; Bubulya, P., Ed.; Tech Publishers: Rijeka, Yugoslavia, 2012.

27. United States Environmental Protection Agency, Office of Radiation and Indoor Air Radiation Protection Division.*Depleted Uranium, Technical Brief*; US EPA: Washington, DC, USA, 2006.

28. WHO. *Health Effects of Depleted Uranium*; A54/19; World Health Organization: Geneva, Switzerland, 2001.

29. Sahoo, S.K.; Matsumoto, M.; Shiraishi, K.; Cuknic, O.; Zunic, Z.S. Dose effect for South Serbians due to ^{238}U in natural drinking water. *Radiat. Prot. Dosimetry* 2007, *127*, 407–420.

30. ATSDR. *Toxicological Profile for Uranium*; U.S. Department of Health and Human Services, Agency for Toxic Substances and Disease Registry, Division of Toxicology and Human Health Sciences, Environmental Toxicology Branch: Atlanta, GA, USA, 2013.

31. Shavrtsbeyn, M.; Tuchinda, P.; Gaitens, J.; Squibb, K.S.; McDiarmid, M.A.; Gaspari, A. Patch testing with uranyl acetate in veterans exposed to depleted uranium during the 1991 Gulf War and the Iraqi conflict. *Dermatitis* 2011, *22*, 33–39.

32. McDiarmid, M.A.; Engelhardt, S.M.; Oliver, M.; Gucer, P.; Wilson, P.D.; Kane, R.; Kabat, M.; Kaup, B.; Anderson, L.; Hoover, D.; *et al*. Biological monitoring and surveillance results of Gulf War I veterans exposed to depleted uranium.*Int. Arch. Occup. Environ. Health A* 2005, *72*, 14–29.

33. Durakovic, A.; Horan, P.; Dietz, L.A.; Zimmerman, I. Estimate of the time zero lung burden of depleted uranium in persian Gulf War veterans by the 24-hour urinary excretion and exponential decay analysis. *Mil. Med.* 2003, *186*, 600–605.

34. Valdéz, M. Estimating the lung burden from exposures to aerosols of depleted uranium. *Radiat. Prot. Dosimetry* 2009,*134*, 23–29.

35. Karpas, Z.; Lorber, A.; Sela, H.; Paz-Tal, O.; Hagag, Y.; Kurttio, P.; Salonen, L. Measurement of the ^{234}U/^{238}U ratio by MC-ICPMS in

drinking water, hair, nails and urine as an indicator of uranium exposure source. *Health Phys.* 2005, *89*, 315–321.
36. Kurttio, P.; Harmoinen, A.; Saha, H.; Solomen, L.; Karpas, Z.; Komulinen, H.; Auvinen, A. Kidney toxicity of ingested uranium from drinking water. *Am. J. Kidney Dis.* 2006, *47*, 972–982.
37. Lind, O.C.; Salbu, B.; Skipperud, L.; Janssens, K.; Jaroszewicz, J.; de Nolf, W. Solid state speciation and potential bioavailability of depleted uranium particles from Kosovo and Kuwait. *J. Environ. Radioact.* 2009, *100*, 301–307.
38. McClain, D.E.; Benson, K.A.; Dalton, T.K.; Ejnik, J.; Emond, C.A.; Hodge, S.J.; Kalinch, J.F.; Landauer, M.A.; Miller, A.C.; Pellmar, T.C.; et al. Biological effects of embedded depleted uranium (DU): Summary of armed forces radiobiology research institute research. *Sci. Total Environ.* 2001, *274*, 115–118.
39. Hooper, F.J.; Squibb, K.S.; Siegel, E.L.; McPhaul, K.; Keogh, J.P. Elevated urine uranium excretion by soldiers with retained uranium shrapnel. *Health Phys.* 1999, *77*, 512–519.
40. Leggett, R.W. Basis for the ICRP's age-specific biokinetic model for uranium. *Health Phys.* 1994, *67*, 589–610.
41. ICRP. Human respiratory tract model for radiological protection. *Ann. ICRP* 1994, *24*, 1–3.
42. ICRP. Age dependent doses to members of the public from intake of radionuclides-Part 3 ingestion dose coefficients.*Ann. ICRP* 1995, *25*, 1.
43. Leggett, R.W.; Harrison, J.D. Fractional absorption of ingested uranium in humans. *Health Phys.* 1995, *68*, 484–498.
44. Leggett, R.W. The behavior and chemical toxicity of U in the kidney: A reassessment. *Health Phys.* 1989, *57*, 365–385.
45. Miller, A.C.; Xu, J.; Stewart, M.; Brooks, K.; Hodge, S.; Shi, L.; Page, N.; McClain, D. Observation of radiation-induced damage in human cells exposed to depleted uranium: Dicentric frequency and neoplastic transformation as endpoints. *Radiat. Prot. Dosimetry* 2002, *99*, 275–278.
46. Barron, E.S.; Muntz, J.A.; Gasvoda, B. Regulatory mechanisms of cellular respiration; The role of cell membranes; Uranium inhibition of cellular respiration. *J. Gen. Physiol.* 1948, *32*, 163–178.
47. McQueney, M.S.; Markham, G.D. Investigation of monovalent cation activation of *S*-adenosylymethionine synthetase using mutagenesis and uranyl inhibitor. *J. Biol. Chem.* 1995, *207*, 18277–18284.
48. Renault, S.; Faiz, H.; Gadet, R.; Ferrier, B.; Martin, G.; Baveral, G.;

Conjard-Duplany, A. Uranyl nitrate inhibits lactate gluconeogenesis in isolated human and mouse renal proximal tubules; A 13C-NMR study. *Toxicol. Appl. Pharmacol.* 2010, *242*, 9–17.

49. Gueguen, Y.; Rouas, C.; Monin, A.; Manens, L.; Stefani, J.; Delissen, O.; Grison, S.; Dublineau, I. Molecular, cellular and tissue impact of depleted uranium on xenobiotic-metabolizing enzymes. *Arch. Toxicol.* 2013.

50. Pititot, F.; Lestaevel, P.; Tourionias, E.; Mazzucco, C.; Jacquinot, S.; Dhieux, B.; Delissen, O.; Toumier, B.B.; Gensdarmes, F.; Beaunier, P.; Dublineau, I. Inhalation of uranium nanoparticles: Respiratory tract deposition and translocation to secondary target organs in rats. *Toxicol. Lett.* 2013, *217*, 217–225.

51. Tirmarche, M.; Baysson, H.; Telle-Amberton, M. Uranium exposure and cancer risk: A review of epidemiological studies. *Rev. Epidemiol. Sante Publique* 2004, *52*, 81–90.

52. Periyakaruppan, A.; Kumar, F.; Sarkar, S.; Sharma, C.S.; Ramesh, G.T. Uranium induces oxidative stress in lung epithelial cells. *Arch. Toxicol.* 2007, *81*, 389–395.

53. Xie, H.; LeCerte, C.; Thompson, W.D.; Wise, J.P. Depleted uranium induces neoplastic transformation in human lung epithelial cells. *Chem. Res. Toxicol.* 2010, *23*, 373–378.

54. Orona, N.S.; Tasat, D.R. Uranyl nitrate-exposed rat alveolar macrophages cell death: Influence of superoxide anion and TNF α mediators. *Toxicol. Appl. Pharmacol.* 2012, *261*, 309–316.

55. Stradling, N.; Hodgson, A.; Ansobrolo, E.; Berard, P.; Etherington, G.; Fell, T.; Rance, E.; le Guen, B. *Industrial Uranium Compounds: Exposure Limits, Assessment of Intake and Toxicity after Inhalation*; National Radiologicl Protection Board: Chilton, UK, 2002; p. 10.

56. Arzuaga, X.; Rieth, S.H.; Bathija, A.; Cooper, G.S. Renal effects of exposure to natural and depleted uranium: A review of the epidemiological and experimental data. *J. Toxicol. Environ. Health B* 2010, *13*, 527–545.

57. Squibb, K.S.; Leggett, R.W.; McDiarmid, M.A. Prediction of renal concentrations of depleted uranium and rdaiayion dose in Gulf War veterans with embedded shrapnel. *Health Phys.* 2005, *89*, 267–273.

58. Zhu, G.; Xiang, X.; Chen, X.; Wang, L.; Hu, H.; Weng, S. Renal dysfunction induced by long-term exposure to depleted uranium in rats. *Arch. Toxicol.* 2009, *83*, 37–46.

59. Zhu, G.; Tan, M.; Li, X.; Xiang, X.; Hu, H.; Zhao, S. Accumulation and distribution of uranium in rats after implantation of depleted uranium

fragments. *J. Radiat. Res.* 2009, *50*, 183–192.

60. Shaki, F.; Pourahmad, J.; Hosseine, M.; Ghazi-Khansari, M. Toxicity of depleted uranium on isolated rat kidney mitochondria. *Res. Pharm. Sci.* 2012.

61. Shaki, F.; Pourahmad, J. Mitochondrial toxicity of depleted uranium: Protection by beta-glucan. *Iran J. Pharm. Res.*2013, *12*, 131–140.

62. Roszell, L.E.; Hahn, F.F.; Lee, R.B.; Parkhurst, M.A. Assessing the renal toxicity of capstone depleted uranium oxides and other uranium compounds. *Health Phys.* 2009, *96*, 343–351.

63. Lestaevel, P.; Romero, E.; Dhieux, B.; Soussan, H.B.; Berradi, H.; Dublineau, I.; Voisin, P.; Gourmelon, P. Different patterns of brain pro-/anti-oxidant activity between depleted and enriched uranium in chronically exposed rats.*Toxicology* 2009, *258*, 1–9.

64. Houpert, P.; Lestaevel, P.; Busby, C.; Paquet, F.; Gourmelon, P. Enriched but not depleted uranium affects central nervous system in long-term exposed rat. *Neuro Biol.* 2005, *6*, 1015–1020.

65. Jiang, G.C.-T.; Aschner, M. eurotoxicity of depleted uranium; Reasons for increased concern. *Biol. Trace Elements* 2006,*110*, 1–17.

66. Jiang, G.C.-T.; Tidwell, K.; McLaughlin, B.E.; Cai, J.; Gupta, R.C.; Milatovic, D.; Nuss, R.; Aschner, M. Neurotoxic potential of depleted uranium-effects in primary cortical neuron cultures and in *Caenorhabditis elegand*. *Toxicological Sci.* 2007, *99*, 553–565.

67. Lestaevel, P.; Airault, F.; Racine, R.; Soussan, H.B.; Dhieux, B.; Delissen, O.; Manens, L.; Aigueperse, J.; Voisin, P.; Souidi, H. Influence of environmental enrichment and depleted uranium on behaviour, cholesterol and acetylcholine in apolioprotein E-defivient mice. *J. Mol. Neurosci.* 2013.

68. Arfsten, D.P.; Still, K.R.; Ritchie, G.D. A review of the effects of uranium and depleted uranium exposure on reproductive and fetal development. *Toxicol. Ind. Health* 2001, *17*, 180–191.

69. Arfsten, D.P.; Bekkedal, M.; Wilfong, E.R.; Rossi, J.; Grasman, K.A.; Healey, L.B.; Rutkiewicz, J.M.; Johnson, E.W.; Thitoff, A.R.; Jung, A.E.; *et al*. Study of the reproductive effects in rats surgically implanted with depleted uranium for up to 90 days. *J. Toxicol. Environ. Health A* 2005, *68*, 967–997.

70. Arfsten, D.P.; Still, K.R.; Wilfong, E.R.; Johnson, E.W.; McInturf, S.M.; Eggers, J.S.; Schaeffer, D.J.; Bekkedal, M.Y. Two-generation reproductive toxicity study of implanted depleted uranium (DU) in CD

rats. *J. Toxicol. Environ. Health* 2009, *72*, 410–427.
71. Miller, A.C.; Stewart, M.; Rivas, R. Perconceptional parental exposure to depleted uranium: Transmission of genetic damage of offspring. *Health Phys.* 2010, *99*, 371–379.
72. Macfarlane, G.J.; Briggs, A.M.; Maconochie, N.; Hotopf, P.; Lunt, M. Incidence of cancer among UK Gulf War veterans: Cohort study. *Brit. Med. J.* 2003, *327*, 1373–1378.
73. Storm, H.H.; Jørgensen, H.O.; Kejs, A.M.; Engholm, G. Depleted uranium and cancer in Danish Balkan veterans deployed 1992–2001. *Eur. J. Cancer* 2006, *42*, 2355–2358.
74. Legorio, S.; Grande, E.; Martina, L. Review of epidemiological studies of cancer risk among Gulf War and Balkans veterans. *Epidemiol. Prev.* 2008, *32*, 145–155.
75. Bogers, R.P.; van Leeuwen, F.E.; Grievink, L.; Schouten, L.J.; Kiememev, L.A.; Schram-Bijkerk, D. Cancer incidence in Dutch Balkan veterans. *Cancer Epidemiol.* 2013, *37*, 550–555.
76. McDiarmid, M.A.; Squibb, K.S.; Engelhardt, S.; Oliver, M.; Gucer, P.; Wilson, P.D.; Kane, R.; Kabet, M.; Kaup, B.; Anderson, L.; *et al*. Surveillance of depleted uranium exposed Gulf War veterans: Health effects observed in an Engaged "friendly fire" cohort. *J. Occ. Environ. Med.* 2001, *43*, 991–1000.
77. McDiarmid, M.A.; Squibb, K.S.; Engelhard, S.M. Biologic monitoring for urinary uranium in Gulf War I veterans. *Health Phys.* 2004, *87*, 51–56.
78. McDiarmid, M.A.; Engelhardt, S.M.; Oliver, M.; Gucer, P.; Wilson, P.D.; Kane, R.; Cernich, A.; Kaup, B.; Anderson, L.; Hoover, D.; *et al*. Health surveillance of Gulf War I veterans expose to depleted uranium: Updating the cohort. *Health Phys.* 2007, *93*, 60–73.
79. McDiarmid, M.A.; Engelhard, S.M.; Dorsey, C.D.; Oliver, M.; Gucer, P.; Wilson, P.D.; Kane, R.; Cernich, A.; Kaup, B.; Anderson, L.; *et al*. Surveillance results of depleted uranium-exposed Gulf War I veterans: Sixteen years of follow up. *J. Toxicol. Environ. Health A* 2009, *72*, 14–29.
80. McDiarmid, M.A.; Engelhardt, S.M.; Dorsey, C.D.; Oliver, M.; Gucer, P.; Gaitens, J.M.; Kane, R.; Cernich, A.; Kaup, B.; Hoover, D.; *et al*. Longitudinal health surveillance in a cohort of Gulf War veterans 18 years after first exposure to depleted uranium. *J. Toxicol. Environ. Health* 2011, *74*, 678–691.
81. Oliver, M.V.B.M.S.; McDiarmid, M.A.; Squibb, K.S.; Tucker, J.D. Long term depleted uranium exposure in Gulf War I veterans does not

induce chromosome aberrations in peripheral blood lymphocytes. *Mutat. Res.* 2011, *720*, 53–57.

82. Bakhmutsky, M.V.; Oliver, M.S.; McDiarmid, M.A.; Squibb, K.S.; Tucker, J.D. Long term depleted uranium exposure in Gulf War I veterans does not cause elevated numbers of micronuclei in peripheral blood lymphocytes. *Mutat. Res.* 2013, *757*, 132–139.

83. Hines, S.E.; Gucer, P.; Klingerman, S.; Breyer, R.; Centeno, J.; Gaithers, J.; Oliver, M.; Engelhardt, S.M.; Squibb, K.S.; McDiarmid, M.A. Pulmonary health effects in Gulf War I service members exposed to depleted uranium. *J. Occup. Environ. Med.* 2013, *55*, 937–944.

84. McDiarmid, M.A.; Gaithers, J.M.; Hines, S.; Breyer, R.; Wong-You-Cheong, J.J.; Engelhardt, S.M.; Oliver, M.; Gucer, P.; Kane, R.; Cernich, A.; et al. The Gulf War depleted uranium cohort at 20 years: Bioassay results and novel approaches to fragment surveillance. *Health Phys.* 2013, *104*, 347–361.

85. Alaani, S.; Tafash, M.; Busby, C.; Hamden, M.; Blaurock-Busch, E. Uranium and other contaminants in hair from the parents of children with congenital abnormalities in Fallujah, Iraq. *Confl. Health* 2011, *5*, 15.

86. Alaani, S.; Fallouji, M.A.A.; Busby, C.; Hamdan, M. Pilot study of congenital anomaly rates at Birth in Fallujah, Iraq, 2010. *J. Islamic Med. Assoc.* 2012, *44*, 8–15.

87. Al-Habithi, T.S.; Al-Diwan, J.K.; Abubakir, A.M.; Saleh, M.; Sabila, P. Birth defects in Iraq and plausibility of environmental exposure: A Review. *Confl. Health* 2012, *6*, 3.

88. Butowinski, A.T.; DeScisciolo, C.; Conlin, A.M.; Ryan, M.A.K.; Sevick, C.J.; Smith, T.C. Birth defects in infants born in 1998–2004 to men and women serving in the U.S. military during the 1990–1991 Gulf War Era. *Birth Defects Res. A* 2012, *94*, 721–728.

89. Busby, C.; Hamdan, M.; Ariabi, E. Cancer, infant mortality and birth sex-ratio in Fallujah, Iraq 2005–2009. *Int. J. Res. Pub. Health* 2010, *7*, 2828–2837.

90. Al-Hamzawi, A.A.; Jaafar, M.S.; Tawfiq, N.F. Uranium concentration in uranium samples of Southern Iraqi leukemia patients using CR-39 track detector. *J. Radioanal. Nuc. Chem.* 2013.

91. Hagopian, A.; Lafta, R.; Hassan, J.; Davis, S.; Mirick, D.; Takaro, T. Trends in childhood leukemia in Basrah, Iraq, 1993–2007. *Am. J. Pub. Health* 2010, *100*, 1081–1087.

92. Alrudainy, L.A.; Hassan, J.G.; Salih, H.M.; Abbas, M.K.; Majeed, A.A.

Time trends and geographical distribution of childhood leukemia in Basrah, Iraq, from 2004 to 2009, Sultan Qaboos. *Univ. Med. J.* 2011, *11*, 215–220.

93. Greiser, E.; Hoffmann, W. Questionable increase of childhood leukemia in Basrah, Iraq. *Am. J. Pub. Health.* 2010, *100*, 1556.

94. Hagopian, A.; Lafta, R.; Hassan, J.; Davis, S.; Mirick, D.; Takaro, T. Hagopian *et al*. Respond. *Am. J. Pub. Health* 2010,*100*, 1557.

95. Fathi, R.A.; Matti, L.Y.; Al-Salih, H.S.; Godbold, D. Environmental pollution by depleted uranium in Iraq with special reference to Mosul and possible effects on cancer and birth defect rates. *Med. Confl. Surviv.* 2013, *29*, 7–25.

96. United Nations Environment Programmme. *Depleted Uranium in Kosovo, Post-Conflict Environmental Assessment*; United Nations Environment Programmme: Nairobi, Kenya, 2001.

97. Nafezi, G.; Gregoric, A.; Vaupotic, J.; Bahtijari, M.; Kugali, M. Radon levels and doses in dwellings in two villages in Kosovo, affected by depleted uranium. *Radiat. Prot. Dosimetry* 2013.

98. United Nations Environment Programmme. *Depleted Uranium in Serbia and Montenegro, Post-Conflict Environmental Assessment in the Republic of Yugoslavia*; United Nations Environment Programmme: Nairobi, Kenya, 2002.

99. United Nations Environment Programmme. *Depleted Uranium in Bosnia and Herzegovina, Post-Conflict Environmental Assessment of Depleted Uranium in Bosnia and Herzegovina*; United Nations Environment Programmme: Nairobi, Kenya, 2003.

100. Handley-Sidhu, S.; Bryan, N.D.; Vaughan, P.J.; Livens, F.R.; Keith-Roach, M.J. Corrosion and transport of depleted uranium in sand-rich environments. *Chempsphere* 2009, *77*, 143–149.

101. Toque, C.; Milodowski, A.E.; Baker, A.C. The corrosion of depleted uranium in terrestrial and marine environments. *J. Environ. Radioact.* 2013.

102. Crean, D.E.; Livens, F.R.; Stennett, M.C.; Grolimund, D.; Borca, C.N.; Hyatt, N.C. Remediation of soil contaminated with particulate depleted uranium by multi stage chemical extraction. *J. Hazard Mater.* 2013.

Chapter 8

SYNTHESES OF SULFO-GLYCODENDRIMERS USING CLICK CHEMISTRY AND THEIR BIOLOGICAL EVALUATION

Yoshiko Miura[1,2], Shunsuke Onogi[2] and Tomohiro Fukuda[2,3]

[1]Department of Chemical Engineering, Graduate School of Engineering, Kyushu University, 744 Motooka, Nishi-ku, Fukuoka 819-0395, Japan

[2]School of Materials Science, Japan Advanced Institute of Science and Technology, 1-1 Asahidai, Nomi, Ishikawa 923-1292, Japan

[3]Department of Applied Chemistry and Chemical Engineering, Toyama National College of Technology, Hongo-campus, 13 Hongo-Machi, Toyama, Toyama 939-8630, Japan

ABSTRACT

A series of novel glycol-clusters containing sulfonated *N*-acetyl-D-glucosamine (GlcNAc) have been synthesized using click chemistry. Three dendrimers with aromatic dendrons were synthesized using chlorination, azidation and click chemistries. The resulting dendrimers were modified with azide-terminated sulfonated GlcNAc using click chemistry. The sulfonated dendrimers showed affinity for proteins, including the lectin wheat germ agglutinin and amyloid beta peptide (1-42). The dendrimers of G1 and G2 in particular showed the largest affinity for the proteins. The addition of the sulfonated GlcNAc dendrimers of G1 and G2 exhibited an inhibition effect on the aggregation of the amyloid beta peptide, reduced the β-sheet conformation, and led to a reduction in the level of nanofiber formation.

INTRODUCTION

Saccharides displayed on the surfaces of cells have been the subject of considerable attention because they can play important roles in living systems [1]. Furthermore, they have been related to a variety of different biological activities, including cell-cell adhesion, protein recognition, pathogen infection, and cancer metastasis. For these reasons, materials capable of interfering with the functions of these saccharides are being considered for use in the fabrication of new biomaterials and the development of new drugs [2]. Saccharide-protein interactions, however, are usually too weak to be used as drugs and biomaterials.

Interestingly, it is well known that saccharide-protein interactions can be amplified by multivalency, otherwise known as "the cluster glycoside effect" [3,4]. Clusters of glycosides can amplify the interaction though multiple saccharide binding interactions to the proteins, and increase the binding probability. A variety of different compounds containing multivalent saccharides have been reported to show strong molecular recognition abilities. Furthermore, several saccharide containing substances have been reported such as artificial glycoprotein conjugates, glycopeptides, saccharide-thin layers and saccharide nanoparticles [5,6,7,8,9,10,11]. Of these saccharide containing substances, polymers with saccharide side chains have attracted the most pronounced level of attention because of the large cluster glycoside effects associated with their saccharide-protein interactions. Throughout the remainder of this paper, these polymers will be referred to as "glycopolymers". These glycopolymers are interesting because they exhibit a strong amplification effect and possess properties making them practical candidates for applications as biomaterials and polymer drug development.

We previously reported the synthesis of a variety of different glycopolymers for lectin recognition, toxin neutralization and cell cultivation [12]. Although these glycopolymers were interesting materials, it was difficult to clarify the detailed mechanism of their interactions, because of the complicated nature of polymer structure resulting from the variable molecular weight and monomer sequence distributions in the copolymers. Of the glycopolymers synthesized and evaluated to date, glycodendrimers possess some of the most interesting characteristics as a consequence of their uniform structure and their extensive molecular recognition ability, with the general expectation that these materials could ultimately be applied as potential polymer drugs [13].

Recently, we reported the synthesis of a sulfonated glycopolymer that interacted with Alzheimer amyloid β (Aβ) and inhibited its aggregation [14,15]. The sulfonated glycopolymer behaved as a mimic of glycosaminoglycans (GAGs) such as heparin and heparan, and interacted with Aβ like GAGs. In the current research, novel glycodendrimers containing a sulfonated saccharide unit were investigated for their ability to inhibit the aggregation of the Aβ (1-42) peptide. Glycodendrimers possess well-defined structures that are generally believed to be more suitable and therefore better suited for use in medicinal applications than liner polymers. The synthesis of the dendrimers in the current report represents one of the key highlights of this research paper because the dendrimers were successfully constructed using click chemistry rather than the tedious multi-step synthetic strategies typically used in the synthesis of these materials [16]. Several dendrimers were synthesized in this way and their inhibitory effects on Aβ aggregation investigated by fluorescence, circular dichroism (CD) spectra, and atomic force microscopy (AFM).

The interactions of these glycodendrimers with wheat germ agglutinin (WGA) were also investigated to clarify the properties of novel cluster glycosides. Since the interactions between the Aβ and GAGs were predominately electrostatic in nature, the synthetic sulfonated glycodendrimer was envisaged to interact with the Aβ in the same way. WGA has a positive net charge in neutral buffer solution [17], and the interaction of the glycodendrimers with WGA was considered to occur though both the molecular recognition of N-acetyl-D-glucosamine (GlcNAc) [18] and electrostatic interactions. The interaction with WGA was investigated in view of the Aβ interaction. The ultimate propose of this study was the fabrication of a glycocluster capable of inhibiting Aβ aggregation.

RESULTS AND DISCUSSION

Synthesis of Dendrimers with 6-Sulfo-GlcNAc

Glycodendrimers of G1 and G2 (Figure 1) were synthesized using click chemistry according to the divergent method depicted in Scheme 1, Scheme 2, Scheme 3, Scheme 4, Scheme 5, Scheme 6 and Scheme 7.

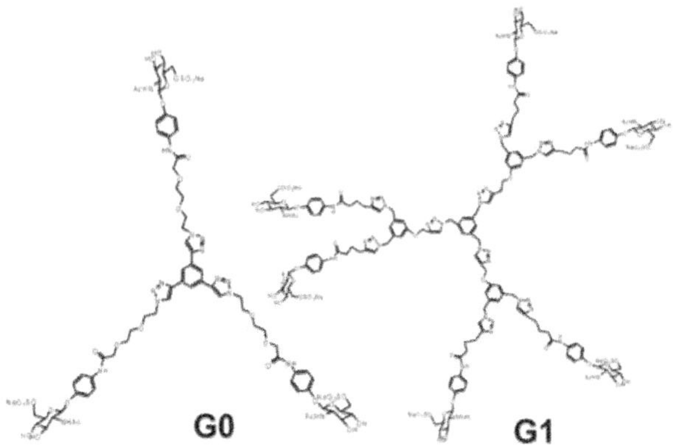

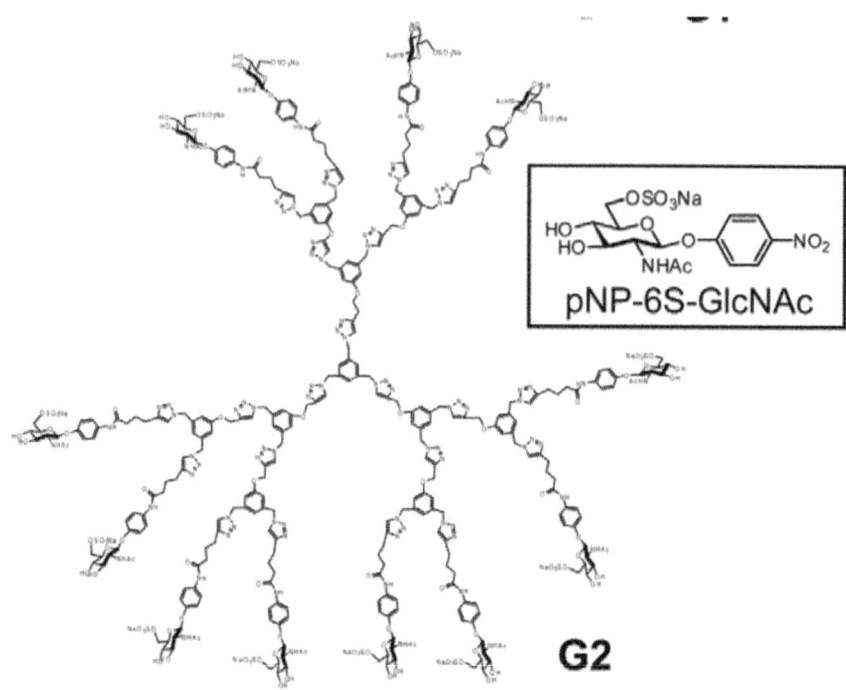

Figure 1: Chemical structures of the saccharide-derivatives.

The trimer of 6-sulfo-GlcNAc (G0) was synthesized via the connection of azide-terminated 6-sulfo-GlcNAc to 1,3,5-triethynylbenzene, using click chemistry [19,20] (Scheme 1 and Scheme 2). *p*-Aminophenyl-6-sulfo-D-GlcNAc was synthesized according to a method previously reported in the literature [14,15] and subsequently connected to *N*-phenylhex-5-ynamide using 2-(7-aza-1*H*-benzotriazole-1-yl)-1,1,3,3-tetramethyl uranium hexafluorophosphate (HATU) and *N*,*N*-diisopropylethylamine (DIEA) (Scheme 3). Starting from 5-hydroxyisophthalic acid, the dendron of 1,3-bis(chloromethyl)-5-(prop-2-yn-1-yloxy)benzene was synthesized by sequential alkylation, hydrogenation and chlorination reactions, with the resulting dendron being attained in high yield (G1-N_3: 93%, G2-N_3: 56%) (Scheme 4 and Scheme 5).

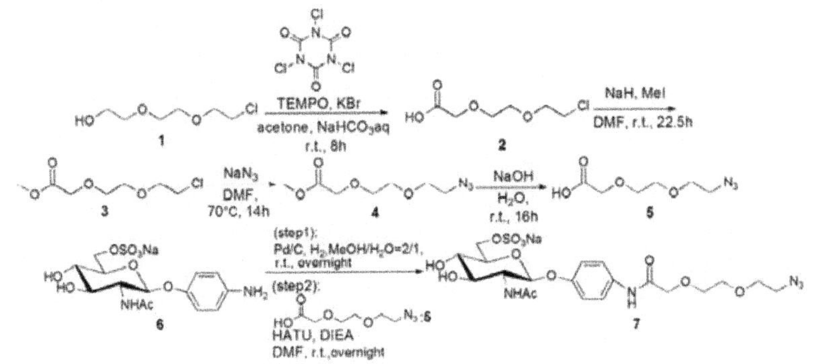

Scheme 1: Syntheses of azide-terminated sulfonated GlcNAc 7.

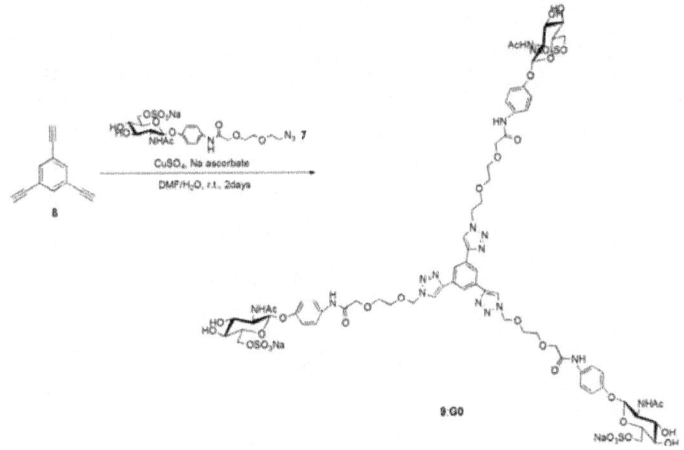

Scheme 2: Synthesis of 6-sulfo-GlcNAc trimer G0: 9.

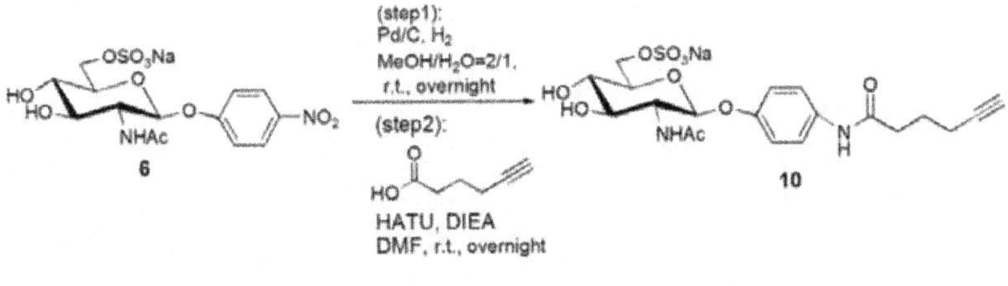

Scheme 3: Synthesis of acetylene-terminated sulfonated GlcNAc 10.

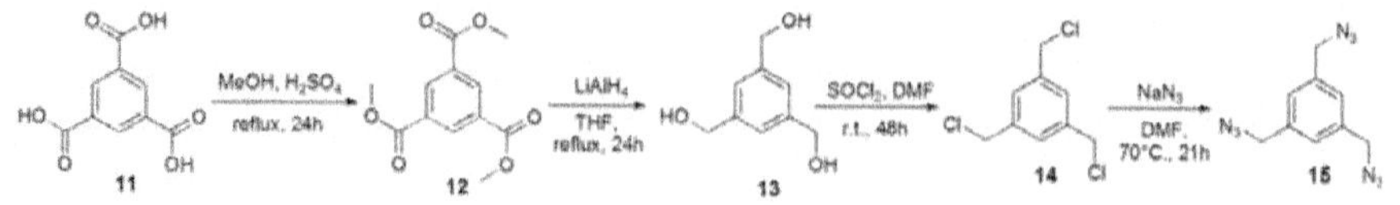

Scheme 4: Syntheses of the dendrimer core 15.

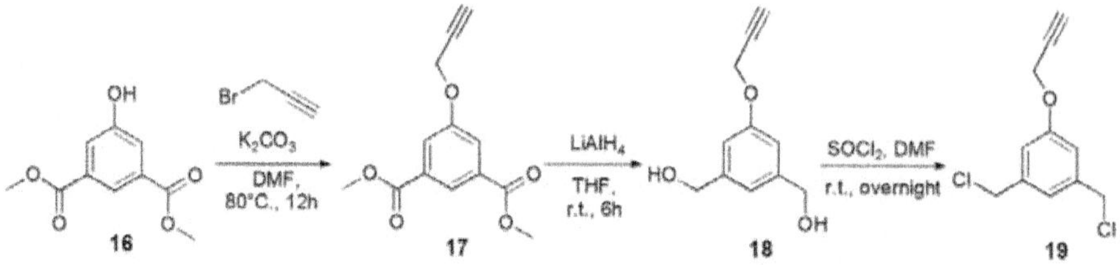

Scheme 5: Syntheses of the dendron 19.

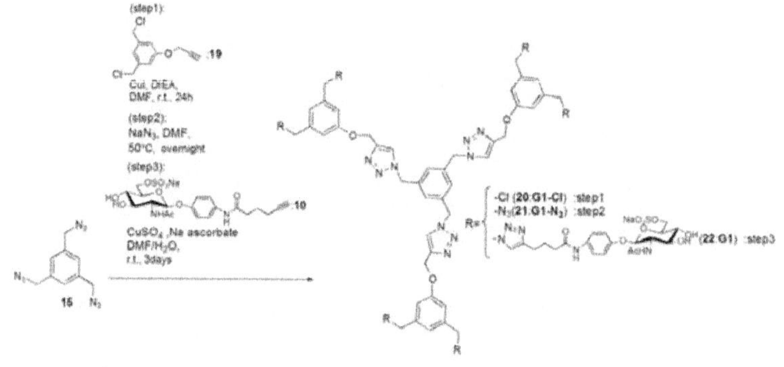

Scheme 6: Syntheses of 6-sulfo-GlcNAc hexamer G1: 22.

Glycodendrimers of G1 and G2 were obtained by click chemistry in a modest yield (G1: 28%, G2: 21%) (Scheme 6 andScheme 7). 1,3,5-Tris(azidomethyl) benzene was used as the core of the dendrimer and was itself derived from 1,3,5-trimesic acid via sequential hydrogenation, chlorination and azidation reactions (Scheme 3). The structures of the compounds were confirmed by ^{1}H and ^{13}C-NMR, electrospray ionization mass spectroscopy (ESI-MS) and matrix-assisted laser desorption ionization and time of flight mass spectroscopy (MALDI-TOF-MS). Detailed syntheses and characterizations are provided in the Experimental Section.

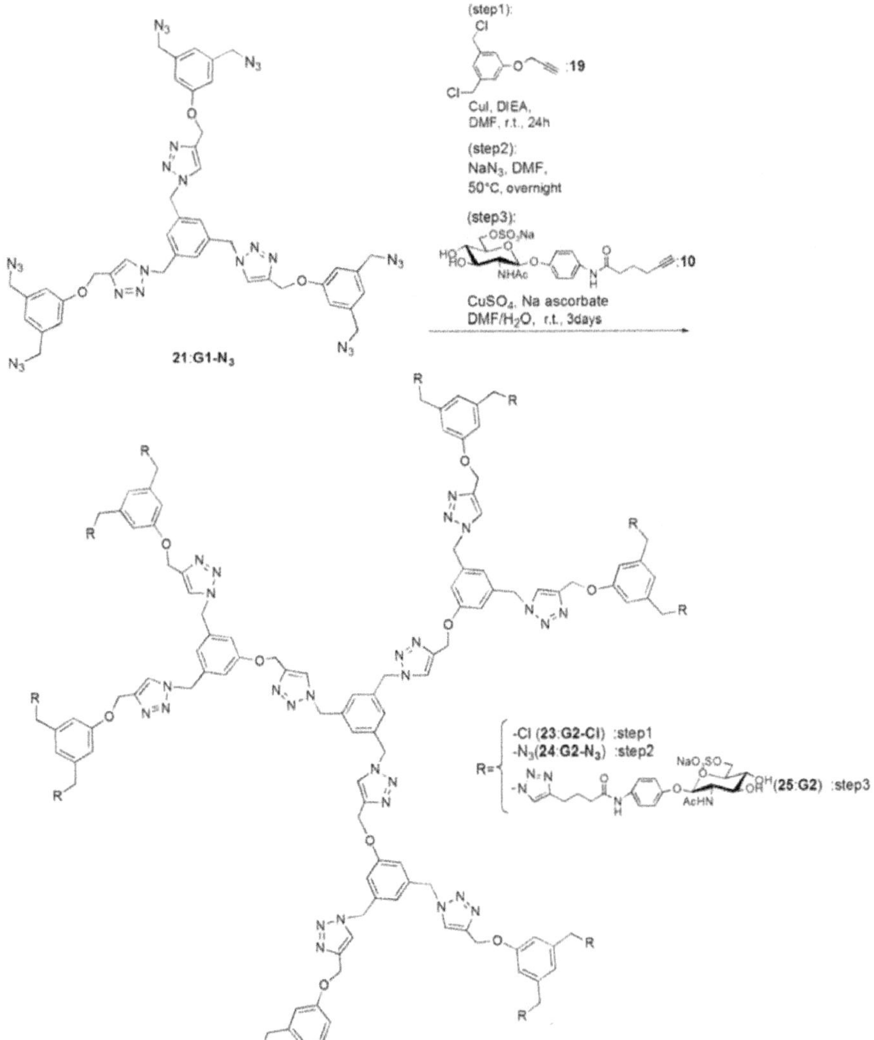

Scheme 7: Syntheses of 6-sulfo-GlcNAc dodecamer G2: 25.

Lectin Recognition Ability

Although the interaction mechanism of WGA is different from that of Aβ, the synthesized glycodendrimers were evaluated in a lectin recognition assay to estimate their biological ability (Figure 2) [19].

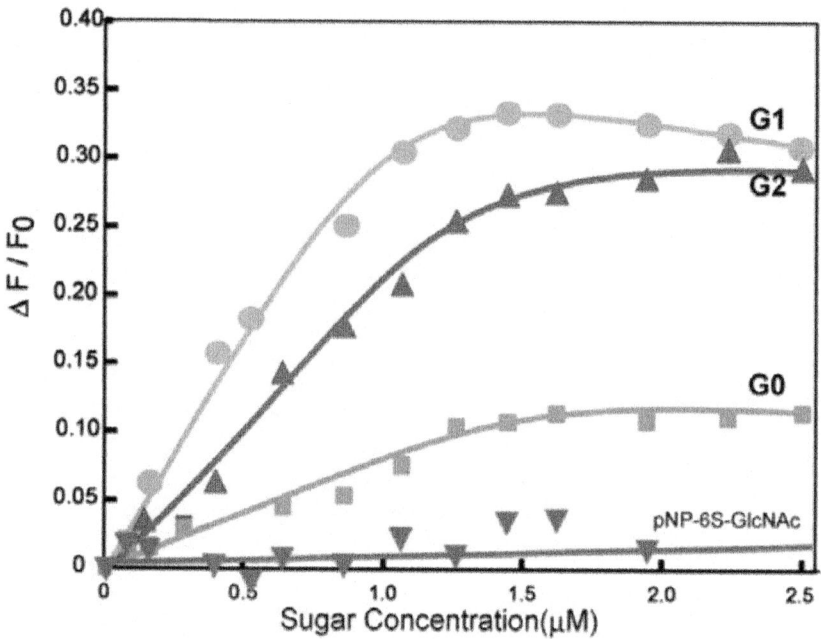

Figure 2: Fluorescence intensity changes of the FITC-WGA with varying sugar concentrations of the sulfonated GlcNAc derivatives.

Fluorescence quenching of the fluorescein isothiocyanate (FITC)-WGA complex occurred upon the addition of the glycodendrimers [21]. The quenching behavior was analyzed using a Scatchard plot. The binding constants (K_a) of G0, G1 and G2 were found to be 6.56×10^5, 5.55×10^6 and 2.12×10^6 (M^{-1}), respectively, with the affinity constants therefore being of the order G1 > G2 >> G0. In contrast to WGA, the monomeric sulfonated GlcNAc did not show the same detectable affinity for lectin, and the addition of the glycodendrimers to FITC-bovine serum albumin did not induce fluorescence quenching. The results showed that the glycodendrimer of G1 and G2 had the ability to amplify the protein-saccharide interactions, indicating that the protein-saccharide interactions were specific to a certain extent. G0 did not show a high level of affinity for WGA, suggesting the potential for weak interactions to proteins including Aβ.

The hydrophobic domains of the dendrimes of G1 and G2 were analyzed by fluorescence of 8-anilino-1-naphthalene sulfonate (ANS) (Figure 3). The solution of G1 showed a larger blue shift in its emission spectra than that of G2, whereas the solution of G0 did not show any blue shift.

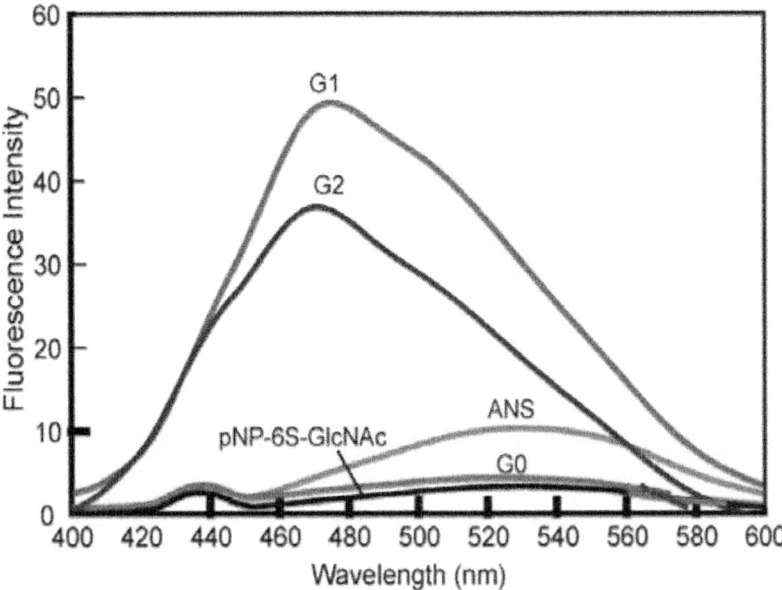

Figure 3: Fluorescence spectra of ANS (20 µM) with sugar derivatives (66.7 µM) in PBS buffer.

The hydrophobicity of dendrimer suggested that they would possess self-assembling properties in solution. Interestingly, the larger fluorescence of G1 *versus* that of G2 indicated the larger self-assembling properties of G1 in the aqueous solution. In addition, the glycodendrimers induced the formation of WGA aggregates. The diameters of WGA-glycodendrimer aggregates in the presence of WGA (1.00 µM) were 752, 496 and 384 nm, respectively, with G1, G2, and G0 (data not shown).

Multivalent GlcNAc generally showed a strong interaction based on the cluster glycoside effect [22], which was similar to the interaction observed for G1 and G2. The positive net charge of WGA also contributed to its interactions with G1 and G2 in an electrostatic manner. The sizes of the dendrimers by dynamic light scattering (DLS) became semi-micro in order in the presence of the WGA because of the cross-linking between the WGA and the dendrimer that occurred as a consequence of molecular recognition. The affinity of G1 was larger than that of G2, although the structure of G2 apparently exhibited a greater degree of multivalency. In addition, considering that the distance between the sugar binding sites was 1.5–5 nm in WGA [23,24], the sugar distance of G2 was better suited to accommodate WGA than that of G1. The affinities of dendrimers, however, showed an opposing trend. Taking the ANS results into account, G1 formed larger glycoside clusters than G2 through self-assembly, which resulted in the larger affinity for WGA. Given that the self-assembly plays an important role in the cluster glycoside effect of dendrimers, the affinity between WGA and dendrimer was considered to be dynamic including the cross linking and statistical re-binding aspects of the process. The results suggested that the glycodendrimers of G1 and G2 provided glycoside clusters suitable for binding WGA and Aβ.

Inhibitory Effect of Glycodendrimer on Aβ (1-42) Aggregation

The interactions of the glycodendrimers with Aβ were investigated. The aggregation properties of Aβ (1-42) were analyzed by fluorescence spectroscopy with thioflavin T (ThT) [14,15]. It is known that ThT has an affinity for amyloid fibril and that the fluorescence of ThT increases when ThT is attached to amyloid fibrils. In the current experiment, the degree of amyloid formation was monitored by the fluorescence of ThT [25]. The level of amyloidosis was examined over an 8 h period and the fluorescence intensity following this period was evaluated as 1.0. The aggregation behaviors with various sugar additives are shown in Figure 4.

The monomeric 6-sulfonated GlcNAc did not exhibit an inhibitory effect on Aβ aggregation, as we previously reported [14,15]. In contrast, the addition of sulfonated glycodendrimer modestly inhibited the aggregation of Aβ. The degree of inhibition was in the order G1 = G2 > G0 >> monomer. The addition of dendrimer inhibited the aggregation by approximately 50% relative to the level of inhibition observed in the absence of the additives.

The dendrimers of G1 and G2 were effective inhibitors of Aβ aggregation, whereas the inhibitory effect of G0 was insufficient. The observed order of the inhibitory effect was similar to that observed for the lectin recognition. The results suggested that G1 and G2 displayed saccharides in a favorable manner for effective protein affinity. The molecular recognition of the saccharides by lectin was affected by the cluster glycoside effect, which was controlled by the spatial arrangement of the saccharides in the dendrimers or the self-assembled dendrimers. The appropriate glycoside clusters for WGA could also become a better inhibitor for Alzheimer amyloid aggregation.

It has been reported that GAGs bind to Aβ via electrostatic interactions with the basic segment 13HHQK16 [26]. Glycopolymers containing sulfonated saccharides were considered to interact with Aβ in the same way via electrostatic interactions [14,15]. Glycodendrimers are also considered to bind to Aβ via the same electrostatic interactions. Although the glycodendrimer G1 showed the largest binding properties, its binding to Aβ was almost identical to that of G2. Although the structures of WGA and Aβ are totally different, it was considered that the larger cluster of the glycosides would show high levels of affinity that were similar to those observed in their interaction with WGA.

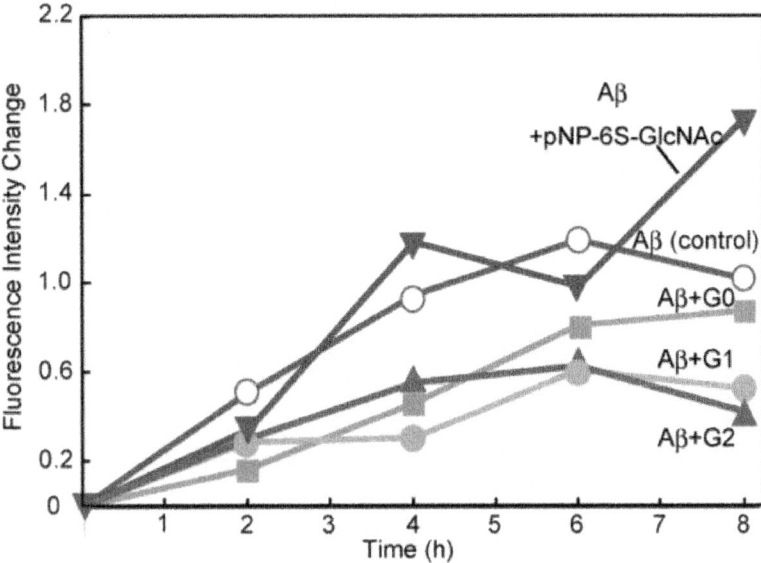

Figure 4: Time course of the fluorescence change in ThT with Aβ (1-42) (20 μM) and sugar derivatives (200 μM) in the phosphate buffer.

Multivalent sulfonated groups have been reported to interact with Aβ with GAGs [26], 6-sulfoGlcNAc polymer [14,15], sulfonated Glc polymer [27], and the polymer with sulfonic acid [28]. It is obvious that the sulfonic acid interact with Aβ via electrostatic interaction. The structural specificity and the mechanism of Aβ interactions are under investigation in our group.

Conformation Analysis of Aβ with CD Spectra

The conformation of Aβ was investigated in the presence of the sulfonated glycodendrimer (Figure 5). In the absence of the sugar additives, Aβ showed the negative cotton effect around 220 nm, suggesting β-sheet conformation [29]. Aβ also showed β-sheet structure in the presence of the monomeric sulfonated saccharide and G1. In contrast, the additions of G1 and G2 led to significant changes in the conformation, with the negative cotton effect observed around 220 nm almost disappearing entirely. The addition of G0 did not lead to a decrease in the negative cotton effect, suggesting that the interaction with Aβ was weak.

CD spectra recorded in the presence of the sulfonated glycodendrimer indicated that its addition led to the inhibition of the aggregation and conformation change of Aβ. Since the amyloidosis relates to changes in the conformation and aggregation, these results were in good agreement with the Th-T experiments. The inhibition of Aβ aggregation was also evaluated with the sulfonated glycopolymer. The results revealed that the sulfonated glycoside clusters inhibited oligomer formation during the early stages of the amyloidosis. It was also considered that the sulfonated glycodendrimer interacted with Aβ and inhibited the formation of the Aβ oligomer [14].

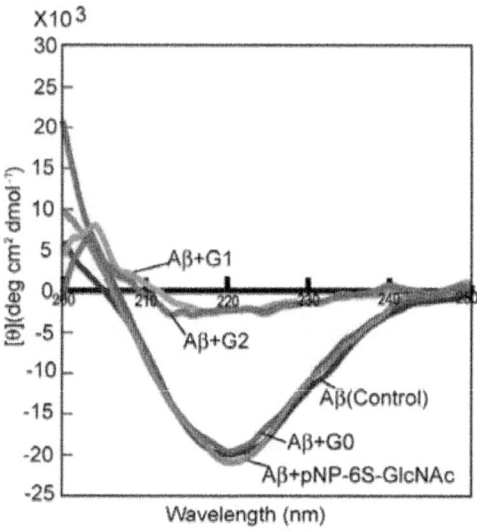

Figure 5: CD spectra for Aβ (1-42) (20 µM) in the presence of the sugar dendrimer additives (200 µM) in the phosphate buffer.

Morphology Observation of Aβ

To develop an understanding of the amyloidosis and the properties of the inhibitors, it was necessary to develop an understanding of the Aβ morphology, because the results from the ThT fluorescence and CD spectra only provided an indirect reflection of the protein amyloidosis. The properties of the amyloid, including its cytotoxicity, were strongly related to the size and shape of the aggregates, and so the AFM measurements were indispensable in evaluating these properties [30]. Aβ spontaneously formed nanofiber upon incubation, with widths of 45-70 nm and length of only a few micrometers (Figure 6a). Aβ also formed nanofibers in the presence of the monomeric saccharide with widths of 35–70 nm, although the fibers were shorter in length at less than 1 µm (Figure 6b). In contrast, the additions of G1 and G2 led to significant changes in the morphology of Aβ. Nanofibers of the Aβ amyloid were not observed at all in the presence of G1 and G2 (Figure 6d,e). Interestingly, the addition of the sulfonated trimer of G0 induced the largest aggregates (Figure 6c).

The morphology of Aβ correlated well with the results of the ThT and CD analyses. Aβ formed nanofibers in the absence of any additives. AFM analysis revealed that the monomeric 6-sulfonated GlcNAc showed the lowest level of Aβ aggregation inhibition, and this result was consistent with the results of the ThT fluorescence and CD spectra. In contrast, the AFM results indicated that the dendrimers of G1 and G2 showed the highest levels of Aβ aggregation inhibition, which were in good agreement with the results of the ThT and CD spectra. The formation of large aggregates was inhibited by the strong interaction with the glycodendrimer.

For G0 (trimer of 6-sulfonated GlcNAc), the AFM results indicated that the material only exerted a minor impact on the inhibition of Aβ aggregation, with the result being consistent with the ThT and CD results. Furthermore, the aggregates in this case were much larger than those formed in the absence of the additives. The shape of the aggregates was different from the nanofiber of the amyloid. The morphology of G0 was measured in the absence of Aβ and the results indicated that G0formed small aggregates (Figure 7) even though the hydrophobicity by ANS fluorescence was small.

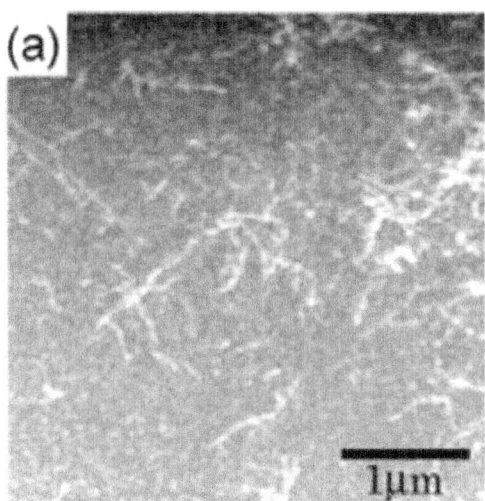

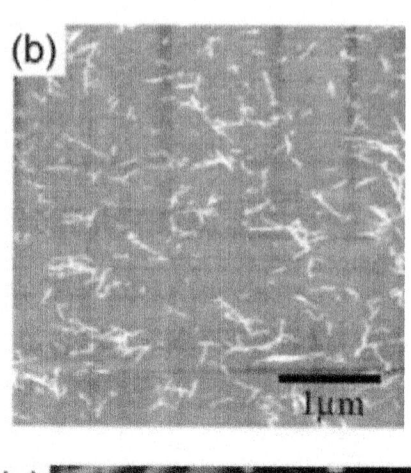

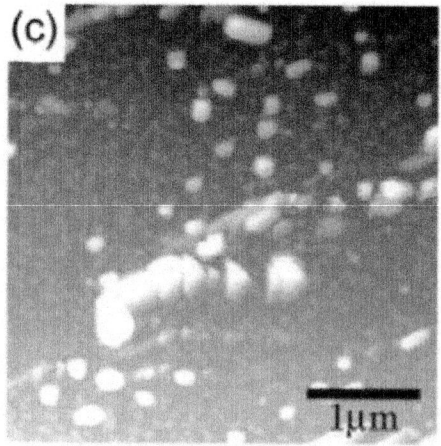

Syntheses of Sulfo-Glycodendrimers Using Click Chemistry and... 245

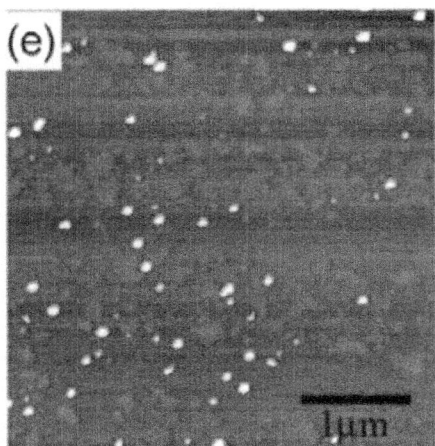

Figure 6: AFM observations of Aβ (1-42) (20 µM) (a) in the absence of an additive, (b) in the presence of pNP-6S-GlcNAc(200 µM), (c) in the presence of G0 (200 µM), (d) in the presence of G1 (200 µM), and (e) in the presence of G2 (200 µM).

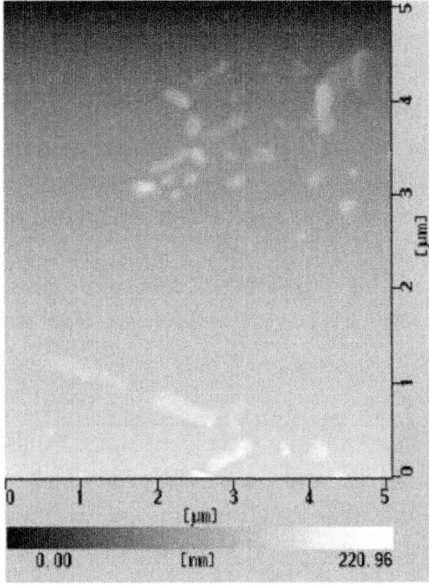

Figure 7: AFM observation of G0 (200 µM).

The morphology of Figure 6c was considered to be representative of a mixture of Aβ and G0 aggregates. Since the interaction of Aβ with G0 was weaker than its interactions with G1 and G2, the nanofiber formation of Aβ aggregates was not hindered. G0 was envisaged to have behaved as a glue, effectively bringing Aβ units together to yield the observed large objects. The morphology in this particular experiment was determined to be mixture of Aβ, Aβ-dendrimer and dendrimer aggregates. In spite of this, it was expected that the cytotoxicity of Aβ would be reduced by the addition of G0because of the resulting reduction in the number of toxic Aβ aggregates.

EXPERIMENTAL

Materials

The following reagents were used as received: N-acetylglucosamine, copper(0) powder, copper(II) sulfate, copper(I)iodide, N,N-dimethylformamide (DMF), chloroform ($CHCl_3$), ethyl acetate (EtOAc), n-hexane, magnesium sulfate ($MgSO_4$), methanol (MeOH), methylene chloride (CH_2Cl_2), potassium bromide (KBr), potassium carbonate (K_2CO_3), 2-propanol, sodium L-ascorbate, sodium azide (NaN_3), sodium bicarbonate ($NaHCO_3$), sodium carbonate (Na_2CO_3) sodium hydride (NaH), sodium hydroxide (NaOH), tetrahydrofuran (THF), thioflavin-T(ThT), thionyl chloride ($SOCl_2$), (Kanto Chemical Co., Inc., Tokyo, Japan), 2-[2-(2-chloroethoxy)ethoxy]ethanol, DIEA, 5-hexynoic acid, propargyl bromide, 2,2,6,6-tetramethylpiperidine 1-oxyl (TEMPO), trichloroisocyanuric acid, 1,3,5-triethynylbenzene, trimesic acid (TCI Co., Tokyo, Japan), amyloid β protein (human, 1-42) (Aβ (1-42)), HATU (Peptide Institute Inc., Osaka, Japan), palladium/charcoal (Pd/C) (Merck & Co., Inc., Whitehouse Station, NJ, USA), and deuterated solvents ($CDCl_3$, CD_3OD, DMF-d_7 and D_2O), dimethyl 5-hydroxyisophtalate, and lithium alminium hydride ($LiAlH_4$) (Sigma-Aldrich, St. Louis, MO, USA). 6-sodium sulfo-N-acetyl-β-D-glucosamine (6) was synthesized according to the previous literature [5,14].

Measurements

The identities of the compounds were confirmed by the following methods. Both ^{1}H-NMR (300 MHz) and ^{13}C-NMR (75 MHz) spectra were recorded using a Varian Gemini 2000 spectrometer and ^{1}H-NMR (500 MHz) spectra was recorded using a Varian UNITY 500plus spectrometer, equipped with a Sun workstation, respectively. The spectra were measured in $CDCl_3$, CD_3OD, DMF-d_7 and D_2O solvents at room temperature. Mass spectra were measured on a MALDI-TOF-MS (Voyager, Applied Biosystems, Foster City, CA, USA) and ESI-MS (LCQ Deca xp, Thermo Fisher Scientific, Waltham, MA, USA). CD spectra were recorded using a JASCO J-720 spectrometer (JASCO, Tokyo, Japan). AFM measurement was carried out using SPA400 instrument (Seiko Instruments Inc., Tokyo, Japan) with a 40 N cantilever. Fluorescence spectra were measured on a JASCO FP-6500 (JASCO) at 25 °C. DLS for the GNPs was determined using a Zetasizer 3000 (Sysmex, Kobe, Japan).

Syntheses

[2-(2-Chloroethoxy)ethoxy]acetic acid (2). [2-(2-Chloroethoxy)ethoxy]ethanol (1, 2.20 g, 13.0 mmol) was dissolved in acetone (30 mL), and 15% $NaHCO_3$ aq. was added to the solution at 0 °C. KBr (0.312 g, 2.60 mmol) and TEMPO (0.040 g, 0.040 mmol) were added to the solution, and trichloroisocyanuric acid (6.10 g, 26.2 mmol) was added dropwise. The solution was allowed to warm up to room temperature, and stirred for 8 h. The reaction was confirmed by TLC (EtOAc–*n*-hexane = 3:1), and 2-propanol (10 mL) was added to quench the reaction. The reaction mixture was filtrated with Celite, and neutralized with sat. Na_2CO_3 aq. The solution was acidified with 1 N HCl, and extracted with $CHCl_3$. The organic phase was dried over $MgSO_4$, and $MgSO_4$ was removed by filtration. The solution was evaporated to yield yellow oil; Yield 1.94 g, 88.6%. ^{1}H-NMR (300 MHz, r.t., $CDCl_3$): δ/ppm 3.58–3.60 (m, 2H, CH_2), 3.64–3.68 (m, 2H, CH_2), 3.69–3.70 (m, 2H, CH_2), 3.71–3.76 (m, 2H, CH_2), 4.18 (s, 2H, CH_2). IR wavenumber [cm^{-1}]: 3470 (OH) 1734 (C=O) 1112 (C-O). ESI-MS (negative): 181.2 [M−H]$^-$.

Methyl 2-[2-(2-chloroethoxy)ethoxy]acetate (3). Compound 2 (0.312 g, 1.78 mmol) was dissolved in DMF (2 mL), and NaH (0.530 g, 2.22 mmol) was added to the solution at 0 °C. After stirring for 30 min, MeI (0.649 g, 3.33 mmol) was added to the solution. The reaction mixture was stirred for 22.5 h at room temperature. The reaction was confirmed by TLC (EtOAc–*n*-hexane = 1:3). The solution was evaporated, and the residue was dissolved in $CHCl_3$. The solution was washed with 1 N HCl, sat. $NaHCO_3$ aq. and brine. The solution was dried over $MgSO_4$, and $MgSO_4$ was removed by filtration. The residue was purified by column chromatography (EtOAc–*n*-hexane = 1:3) to yield colorless oil; Yield 0.242 g, 74.4%. ^{1}H-NMR (300 MHz, r.t., $CDCl_3$): δ/ppm 3.62 (t, 2H, J = 6 Hz, J = 6 Hz, CH_2), 3.71–3.72 (m, 2H, CH_2), 3.72–3.74 (m, 2H, CH_2), 3.73 (s, 3H, CH_3), 3.74–3.76 (m, 2H, CH_2), 4.16 (s, 2H, CH_2). ESI-MS (positive): 219.1 [M+Na]$^+$.

Methyl 2-[2-(2-azidoethoxy)ethoxy]acetate (4). Compound 3 (0.242 g, 1.31 mmol) was dissolved in DMF (15 mL). NaN_3 (0.511 g, 7.86 mmol) was added to the solution, and stirred for 14 h at 70 °C. The reaction was confirmed by TLC (EtOAc–*n*-hexane = 1:3). NaN_3 was removed by filtration, and the solution was evaporated. The residue was dissolved in $CHCl_3$, and the solution was washed with 1 N HCl, sat. $NaHCO_3$ aq. and brine. The organic phase was dried over $MgSO_4$, and $MgSO_4$ was removed by filtration. The solution was evaporated to yield colorless oil; Yield 0.220 g, 87.8%. ^{1}H-NMR (300 MHz, r.t., CDCl$_3$): δ/ppm 3.38 (t, 2H, J = 5 Hz, J = 5 Hz, CH$_2$), 3.65–3.67 (m, 2H, CH$_2$), 3.67–3.67 (m, 2H, CH$_2$), 3.70–3.72 (m, 2H, CH$_2$), 3.73 (s, 3H, CH$_3$), 4.16 (s, 2H, CH$_2$). ESI-MS (positive) 226.1 [M+Na]$^+$.

[2-(2-Azidoethoxy)ethoxy]acetic acid (5). Compound 4 (0.220 g, 1.15 mmol) was dissolved in 1 N NaOHaq (15 mL), and the solution was stirred for 16 h. The reaction was confirmed by TLC (EtOAc–*n*-hexane = 1:3). The solution was acidified with 1 N HCl, and extracted with CHCl$_3$. The solution was dried over MgSO$_4$, and MgSO$_4$ was removed by filtration. The solution was evaporated to yield colorless oil; Yield 0.172g, 84.4%. ^{1}H-NMR (300 MHz, r.t., CDCl$_3$): δ/ppm 3.41 (t, 2H, J = 4.8 Hz, J = 5.1 Hz, CH$_2$), 3.67–3.69 (m, 2H, CH$_2$), 3.70–3.71 (m, 2H, CH$_2$), 3.75–3.78 (m, 2H, CH$_2$), 4.16 (s, 2H, CH$_2$). ESI-MS (negative): 188.6 [M−H]$^-$.

p-N-[2-(2-Azidoethoxy)ethoxy]amidophenyl-2-acetamido-2-deoxy-6-sulfo- -D-glucopyranoside (7). Compound 6 was synthesized by following the previous method [5,14]. Next 6 (0.142 g, 0.320 mmol) was dissolved in MeOH–H$_2$O = 2:1 (10 mL). Pd/C (20 mg) was added to the solution, and the solution was stirred overnight under H$_2$. Pd/C was removed by filtration, and the solution was evaporated. The residue was dissolved in DMF (5 mL), and 5 (86.2 mg, 0.456 mmol) was added to the solution. HATU (185 mg, 0.487 mmol) and DIEA (83 μL, 0.487 mmol) were added to the solution, and the solution was stirred overnight. The solution was evaporated, and the residue was purified by column chromatography (H$_2$O–MeOH = 5:1) to yield a brown solid; Yield 0.160 g, 83.2%. ^{1}H-NMR (500 MHz, r.t., D$_2$O): δ/ppm 2.09 (s, 3H, OCOC$\underline{H}_3$), 3.36–3.39 (m, 2H, CH$_2$), 3.48 (t, 1H, J_{H3-H4} = 9.0 Hz, J_{H4-H5} = 9.5 Hz, H4), 3.54 (t, 1H, J_{H2-H3} = 10.0 Hz, J_{H3-H4} = 9.5 Hz, H3), 3.57–3.62 (m, 2H, CH$_2$), 3.66–3.68 (m, 2H, CH$_2$), 3.71–3.73 (m, 2H, CH$_2$), 3.74–3.75 (m, 1H, H5), 3.88 (t, 1H, J_{H1-H2} = 9.5 Hz, J_{H2-H3} = 9.5 Hz, H2), 4.11 (s, 2H, CH$_2$), 4.13 (dd, 1H, $J_{H5-H6proR}$ = 4.0 Hz, $J_{H6proR-H6proS}$ = 6.0 Hz, H6proR), 4.28 (d, 1H, J = 10 Hz H6proS), 5.01 (d, J = 8.5 Hz, H1), 6.98(d, 2H, J_{o-m}=7.0 Hz, Ph-o), 7.27 (d, 2H, J_{o-m} = 9.0 Hz, Ph-m). ESI-MS (negative): 421.1 [M−Na]$^-$.

pNP 6-Sulfo-GlcNAc trimer (G0:9). Compound 7 (91.8 mg, 0.157 mmol) and 1,3,5-triethylbenzene (8) (5.12 mg, 0.0341 mmol) were dissolved in DMF (20 mL). To the solution, sodium ascorbate (12.2 mg, 0.6 eq.) and CuSO$_4$ (4.91 mg, 0.3 eq.) was added. The solution was stirred for 2 days, and the reaction was confirmed by reversed phase TLC (H$_2$O–MeOH = 5:1). Copper was removed by centrifugation, and the residue was purified by column chromatography (H$_2$O–MeOH = 5:1) to yield colorless solid; Yield 0.0320g, 49.2%. ^{1}H-NMR (300 MHz, r.t., D$_2$O):δ/ppm 1.86(s, 9H, CH$_3$), 3.41–3.47 (overlap, 6H, CH$_2$), 3.41–3.47 (overlap, 3H, H4), 3.52–3.57 (overlap, 3H, H3), 3.52–3.57 (overlap,

12H, CH$_2$ × 2), 3.66 (broad, 3H, 5H), 3.66 (broad, 6H, CH$_2$), 3.75–3.77 (overlap, 3H, H2), 3.77 (s, 6H, CH$_2$), 3.98 (s, 6H, NH), 4.05–4.07 (m, 3H, H6proR), 4.16 (d, 3H, J = 11.5 Hz H6proS), 4.61–4.62 (overlap, 3H, H1), 6.47 (d, 6H, J_{o-m} = 9.0 Hz, Ph-o), 6.77 (d, 6H, J_{o-m} = 8.5 Hz, Ph-m), 7.53 (s, 3H, Ar), 8.18 (s, 3H, C=CH). ^{13}C-NMR (75 MHz, r.t., D$_2$O): δ/ppm 21.6, 49.5, 53.7, 54.8, 66.2, 67.8, 68.7, 68.8, 69.2, 72.9, 73.3, 99.3, 116.1, 121.2, 121.7, 122.6, 130.0, 130.6, 153.2, 169.2, 174.0. ESI-MS: m/z 976.1 [M+2Na]$^{2+}$, (LCMS, negative) m/z 822.9 [M−3Na]$^{3-}$.

p-N-(5-Hexynoic)amidophenyl-2-acetoamido-2-deoxy-6-sulfo- -D-glucopyranoside (10). Compound 6 was dissolved in MeOH–H$_2$O = 2:1. Pd/C was added to the solution, and the solution was stirred overnight under H$_2$. Pd/C was removed by filtration, and the solution was evaporated. The residue, *p*-aminophenyl-2-acetoamido-2-deoxy-6-sulfo-β-D-glucopyranoside (0.353 g, 0.853 mmol) was dissolved in DMF (20 mL), and 5-hexynoic acid (0.143 g, 1.28 mmol) was added to the solution at 0 °C. HATU (0.486 g, 1.28 mmol) and DIEA (217 μL, 1.28 mmol) was added to the solution, and the solution was stirred overnight. DMF was evaporated, and the residue was purified by reversed phase column chromatography (H$_2$O–MeOH = from 5:1 to 2:1) to yield white solid; Yield 0.366g, 81.6%. ^{1}H-NMR (500 MHz, r.t., D$_2$O): δ/ppm 1.74 (t, J = 7.0Hz, 2H, CH$_2$), 1.90(s, 3H, OCOC$\underline{H}_3$), 2.15–2.18 (m, 2H, CH$_2$), 2.25 (t, J = 2.5Hz, 1H, C≡CH), 2.36–2.39 (m, 2H, CH$_2$), 3.47 (t, 1H, J_{H3-H4} = 9.5 Hz, J_{H4-H5} = 9.5 Hz, H4), 3.52–3.55 (m, 1H, H3), 3.70–3.73 (m, 1H, H5), 3.87 (t, 1H, J_{H1-H2} = 8.5Hz, J_{H2-H3} = 10 Hz, H2), 4.11 (dd, 1H, $J_{H5-H6proR}$ = 5.5Hz, $J_{H6proR-H6proS}$ = 5.0 Hz, H6proR), 4.25 (d, 1H, J = 11.5 Hz, H6proS), 4.99 (d, J = 8.5 Hz, H1), 6.93 (d, 2H, J_{o-m} = 9.5 Hz, Ph-o), 7.27 (d, 2H, J_{o-m} = 9.5 Hz, Ph-m).

Trimethyl 1,3,5-benzenetricarboxylate (12). Trimesic acid (11, 8.00 g, 38.1 mmol) was dissolved in MeOH (140 mL), and conc. H$_2$SO$_4$ (2 mL) was added to the reaction mixture. The solution was refluxed for 24 h, and the solution was evaporated. The residue was dissolved in CHCl$_3$, and washed with sat. Na$_2$CO$_3$aq. The solution was dried with MgSO$_4$, and removed by filtration. The solution was evaporated to obtain white solid; Yield 9.09 g, 94.7%. ^{1}H-NMR (300 MHz, r.t., CDCl$_3$): δ/ppm 3.96 (s, 9H, OMe), 8.83 (s, 3H, Ar).

1,3,5-Trihydroxymethylbenzene (13). Compound 12 (1.00 g, 3.97 mmol) was dissolved in THF (70 mL). LiAlH$_4$ (35.7 mmol) was added to the solution, and the solution was refluxed for 24 h. The reaction was confirmed by TLC (EtOAc–*n*-hexane=1:1), and water was added to the reaction in order to quench the reaction. The solution was dried over MgSO$_4$ and evaporated to yield white solid; Yield 0.571 g, 85.6%. ^{1}H-NMR (300 MHz, r.t., CDCl$_3$): δ/ppm 4.60 (s, 6H, CH$_2$), 7.25 (s, 3H, Ar).

1,3,5-Tribenzylchloride (14). Compound 13 (0.200 g, 1.19 mmol) was dissolved with $SOCl_2$ (5 mL) at 0 °C and to the solution the catalytic amount of DMF was added. The solution was stirred for 48 h at room temperature. The reaction was confirmed by TLC (EtOAc–n-hexane = 1:1). The reaction mixture was poured into ice-water, and extracted with $CHCl_3$. The organic phase was washed with water, and dried over $MgSO_4$. The solution was evaporated to yield white solid; Yield 0.241 g, 93.9%. ^{1}H-NMR (300 MHz, r.t., $CDCl_3$): δ/ppm 4.57 (s, 6H, CH_2), 7.36 (s, 3H, Ar).

1,3,5-Tribenzylazide (15). Compound 14 (0.214 g, 0.964 mmol) was dissolved in DMF (10 mL). NaN_3 (1.13 g, 17.4 mmol, 18 eq.) was added to the solution, and the solution was stirred at 70 °C for 21 h. The solution was evaporated, and the residue was dissolved with chloroform. The solution was washed with 1 N HCl aq., sat. $NaHCO_3$, and brine. The solution was dried over $MgSO_4$. The solution was evaporated to yield colorless oil; Yield 0.218 g, 93.0%. ^{1}H-NMR (300 MHz, r.t., $CDCl_3$): δ/ppm 4.38 (s, 6H, CH_2), 7.23 (s, 3H, Ar).

1-(3-Acetyl-5-prop-2-ynyloxy-phenyl)ethanone (17). Dimethyl 5-hydroxyisophtalate (16, 5.01 g, 23.8 mmol) and K_2CO_3 (93 g, 35.7 mmol, 1.5 eq.) were dissolved in DMF (100 mL), and the solution was degassed with bubbling of N_2. The solution was heated up to 80 °C, and propargyl bromide (2.80 mL, 35.7 mmol, 1.5 eq.) was added under N_2. The solution was stirred for overnight. The solution was allowed to be cooled to room temperature. K_2CO_3 was removed by filtration, and the solution was evaporated. The residue was recrystallized in EtOH to yield white solid; Yield: 4.52 g, 76.5%. ^{1}H-NMR (300 MHz, r.t., $CDCl_3$): δ/ppm 2.53 (t, J = 2.1 Hz, J = 2.4 Hz, 1H, C≡CH), 3.92 (s, 6H, CH_3), 4.76 (d, J = 2.4 Hz, 2H, CH_2), 7.80 (d, J = 1.5 Hz, 2H, ArH), 8.30 (t, J = 1.5 Hz, J = 1.2 Hz, 1H, ArH).

(3-Hydroxymethyl-5-prop-2-ynyloxy-phenyl)methanol (18). Compound 17 (0.500 g, 2.01 mmol) was dissolved in THF (35.0 mL), and degassed. $LiAlH_4$ (12.1 mmol) was added to the solution. The reaction mixture was stirred for 6 h, and the reaction was confirmed by TLC (EtOAc–n-hexane = 1:1). Then, water was added to the solution in order to quench the reaction. The solution was dried over $MgSO_4$, and $MgSO_4$ was removed by filtration. The solution was evaporated to yield white solid; Yield 0.425 g, 99%. ^{1}H-NMR (300 MHz, r.t., CD_3OD): δ/ppm 2.91 (t, J = 2.7 Hz, J = 2.4 Hz, 1H, C≡CH), 4.57 (s, 4H, CH_2OH), 4.72 (d, J = 2.4 Hz, 2H, CH_2), 6.89 (s, 2H, ArH), 8.30 (d, J = 0.6 Hz, 1H, ArH).

1,3-Bischloromethyl-5-prop-2-ynyloxy-benzene (19). Compound 18 (0.425 g, 2.21 mmol) was dissolved in CH_2Cl_2, and $SOCl_2$ was added to the solution with catalytic amount of DMF at 0 °C. The reaction mixture was stirred for overnight, and the reaction was confirmed by TLC (EtOAc–n-hexane = 1:1). The reaction mixture was poured into ice-water, and extracted with $CHCl_3$. The organic phase was washed with water, and dried over $MgSO_4$. $MgSO_4$ was removed by filtration, and the solution was evaporated to yield white solid; Yield. 0.462 g, 91.2%. ^{1}H-NMR (300 MHz, r.t., $CDCl_3$): δ/ppm 2.52 (t, J = 2.4 Hz, J = 2.4 Hz, 1H, C≡CH), 4.53 (s, 4H, CH_2), 4.70 (d, J = 2.4 Hz, 2H, CH_2), 6.95 (d, J = 1.5 Hz, 2H, ArH), 7.02 (t, J = 0.6 Hz, J = 0.6 Hz, 1H, ArH).

G1-Cl (20). Compounds 15 (95.0 mg, 0.391 mmol) and 19 (0.420 g, 1.83 mmol) was dissolved in DMF (3 mL). To the reaction mixture, CuI (111 mg, 0.586 mmol) was added and suspended. DIEA (100 mL, 0.588 μmol) was added to the solution. The solution was stirred for 24 h. The reaction mixture was evaporated, and the residue was purified by column chromatography ($CHCl_3$–MeOH = 60:1) to yield white solid; Yield 0.268 g, 73.7%. ^{1}H-NMR (300 MHz, r.t., $CDCl_3$): δ/ppm 4.50 (s, 12H, CH_2), 5.20 (s, 6H, CH_2), 5.46 (s, 6H, CH_2), 6.94 (s, 6H, ArH), 6.99 (s, 3H, ArH), 7.10 (s, 3H, ArH), 7.54 (s, 3H, C=CH).

G1-N_3 (21). Compound 20 (0.268 g, 0.288 mmol) was dissolved in DMF (10 mL). NaN_3 (3.63 mmol) was added to the solution, and the solution was stirred at 50 °C for overnight. The solution was evaporated, and the residue was dissolved in $CHCl_3$. The organic phase was washed with water for three times. The solution was dried with $MgSO_4$, and $MgSO_4$ was removed by filtration. The residue was purified by column chromatography ($CHCl_3$–MeOH = 100:1) to yield white solid; Yield 0.260 g, 93.1%. ^{1}H-NMR (300 MHz, r.t., $CDCl_3$): δ/ppm 4.29 (s, 12H, CH_2), 5.21 (s, 6H, CH_2), 5.45 (s, 6H, CH_2), 6.85 (s, 3H, ArH), 6.88 (s, 6H, ArH), 7.13 (s, 3H, ArH), 7.55 (s, 3H, C=CH). ^{1}H-NMR (300 MHz, r.t., DMF-d_7): δ/ppm 4.50 (s, 12H, CH_2) 5.23 (s, 6H, CH_2), 5.71 (s, 6H, CH_2), 7.05 (s, 3H, ArH), 7.10 (s, 6H, ArH), 7.45 (s, 3H, ArH), 8.39 (s, 3H, C=CH). ^{13}C-NMR (75 MHz, r.t., $CDCl_3$): δ/ppm 53.4, 54.4, 62.1, 114.3, 120.6, 122.9, 127.6, 136.8, 137.7, 144.4, 158.8. ESI-MS: m/z 969.9.0 [M+H]$^+$, 992.0 [M+Na]$^+$.

pNP 6-Sulfo-GlcNAc hexamer(G1:22). Compounds 21 (6.15 mg, 6.34 μmol) and 10 were dissolved in DMF (400 μL), and an aqueous solution of sodium ascorbate (6.02 mg, 0.0304 mmol) and $CuSO_4$ (2.40 mg, 0.0150 mmol) was added to the solution. The reaction mixture was stirred for 3 days. The solvent was evaporated and the residue was purified by reversed phase chromatography (H_2O–MeOH = 2:1) to yield a white solid; Yield 7.70 mg, 28.2%. ^{1}H-NMR (300 MHz, r.t., DMF-d_7): δ/ppm 1.26–1.28 (m, 12H, CH_2), 1.88 (s, 18H, OCOCH_3), 1.96 (t, J = 7.5Hz, J = 7.5 Hz, 12H, CH_2), 2.41 (t, J = 7.5 Hz, J = 7.8 Hz, 12H, CH_2), 3.92 (dd, J_{H1-H2} = 9.5Hz, J_{H2-H3} = 9.0 Hz, 6H, H2), 4.03–4.06 (m, 6H, H5), 4.07–4.11 (m, 6H, H3), 4.19–4.26 (m, 6H, H4),

5.02 (d, J_{H1-H2} = 9.0 Hz, 6H, H1), 5.16 (s, 6H, CH_2), 5.29 (broad, 6H, H6), 5.36 (broad, 6H, H6), 5.57 (s, 12H, CH_2), 5.66 (s, 6H, CH_2), 6.91 (s, 3H, ArH), 6.94 (d, 12H, J_{om} = 9.0Hz, Ph-o), 7.03 (s, 6H, ArH), 7.44 (s, 3H, ArH), 7.58 (d, 12H, J_{o-m} = 9.0Hz, Ph-m), 7.94 (s, 6H, C=CH), 8.40 (s, 3H, C=CH). ^{13}C-NMR (75 MHz, r.t., DMF-d_7): δ/ppm 23.5, 28.2, 30.6, 31.0, 47.9, 58.4, 59.8, 61.8, 67.1, 71.7, 76.8, 80.4, 81.2, 106.0, 115.2, 119.8, 122.5, 126.0, 130.6, 133.6, 140.1, 143.1, 144.2, 159.4, 164.6, 175.5, 176.4. ESI-MS: m/z 822.9 [M+H+4Na]$^{5+}$.

G2-Cl (23). G1-N_3 (21, 0.0597 g, 0.0616 mmol) and 19 (0.114 g, 0.498 mmol) was dissolved in DMF (750 μL). CuI (21.0 mg, 0.110 mmol) was added to the solution and suspended. DIEA was added to the solution, and the solution was stirred for 24 h at room temperature. The reaction was confirmed by TLC ($CHCl_3$–MeOH = 10:1), and the residue was purified by column chromatography ($CHCl_3$–MeOH = 40:1) to yield a white solid; Yield 43.5 mg, 30.1%. ^{1}H-NMR (300 MHz, r.t., $CDCl_3$): δ/ppm 4.45 (s, 24H, CH_2), 4.99 (s, 6H, CH_2), 5.07 (s, 12H, CH_2), 5.12 (s, 6H, CH_2), 5.37 (s, 12H, CH_2), 6.73 (s, 3H, ArH), 6.75 (s, 6H, ArH), 6.88 (s, 12H, ArH), 6.94 (s, 6H, ArH), 7.05 (s, 3H, ArH), 7.55 (s, 3H, C=CH), 7.62 (s, 6H, C=CH).

G2-N_3 (24). 23 (0.0435 g, 0.0186 mmol) was dissolved in DMF (10 mL). NaN_3 (0.0201 g, 0.334 mmol) was added to the solution, and stirred at 50 °C for overnight. The solution was evaporated and the residue was dissolved in $CHCl_3$. The organic phase was washed with water three times. The organic phase was dried over $MgSO_4$, and $MgSO_4$ was filtrated. The residue was purified by column chromatography ($CHCl_3$–MeOH = 40:1) to yield a white solid; Yield 0.253 g, 56.3%. ^{1}H-NMR (300 MHz, r.t., $CDCl_3$): δ/ppm 4.27 (s, 24H, CH_2), 4.5.04 (s, 6H, CH_2), 5.16 (s, 12H, CH_2), 5.17 (s, 6H, CH_2), 5.40 (s, 12H, CH_2), 6.76 (s, 3H, ArH), 6.78 (s, 6H, ArH), 6.83 (s, 6H, ArH), 6.86 (s, 12H, ArH), 7.08 (s, 3H, ArH), 7.54 (s, 3H, C=CH), 7.61 (s, 6H, C=CH). ^{1}H-NMR (300 MHz, r.t., DMF-d_7): δ/ppm 4.48 (s, 24H, CH_2), 5.15 (s, 6H, CH_2), 5.24 (s, 12H, CH_2), 5.66 (s, 6H, CH_2), 5.69 (s, 12H, CH_2), 7.03 (s, 6H, ArH), 7.06 (s, 9H, ArH), 7.09 (s, 12H, ArH), 7.44 (s, 3H, ArH), 8.35 (s, 3H, C=CH), 8.41 (s, 6H, C=CH). ^{13}C-NMR (75 MHz, r.t., $CDCl_3$): δ/ppm 53.3, 53.6, 54.3, 61.7, 62.0, 114.2, 114.6, 120.0, 120.6, 123.2, 123.5, 127.7, 136.7, 137.2, 137.7, 143.6, 144.1, 158.7, 159.0. MALDI-TOF-MS: m/z 2447.2 [M+Na]$^+$.

pNP 6-Sulfo-GlcNAc dodecamer (G2:25). Compounds 24 (8.00 mg, 3.30 μmol) and 10 were dissolved in DMF (350 μL). The aqueous solution (83 μL) of sodium ascorbate (6.28 mg, 31.7 mmol) and $CuSO_4$ (2.52 mg, 0.0158 mmol) was added to the solution. The reaction mixture was stirred for 3 days. The reaction was confirmed by TLC (H_2O–MeOH = 3:1). The solution was evaporated and the residue was purified by reversed phase column chromatography (H$_2$O–MeOH = 3:1) to yield a white solid; Yield 6.30 mg, 21.1%. ^{1}H-NMR (500 MHz, r.t., DMF-d_7): δ/ppm 1.26–1.28 (m, 24H, CH$_2$), 1.88 (s, 36H, OCOCH$_3$), 1.94–199 (m, 24H, CH$_2$), 2.40 (t, J = 7.5 Hz, J = 7.0 Hz, 24H, CH$_2$), 3.78 (broad, 12H, H5), 3.87–3.90 (m, 12H, H2), 4.06–4.10 (m, 12H, H3), 4.23–4.26 m, 12H, H4), 4.25 (m, 6H, H4), 5.01 (d, $J_{\text{H1-H2}}$ = 8.5 Hz, 12H, H1), 5.16–5.18 (m, 12H, H6), 5.22–5.24 (broad, 6H, CH$_2$), 5.31–5.34 (m, 6H, H6), 5.36 (broad, 6H, H6'), 5.53 (s, 24H, CH$_2$), 5.56 (s, 12H, CH$_2$), 5.58 (s, 12H, CH$_2$), 5.61–5.66 (broad, 6H, CH$_2$), 6.88 (s, 6H, ArH), 6.96–6.98 (overlap, 24H, Ph-o), 6.96–6.98 (overlap, 9H, ArH), 7.03(s, 12H, ArH), 7.44 (broad, 3H, ArH), 7.58 (d, 2H, $J_{\text{o-m}}$ = 8.1 Hz, Ph-m), 7.94 (s, 12H, C=CH), 8.35 (s, 3H, C=CH), 8.35 (s, 6H, C=CH). ^{13}C-NMR (75 MHz, r.t., DMF-d_7): δ/ppm 22.6, 28.2, 30.6, 31.1, 58.4, 61.3, 61.8, 67.1, 71.7, 76.8, 80.3, 81.2, 82.3, 91.5, 106.0, 115.4, 120.0, 122.5, 126.0, 127.8, 130.0, 140.1, 143.2, 143.4, 144.2, 148.7, 155.4, 156.6, 156.6, 164.6, 164.7, 176.4, 181.0. ESI-MS m/z 1238.0 [M+H+6Na]$^{7+}$.

Lectin Recognition Assay

Each sugar dendrimer was dissolved in PBS (pH 7.4) at a concentration of 8.00 µM. The lectin (FITC-WGA) was also dissolved in PBS (pH 7.4) at a concentration of 0.100 µM. To a solution of WGA (300 µL) were sequentially added aliquots of the sugar dendrimer solutions (3, 3, 5, 5, 5, 5, 10, 10, 10, 10, 10, 20, 20 and 20 µL) at 25 °C. Fluorescent measurements were taken 5 min after the addition of each aliquot. Fluorescence spectra of the FITC were measured with an excitation wavelength at 490 nm and an emission wavelength of 510 nm. The PBS buffer without a sugar dendrimer was also added in the same way as a reference. Changes in the fluorescence of the sugar-protein interaction were calculated by subtracting two of fluorescence intensities (*i.e.*, fluorescence change by sugar protein interaction = fluorescence change by addition of the sugar dendrimer solution–fluorescence change by the addition of the reference PBS solution). Changes in the fluorescence were plotted according to the following formula [19,20]:

$$\frac{[\text{Sugar}]F_0}{\Delta F} = \frac{[\text{Sugar}]F_0}{\Delta F_{\text{max}}} + \frac{F_0}{\Delta F_{\text{max}} \cdot K_a}$$

where [Sugar], K_a, F_0, ΔF, and ΔF_{max} represent the sugar concentration of the dendrimer solution (M), the association constant (M^{-1}), the initial fluorescence intensity, the fluorescence change, and the maximum fluorescence change, respectively.

In Vitro *Amyloid Formation of Aβ (1-42) and* **ThT Fluorescence Assay [14]**

Aβ (1-42) was dissolved in a 0.02% ammonia solution at a concentration of 200 μM. Any aggregates formed were removed by centrifugation using a CS120FX centrifuge (Hitachi, Tokyo, Japan) for 3 h at 16,000 g and 4 °C. The supernatant was mixed with phosphate buffer (20 mM phosphate buffer, pH 7.4, 100 mM NaCl) to a final peptide concentration of 20 μM. The peptide solution was incubated with each sugar additive at 37 °C. The concentration of the sugar additives was 200 μM. Amyloid fibril formation was evaluated by the fluorescence emission of ThT. Following a period of incubation at 37 °C, a 5 μL solution of Aβ in buffer was added to 300 μL of the ThT solution (50 mM in the same phosphate buffer). Following a period of 10 s, the fluorescence intensity was measured at an excitation wavelength of 450 nm and an emission wavelength of 482 nm.

AFM Measurements

The Aβ peptide (1-42) was incubated in phosphate buffer (20 mM phosphate buffer, 100 mM NaCl) at 37 °C for 12 h, with the peptide and the sugar additives being added at a concentration of 20 μM. After the incubation of the sample, a 5 μL portion of the sample solution was placed on freshly cleaved mica and dried under N2. The mica substrates were then washed with 100 μL of MilliQ water.

CONCLUSIONS

A series of novel glycodendrimers containing sulfonated GlcNAc have been successfully synthesized using click chemistry. The glycodendrimers of G1 and G2 formed large glycoside clusters and exhibited strong affinities to proteins of lectin and Aβ (1-42) as a consequence of the multivalent effect. G1 showed the highest level of affinity to proteins because of the self-assembling properties of the dendrimer. G1 and G2 both strongly inhibited Aβ aggregation and β-sheet conformation.

Click chemistry was used in the current paper to successfully synthesize the glycodendrimers. The procedure was facile and proceeded without the need for protecting group chemistry. Work focused on the fabrication of functional cluster of glycosides with saccharides suitable for application in pathogens is currently underway in our laboratories.

ACKNOWLEDGMENTS

This work was supported by a Grand-in-Aid for Scientific Research on

Innovative Areas of Soft-Interface (20106003) and by Grand-in-Aid for Young Scientists (A) (23685027).

REFERENCES

1. Varki, A. Biological roles of oligosaccharides: All of the theories are correct. *Glycobiology* 1993, *3*, 97–130.
2. Baskaran, S.; Grande, D.; Sun, X.L.; Yayon, A.; Chaikof, E.L. Glycosaminoglycan-mimetic biomaterials. 3. Glycopolymers prepared from alkene-derivatized mono-and disaccharide-based glycomonomers. *Bioconjug. Chem.* 2002, *13*, 1309–1313.
3. Mammen, M.; Choi, S.K.; Whitesides, G.M. Polyvalent interactions in biological systems: Implications for design and use of multivalent ligands and inhibitors. *Angew. Chem. Int. Ed.* 1998, *37*, 2754–2794.
4. Lee, Y.C.; Lee, R.T. Carbohydrate-protein interactions: Basis of glycobiology. *Acc. Chem. Res.* 1995, *28*, 321–327.
5. Fukuda, T.; Matsumoto, E.; Onogi, S.; Miura, Y. Aggregation of Alzheimer amyloid beta peptide (1-42) on the multivalent sulfonated sugar interface. *Bioconjug. Chem.* 2010, *21*, 1079–1086.
6. Ko, K.-S.; Jaipuri, F.A.; Pohl, N.L. Fluorous-based carbohydrate microarrays. *J. Am. Chem. Soc.* 2005, *127*, 13162–13163.
7. Liang, P.H.; Wu, C.Y.; Greenberg, W.; Wong, C.H. Glycan arrays: Biological and medical applications. *Curr. Opin. Chem. Biol.* 2008, *12*, 86–92.
8. Miura, Y.; Sasao, Y.; Dohi, H.; Nishida, Y.; Kobayashi, K. Self-assembled monolayers of globotriaosylceramide (Gb3) mimics: Surface-specific affinity with Shiga toxins. *Anal. Biochem.* 2002, *310*, 27–35.
9. Nagahori, N.; Abe, M.; Nishimura, S.I. Structural and functional glycosphingolipidomics by glycoblotting with aminooxy-functionalized gold nanoparticle. *Biochemistry* 2009, *48*, 583–594.
10. Osaki, F.; Kanamori, T.; Sando, S.; Sera, T.; Aoyama, Y. A quantum dot conjugated sugar ball and its cellular uptake. On the size effects of endocytosis in the subviral region. *J. Am. Chem. Soc.* 2004, *126*, 6520–6521.
11. Otsuka, H.; Akiyama, Y.; Nagasaki, Y.; Kataoka, K. Quantitative and reversible lectin-induced association of gold nanoparticles modified with α-lactosyl-ω-mercapto-poly (ethylene glycol). *J. Am. Chem. Soc.* 2001, *123*, 8226–8230.
12. Miura, Y. Synthesis and biological application of glycopolymers. *J.*

Polym. Sci. A1 2007, *45*, 5031–5036.
13. Chabre, Y.M.; Roy, R. Recent trends in glycodendrimer syntheses and application. *Curr. Top. Med. Chem.* 2008, *8*, 1237–1285.
14. Miura, Y.; Yasuda, K.; Yamamoto, K.; Koike, M.; Nishida, Y.; Kobayashi, K. Inhibition of Alzheimer amyloid aggregation with sulfonated glycopolymers. *Biomacromolecules* 2007, *8*, 2129–2134.
15. Miura, Y.; Mizuno, H. Interaction analyses of Amyloid b peptide (1-40) with glycosaminoglycan model polymers.*Bull. Chem. Soc. Jpn.* 2010, *83*, 1004–1009.
16. Rostovtsev, V.V.; Green, L.G.; Fokin, V.V.; Sharpless, K.B. A stepwise huisgen cycloaddition process: Copper(I)-catalyzed regioselective "ligation" of azides and terminal alkynes. *Angew. Chem. Int. Ed.* 2002, *41*, 2596–2599.
17. Rice, R.H.; Etzler, M.E. Chemical modification and hybridization of wheat germ agglutinins. *Biochemistry* 1975, *14*, 4093–4099.
18. Nagata, Y.; Burger, M.M. Wheat germ agglutinin: Molecular characteristic and specificity for sugar binding. *J. Biol. Chem.* 1974, *249*, 3116–3122.
19. Wu, P.; Malkoch, M.; Hunt, J.N.; Vestberg, R.; Kaltgrad, E.; Finn, M.G.; Fokin, V.V.; Sharpless, K.B.; Hawker, C.J. Multivalent, Bifunctional dendrimers prepared by click chemistry. *Chem. Commun.* 2005, 5775–5777.
20. Miura, Y.; Yamauchi, T.; Sato, H.; Fukuda, T. The self-assembled monolayer of saccharide via click chemistry: Formation and protein recognition. *Thin Solid Films* 2008, *516*, 2443–2449.
21. Miura, Y.; Ikeda, T.; Kobayashi, K. Chemoenzymatically synthesized glycoconjugate polymers. *Biomacromolecules*2003, *4*, 410–415.
22. Zanini, D.; Roy, R. Chemoenzymatic synthesis and lectin binding properties of dendritic *N*-acetyllactosamine.*Bioconjugate Chem.* 1997, *8*, 187–192.
23. Schwefel, D.; Maierhofer, C.; Beck, J.G.; Seeberger, S.; Diederichs, K.; Möller, H.M.; Welte, W.; Wittmann, V. Structural basis of multivalent binding to wheat germ agglutinin. *J. Am. Chem. Soc.* 2010, *132*, 8704–8719.
24. Wright, C.S. 2.2Å Resolution structure analysis of two refined *N*-acetylneuraminyl-lactose-wheat germ agglutinin isolectin complexes. *J. Mol. Biol.* 1990, *215*, 635–651.
25. Levine, H., III. Thioglavine T interaction with synthetic Alzheimer's disease β-amyloid peptides: Detection of amyloid aggregation in

solution. *Prot. Sci.* 1993, *2*, 404–410.
26. Giulian, D.; Haverkamp, L.J.; Yu, J.; Kazanskaia, A.; Kirkpatrick, J.; Roher, A.E. The HHQK domain of β-amyloid provies a structural basis for the immunopathology of Alzheimer's disease. *J. Biol. Chem.* 1998, *273*, 29719–29726.
27. Kitano, H.; Saito, D.; Kamada, T.; Gmmei-Ide, M. Binding of b-amyloid to sulfated sugar residues in a polymer brush.*Collid. Surf. B. Biointerface* 2001, *93*, 9–225.
28. Fukuda, T.; Kawamura, M.; Mizuno, H.; Miura, Y. Glycosaminoglycan Model Polymers with poly(γ-glutamate) backbone to inhibit aggregation of β-amyloid peptide. *Polym. J.* 2012.
29. Yankner, B.A.; Duffy, L.K.; Kirschner, D.A. Neurotropic and neurotoxic effects of amyloid beta protein: Reversal by tachykinin neuropeptides. *Science* 1990, *250*, 279–282.
30. Roychaudhuri, R.; Yang, M.; Hoshi, M.; Teplow, D.B. Amyloid b-protein assembly and Alzheimer disease. *J. Biol. Chem.* 2009, *284*, 4749–4753.

Chapter 9

BIOLOGICAL CHEMISTRY OF REACTIVE OXYGEN AND NITROGEN AND RADIATION-INDUCED SIGNAL TRANSDUCTION MECHANISMS

Ross B Mikkelsen[1] and Peter Wardman[2]

[1]Department of Radiation Oncology, Virginia Commonwealth University, 401 College Street, Richmond, VA 23298, USA
[2]Gray Cancer Institute, Mount Vernon Hospital, Northwood, Middlesex HA6 2JR, UK

ABSTRACT

In the past few years, nuclear DNA damage-sensing mechanisms activated by ionizing radiation have been identified, including ATM/ATR and the DNA-dependent protein kinase. Less is known about sensing mechanisms for cytoplasmic ionization events and how these events influence nuclear processes. Several studies have demonstrated the importance of cytoplasmic signaling pathways in cytoprotection and mutagenesis. For cytoplasmic signaling, radiation-stimulated reactive oxygen species (ROS) and reactive nitrogen species (RNS) are essential activators of these pathways. This review summarizes recent studies on the chemistry of radiation-induced ROS/RNS generation and emphasizes interactions between ROS and RNS and the relative roles of cellular ROS/RNS generators as amplifiers of the initial ionization events. Cellular mechanisms for regulating ROS/RNS levels are discussed. The mechanisms by which cells sense ROS/RNS are examined in terms of how ROS/RNS modify protein structure and function, for example, interactions with metal–thiol clusters, protein tyrosine nitration, protein cysteine oxidation, S-thiolation and S-nitrosylation. We propose that radiation-induced ROS are the initiators and that nitric oxide ($NO^{\bullet}$) or derivatives are the effectors activating these signal transduction pathways. In responding to cellular ionization events, the cell converts an oxidative signal to a nitrosative one because ROS are too reactive and unspecific in their reactions for regulatory purposes and the cell is equipped to precisely modulate NO$^\bullet$ levels.

INTRODUCTION

Exposure of cells to clinically relevant doses of ionizing radiation causes significant nuclear DNA damage: 1000 single-strand breaks, 40 double-strand breaks and 3000 damaged bases per gray (Gy) (Ward, 1994). Sensing mechanisms within the nucleus detect this damage and initiate signal transduction pathways, resulting in activation of cell cycle checkpoints and DNA damage repair. In the past few years, components of these nuclear DNA damage-sensing mechanisms, including ATM/ATR, DNA-dependent protein kinase, and CHK1/2 kinases and their homologs from yeast to man, have been identified (Durocher and Jackson, 2001; Khanna et al., 2001; Shiloh and Kastan, 2001).

Less is known about the consequences of cytoplasmic irradiation and how cytoplasmic ionization events influence nuclear processes. Nonetheless, there is a significant body of published information indicating the importance of radiation-stimulated cytoplasmic signaling in terms of cytoprotection and mutagenesis. Ionizing radiation activates cytoplasmic signal transduction pathways involved in cell proliferation and antiapoptotic mechanisms, including growth factor receptors, changes in cytoplasmic Ca^{2+} levels and stress-response kinases (e.g.Hallahan et al., 1991; Haimovitz-Friedman et al., 1994; Stephenson et al., 1994;Todd and Mikkelsen, 1994; Kasid et al., 1996; Kavanagh et al., 1998; Tuttle et al., 1998; Schmidt-Ullrich et al., 2000). Alpha-particle microbeam irradiation of cytoplasm elicits a spectrum of mutations different from that obtained with nuclear radiation and reflective of the type of mutations spontaneously produced by endogenous metabolism (Wu et al., 1999).

An important question that arises is how the few primary ionization events produced at clinically relevant doses (approximately 2000/Gy/cell) are amplified to account for the rapid and robust activation of cellular signal transduction pathways (Ward, 1994). Even considering secondary free radical products resulting from the initial ionization, the calculated amount of reactive oxygen species (ROS) generated is relatively insignificant compared to the amount produced by metabolism (Ward, 1994; see below). However, recent studies indicate that cells are endowed with cytoplasmic amplification mechanisms involving ROS/RNS (reactive nitrogen species) and responsive to low doses of ionizing radiation (e.g. Clutton et al., 1996; Narayanan et al., 1997; Morales et al., 1998; Leach et al., 2001, 2002). These mechanisms appear to be part of general cellular response pathways to oxidative stress (e.g. Ichas and Mazat, 1998; Marshall et al., 2000; Zorov et al., 2000; Droge, 2002; Paxinou et al., 2001). Our intent in this review is to discuss the overall chemistry of free radical production in terms of what ROS/RNS are produced by cellular metabolism and following a radiation exposure, describe possible cellular radiation-induced ROS-sensing mechanisms, and present possible mechanisms emphasizing regulatory protein modification by which an initial radiation-induced ionization event is transduced into a cellular response. Reaction rates, relative intracellular concentrations of ROS/RNS and their subcellular localization are assessed in evaluating biological relevance. What emerges from this analysis is the concept that although ROS are the initial reactants produced from an ionization event, RNS are the actual effectors/activators of redox-dependent cellular signal transduction pathways.

OXIDATION AND NITRATION INITIATED BY SUPEROXIDE ($^\bullet O_2^-$) AND NITRIC OXIDE (NO$^\bullet$)

Although the overall chemical biology of oxidative and nitrosative stress remains complex and ill defined, many individual chemical reactions are quite well characterized. We review here the main reactions, focusing on protein oxidation/nitration and the potential basis for signaling pathways involving short-lived free radicals.

'Oxidative stress' is a term associated with both enhanced production of ROS and reduced efficacy of protection by antioxidant enzymes or low molecular weight antioxidants. After the discovery of NO$^\bullet$ as a biological entity (Arnold et al., 1977; Palmer et al., 1987), and especially the realization (Beckman et al., 1990) that superoxide radicals ($^\bullet O_2^-$) and NO$^\bullet$ form more powerful or 'cloaked' oxidants (Koppenol et al., 1992) via ONOO$^-$ formation, oxidative stress has become unavoidably linked to 'nitrosative stress' and RNS. A further complication is the involvement of carbon dioxide as a major potential modifier of nitrosative stress, via its interaction with ONOO$^-$ to produce carbonate radicals, which had previously not been considered in a biological context (Lymar and Hurst, 1995). The key species associated with oxidative and nitrosative stress and the main routes of interaction between the species are shown in Figure 1.

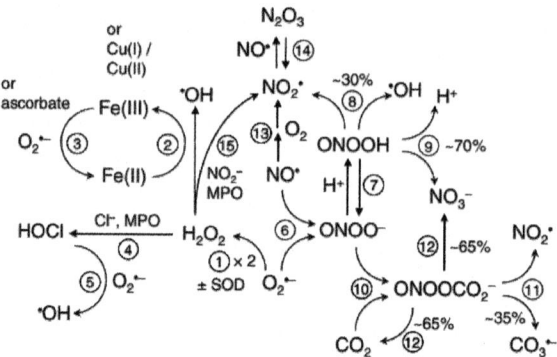

Figure 1: Main pathways for the formation of reactive oxidants from $^\bullet O_2^-$ and NO$^\bullet$ radicals.

Prior to 1990, oxidative stress was exclusively focused on the production of $^\bullet O_2^-$ and more reactive oxidants derived from $^\bullet O_2^-$ involving redox catalysts (Halliwell and Gutteridge, 1999). While $^\bullet O_2^-$ radicals disproportionate quite rapidly to form hydrogen peroxide in a pH-dependent reaction (Bielski *et al.*, 1985), superoxide dismutase (SOD) efficiently catalyzes the reaction (Fridovich, 1995) (Figure 1, reaction 1) and thus is a major controller of the steady-state concentration of $^\bullet O_2^-$. Hydrogen peroxide presents a potential oxidative challenge, although it is efficiently deactivated by glutathione peroxidase and catalase. However, in the presence of trace redox metals (principally iron and copper), production of highly reactive hydroxyl radicals via Fenton chemistry (Figure 1, reaction 2) can enhance damage. Note that only catalytic levels of redox metal are required for this pathway, since $^\bullet O_2^-$ itself, or alternative reductants such as ascorbate, 'recycle' the oxidized metal to the active, reduced form (reaction 3). A recent study demonstrated colocalization of a local Fenton reaction (or at least, oxidation of dihydrorhodamine; see below) and iron-containing mass-dense particles in hepatocytes (Kietzmann *et al.*, 2000).

Another catalyzed transformation of hydrogen peroxide to a more reactive oxidant involves chloride and myeloperoxidase to form hypochlorous acid/hypochlorite (HOCl/ClO$^-$, reaction 4). The latter species are both oxidants and chlorinating agents (Winterbourn and Kettle, 2000); HOCl (but not ClO$^-$; Long and Bielski, 1980), pK_a HOCl=7.5), reacts rapidly with $^\bullet O_2^-$ to form hydroxyl radicals (Candeias *et al.*, 1993) (reaction 5). Owing to the high reactivity of HOCl, its steady-state concentration is likely to be generally low. Therefore, except in activated phagocytes (where high concentrations of $^\bullet O_2^-$ are also formed), reaction 5 seems unlikely to be a significant source of $^\bullet$OH radicals.

Hydroxyl radicals are highly reactive (Ross et al., 1998) – 'promiscuous' is a term often seen – but high reactivity implies low selectivity and diffusion distance, and inability to act as a cellular messenger. NO$^\bullet$ is chemically rather unreactive toward most cellular constituents except for the heme centers that are its prime biological targets in controlling vascular tone (Ignarro, 2000), but NO$^\bullet$ reacts with $^\bullet$O$_2^-$ on every collision (Nauser and Koppenol, 2002) to form ONOO$^-$ (Figure 1, reaction 6). This ‹diffusion-controlled› reaction will usually not outcompete the somewhat slower SOD-catalyzed dismutation, reaction 1. This is because the relative importance of the two reactions as ‹sinks› for $^\bullet$O$_2^-$ depends also on the steady-state concentrations of NO$^\bullet$ and SOD. Leaving aside cellular heterogeneity, SOD levels are usually several micromolar (Fridovich, 1995;Halliwell and Gutteridge, 1999) whereas steady-state NO$^\bullet$ levels are probably considerably lower except under some pathological conditions (Ignarro, 2000). The concentrations of NO$^\bullet$ needed to saturate heme targets are nearer nanomolar (Bellamy et al., 2002). However, particularly under conditions where NO$^\bullet$ synthases (NOS) are stimulated, ONOO$^-$ formation (reaction 6) may become important. This may explain the enhanced tyrosine nitration of the mitochondrial enzyme, Mn-SOD, following radiation exposure of tissue culture cells (Leach et al., 2002). Other mechanisms involving myeloperoxidase and discussed below may need to be invoked to explain the enhanced tyrosine nitration of Mn-SOD observed *in vivo* in rejected kidney allografts (MacMillan-Crow et al., 1996).

Peroxynitrite exists in fast, dynamic equilibrium with its conjugate acid (ONOOH) under physiological conditions (pK_a ONOOH=6.8, Figure 1, equilibrium 7) (Radi et al., 2000). In the absence of carbon dioxide, ONOO$^-$/ONOOH decays via the protonated form, reactions 8 and 9 (ONOO$^-$ is stable at high pH). The minor pathway 8 produces reactive NO$_2^\bullet$ and $^\bullet$OH radicals, while a majority of ONOOH decomposes harmlessly to nitrate, reaction 9. Nitrogen dioxide (NO$_2^\bullet$) is *much* more reactive than NO$^\bullet$ toward common cellular constituents such as thiols (Wardman, 1998; Kirsch et al., 2002). The effective half-life of ONOO$^-$/ONOOH at pH 7.4, 37 °C is ~0.7 s in the absence of other reactants (Radi et al., 2000). This half-life is important because it enables us to assess the importance of alternative pathways for ONOO$^-$ decomposition. In particular, carbon dioxide reacts with ONOO$^-$ sufficiently fast that under physiological levels of carbon dioxide, reaction 10 to form an unstable intermediate (ONOOCO$_2^-$), has a half-life of only ~0.1 s (Lymar and Hurst, 1995), and is thus the major route for ONOO$^-$ disappearance *in vivo*. Whether experimental models accurately mimic ONOO$^-$ decay pathways in biology thus depends, for example, on whether bicarbonate/carbon dioxide buffer systems are used.

The importance of carbon dioxide as a new (or at least, unexpected) reaction partner in free radical biology is that the carbon dioxide/ONOO⁻ adduct formed in reaction 10 is very short-lived (<3 ms; Lymar *et al.*, 1996), and a substantial fraction decomposes (reaction 11) to yield $NO_2^{\bullet}$ and the carbonate radical, $^{\bullet}CO_3^-$, recently described (Augusto *et al.*, 2002) as 'two emerging radicals in biology'. About two-thirds of the adduct $ONOOCO_2^-$ decomposes to NO_3^- (nitrate) and carbon dioxide (reaction 12). Carbonate radical was previously only considered as a biologically important species via reaction of $^{\bullet}OH$ with bicarbonate/carbonate in competition with many other, more reactive cellular constituents; however, the reactivity of $^{\bullet}CO_3^-$ has been quite well established using radiation and photochemical methods (Ross *et al.*, 1998). It is often a rather more reactive one-electron oxidant than $NO_2^{\bullet}$ (Ross *et al.*, 1998; Augusto *et al.*, 2002).

$NO_2^{\bullet}$ can be produced via $NO^{\bullet}$ not involving the $^{\bullet}O_2^-$ pathway to ONOO⁻, reaction 6. In aerobic systems, reaction between $NO^{\bullet}$ and oxygen very probably involves $NO_2^{\bullet}$ as an intermediate, but the reaction is complex (see Wardman, 1998), and the rate is proportional to the *square* of the NO$^{\bullet}$ concentration as well as the oxygen concentration. Thus, formation of $NO_2^{\bullet}$ via reaction 13 is far less likely when the steady-state concentrations of $NO^{\bullet}$ are sub-micromolar and at physiological oxygen tensions than in many laboratory experiments where the steady-state $NO^{\bullet}$ concentrations might rise to tens of micromolar and the oxygen tension is around an order of magnitude higher than in tissues. The corollary is that numerous laboratory studies that have utilized convenient chemical sources of $NO^{\bullet}$, such as the more rapidly decomposing 'NONOates' (Fitzhugh and Keefer, 2000), in solutions equilibrated with atmospheric air (Schmidt *et al.*, 1997) may well be misinterpreted because of the generation of significant quantities of $NO_2^{\bullet}$ and the consequent depletion of cellular thiols and other antioxidants, as well as much more protein nitration than might arise via the $^{\bullet}O_2^-/NO^{\bullet}$ pathway (Wardman, 1998). A further consequence of working with nonphysiological concentrations of both $NO^{\bullet}$ and oxygen is that the resulting nonphysiological $NO_2^{\bullet}$ concentrations can lead to N_2O_3 formation by equilibrium 14. The latter is a potent nitrosating agent (Williams, 1988), but whether it is formed to a significant extent in biology, except perhaps in hydrophobic regions of protein- or lipid-rich environments (e.g. membranes), has been questioned (Ford *et al.*, 2002a). Both $NO^{\bullet}$ and $NO_2^{\bullet}$ partition from aqueous to lipid-rich compartments, so reactions 13 and 14 are favored in lipophilic regions (Liu *et al.*, 1998), and N_2O_3 formation is plausible at the membrane/cytosol interface (Ramachandran *et al.*, 2001). Much the same considerations apply to extrapolating results from experiments involving the bolus addition of ONOO⁻. Williams (1997) concluded that 'there is no experimental evidence... that

ONOOH or ONOO⁻ can act as direct electrophilic nitrosating agents', pointing to N_2O_3 (via NO•+NO_2•, as above) as 'a more likely possibility'. However, direct reaction of ONOO⁻ with some cellular targets is possible (Arteel *et al.*, 1999; Crow, 2000; see below).

NO_2• can be formed in a pathway involving neither ONOO⁻ as an intermediate nor atmospheric oxidation. Nitrite (NO_2•) is a substrate for peroxidases, especially myeloperoxidase (Burner *et al.*, 2000), and protein nitration via NO_2• generated from the peroxidase-catalyzed oxidation of NO_2^- by H_2O_2 as cofactor is another important pathway (Eiserich *et al.*, 1988; Baldus *et al.*, 2002) (Figure 1, reaction 15). Spatial mapping showed that myeloperoxidase immunoreactivity strongly colocalized with nitrotyrosine formation (Baldus *et al.*, 2002). Peroxynitrite and myeloperoxidase 'leave the same footprint in protein nitration' (Kettle *et al.*, 1997). Kinetic arguments show that this is much more likely to involve NO_2• generated from NO_2^- by myeloperoxidase than reaction with HOCl (Kettle *et al.*, 1997).

Figure 1 is not exhaustive in its mapping of reaction pathways: radical–radical reactions are often facile, and several of relevance have been characterized (Ross *et al.*, 1998). Thus, NO_2•, as well as NO•, is reactive toward •O_2^-, producing peroxynitrate (ONOOO⁻) rather than peroxynitrite (ONOO⁻). However, radical–radical reactions involving NO_2• and •OH, for example, are unlikely to be of biological importance because of the extremely low steady-state concentrations of such reactive, and therefore short-lived, radicals. Figure 1 also does not show the routes of formation of •O_2^- and NO•, which are outlined briefly below in considering cellular homeostasis of these species.

PROTEIN NITRATION: 'HOW THE RADICALS DO THE JOB' (GOLDSTEIN *ET AL.*, 2000)

Tyrosine is a common target of nitrosative stress: 3-nitrotyrosine residues have been detected in numerous human diseases and animal models of disease and after cell irradiation (Ischiropoulos, 1998; Leach *et al.*, 2002). Figure 2 illustrates the key features of how 3-nitrotyrosine can be formed via radical pathways (Lymar *et al.*, 1996; Goldstein *et al.*, 2000; Hodges *et al.*, 2000; Lymar, 2001; Radi *et al.*, 2001). As with the use of fluorescent probes for ROS/RNS (see below), it is often overlooked that even though a target or probe eventually produces a stable ('spin paired') molecule from a radical, the reaction must involve a radical intermediate. It is equally important to recognize that the first step – the formation of a phenoxyl or tyrosyl radical – can be accomplished by diverse one-electron oxidants. These may be simple free radicals, as shown, but radicals can be formed via other routes during enzyme catalysis. Tyrosyl radicals are long lived in proteins because of reduced capacity for radical–radical reactions; the radical in ribonucleotide reductase is stable for days (Hoganson *et al.*, 1996). The enzyme prostaglandin H synthase-2 forms a tyrosyl radical during catalysis of prostaglandin formation (Gunther *et al.*, 2002). Reaction of this tyrosyl radical with NO$^\bullet$ leads to tyrosine nitration but does not involve formation of the tyrosyl radical by $^\bullet$OH, NO$_2^\bullet$ or $^\bullet$CO$_3^-$.

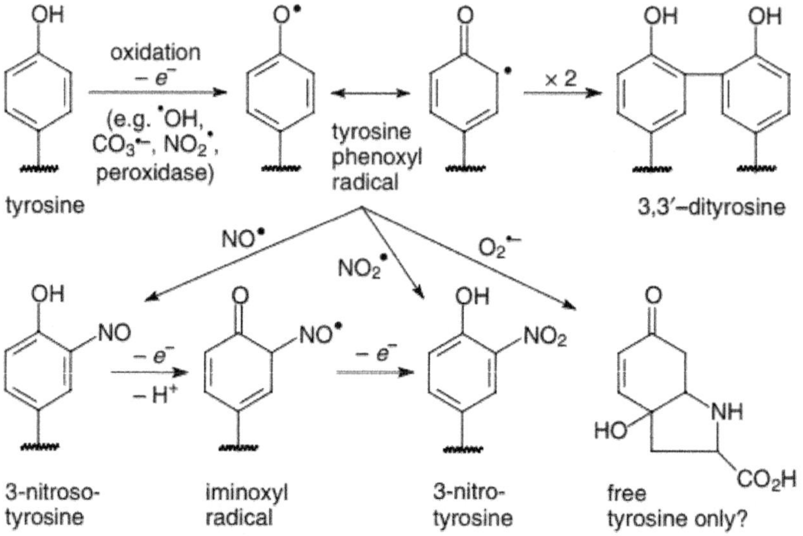

Figure 2: Reaction pathways in tyrosine oxidation/nitration.

In Figure 2, we see that 3-nitrotyrosine can be generated from the phenoxyl radical by either NO$_2^\bullet$ or NO$^\bullet$. The latter pathway involves the intermediate formation of nitrosotyrosine, which is oxidized (e.g. by peroxidase compound I or II intermediates) to an iminoxyl radical and thence to nitrotyrosine (Gunther *et al.*, 2002). Rate constants for reaction of both NO$^\bullet$ and NO$_2^\bullet$ with the tyrosine phenoxyl radical are equally high (Prütz *et al.*, 1985; Eiserich *et al.*, 1995). Generating NO$^\bullet$ and $^\bullet$O$_2^-$ simultaneously in varying ratios or adding ONOO$^-$ to tyrosine yields nitrotyrosine and dityrosine but "all the experimental results can be explained in terms of free radical chemistry" (Hodges *et al.*, 2000), that is, without direct reaction of tyrosine with ONOO$^-$ or its CO$_2$ adduct.

A key factor in the enhancement of tyrosine nitration by CO$_2$ is the ~100-fold higher reactivity of the carbonate ($^\bullet$CO$_3^-$) radical (compared with NO$_2^\bullet$) toward tyrosine at physiological pH (Ross *et al.*, 1998). Thus, the first step in the two-stage nitration process is accomplished faster in the presence of

CO_2 via $^\bullet CO_3^-$ production from $^\bullet O_2^-/NO^\bullet/CO_2$ (Figure 1, reactions 6, 10, 11). Too high a flux of $^\bullet O_2^-$ can diminish tyrosine nitration in model experiments, because the phenoxyl radical reacts rapidly with $^\bullet O_2^-$ (Jin et al., 1993). The rate constant for reaction of $^\bullet O_2^-$ with $NO_2^\bullet$ to form peroxynitrate (ONOOO-) is also high (Ross et al., 1998), but the steady-state concentration of the latter is too low in biology for this radical–radical reaction to be important. It may be a feature in model experiments with oxidation of leuco dyes (Jourd'heuil et al., 2001) or vicinal diamine probes for $NO^\bullet$ (Espey et al., 2002) (see below) when similar considerations apply.

In Figure 2, two decay pathways for tyrosyl radicals not involving $NO^\bullet$ or $NO_2^\bullet$ are also shown. These can be viewed as arising from the resonance form of the radical shown to the right of the phenoxyl radical. It might be expected that radical–radical coupling to form fluorescent dityrosine, while facile with free tyrosine (Prütz et al., 1983), is less likely to occur in proteins. However, protein–protein crosslinks have been observed via radical-induced tyrosine oxidation (Hashimoto et al., 1982). Whether the product shown in Figure 2, arising from the fast reaction between tyrosyl and $^\bullet O_2^-$ radicals (Jin et al., 1993), is formed with tyrosine residues in proteins is not known.

PROBLEMS IN THE USE OF FLUORESCENT PROBES FOR REACTIVE OXYGEN/NITROGEN SPECIES

Probably, thousands of studies have used molecular probes for reactive intermediates in oxidative and nitrosative stress, probes that typically show weak fluorescence in their reduced form but fluoresce strongly on oxidation. While of undoubted value, the interpretation of such experiments must be approached with caution. Figure 3 presents simplified schemes for formation of the oxidized forms of two common types of probe, as well as the basis for several probes reactive toward $NO^\bullet$ (or at least, reactive once activated). Reduced (dihydro) fluoresceins, or fluorescins, are the most common probes, especially the dichlorinated form, $DCFH_2$ (Ischiropoulos et al., 1999). As are most phenols, they are rapidly oxidized by $^\bullet OH$, $NO_2^\bullet$ and $^\bullet CO_3^-$ radicals (Wardman et al., 2002), forming phenoxyl radicals that may disproportionate to form an oxidized quinoid structure and parent dye. The extensive resonance in the quinoid structure (not shown) results in strong fluorescence. Obviously, $DCFH_2$ or similar molecules are good probes for *nonspecific* free radical oxidants.

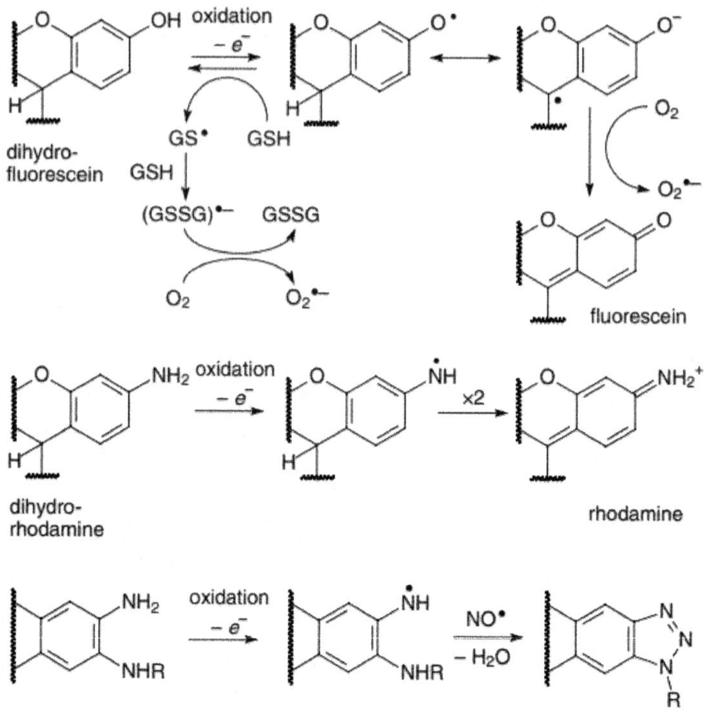

Figure 3: Chemical pathways in the detection of reactive oxygen and nitrogen species using common fluorescent probes. Dihydrofluoresceins and dihydrorhodamines are general probes for oxidizing species; the intermediate free radicals are thiol reactive and can initiate a chain reaction generating further $^{\bullet}O_2^-$ as shown. Vicinal diamines are probes for nitric oxide, but require prior oxidation to a reactive intermediate

However, there are major problems with their use. The first is that the oxidation is very slow with H_2O_2, but fast in the presence of peroxidase and similar catalysts (Ischiropoulos et al., 1999). In this class, the most important is cytochrome c, which catalyzes the oxidation of $DCFH_2$ at concentrations as low as a few nanomolar in a model system (Burkitt and Wardman, 2001). At a constant flux of $^{\bullet}O_2^-$ radicals from xanthine/xanthine oxidase, the fluorescence signal was proportional to the cytochrome c concentration in the 1–25 nm range; with 10 nm cytochrome c, fluorescence was almost invariant over an ~20-fold range of $^{\bullet}O_2^-$ production (Burkitt and Wardman, 2001). The importance of these observations is that cytochrome c is released from mitochondria to cytosol during apoptosis (Yang et al., 1997). Hence, apparent modulation of oxidative stress suggested by changes in fluorescence of $DCFH_2$ might reflect apoptotic pathways rather than real changes in ROS or

RNS. Another problem is the generation of $^{\bullet}O_2^-$ radicals, either by the radical intermediate reducing oxygen (Rota et al., 1999a), or the peroxidase-catalysed formation of a phenoxyl radical center on the remaining phenol group in the oxidized dye (not shown in Figure 3), reacting with glutathione (GSH) to form thiyl radicals (Rota et al., 1999b). The latter can reduce oxygen via the well-known 'redox switch' involving thiol conjugation as shown (Wardman, 1999). It is important to note that the reactions involving radicals with either oxygen or thiols need not be thermodynamically favored to proceed: removal of products from an unfavorable equilibrium can rapidly drive an overall reaction (Wilson et al., 1986). The oxidized, fluorescent dye can also be photo-reduced, a potential problem in measurement (Marchesi et al., 1999). While it might be regarded that the 'self-amplification' of signal resulting from dye radical chemistry is a bonus, quantitative interpretation is fraught with difficulty.

The same factors are probably a feature of the chemistry of the rhodamine dyes that are less well characterized (Figure 2). Whether the radicals disproportionate or react with oxygen to form the fluorescent, oxidized form will depend on kinetic and redox properties. Other difficulties that arise with these dyes include photo-oxidation. For dyes that are esterified and become sequestered upon action of cytoplasmic esterases, dye leakage from the cell and normalization of intracellular dye concentrations represent additional problems (e.g. Leach et al., 2002).

Peroxidase chemistry is also a factor in the use of vicinal diamines to identify $NO^{\bullet}$ production in biology. 1,2,-Phenylenediamine, 2,3-diaminonaphthalene and the newer diaminofluorescein dyes (e.g. DAF-2; Kojima et al., 1998) or their rhodamine analoges (Kojima et al., 2001) all require prior activation by nonspecific oxidants before reaction with $NO^{\bullet}$. Thus, all these share a common reaction scheme in which a fluorescent triazole is eventually formed from $NO^{\bullet}$, but requires an obligate anilyl radical intermediate (Figure 3) (Uppu and Pryor, 1999). This intermediate can obviously be formed from a wide variety of radical and enzyme-based oxidants, the levels of which will modulate the signal (Jourd'heuil, 2002). With $NO^{\bullet}$ in aerobic environments, aerobic oxidation of $NO^{\bullet}$ to $NO_2^{\bullet}$ will provide such an oxidant. However, relying on this for quantitative measurements in biological systems introduces uncertainties because the levels of competing reactants for $NO_2^{\bullet}$, particularly thiols, urate and ascorbate, might not be constant (Ford et al., 2002a).

It is noteworthy that other probes suffer from different problems: the EPR-detectable probe carboxy-'PTIO' (an imidazoline-3-oxide-1-oxyl derivative) reacts with $NO^{\bullet}$, but it is thought that $NO_2^{\bullet}$ is produced (Hogg et al., 1995; Pfeiffer et al., 1997), which oxidizes cellular thiols in microseconds (Ford et al., 2002a). This probe was used in a study of an $NO^{\bullet}$-mediated bystander response (Matsumoto *et al.*, 2001). In summary, the different chemical probes used to detect ROS/RNS each have individual problems that make interpretation of results obtained with their use potentially ambiguous. Until better, more specific probes are developed, other correlative approaches must be applied to more definitively identify specific ROS/RNS. For example, measurements of protein nitrotyrosine and NOS activities can be used to distinguish between RNS from ROS formation.

LIFETIMES OF RADICALS AND DIFFUSION DISTANCES

An estimate for the typical diffusion distance of a species can be made if the lifetime of the species can be calculated. The root-mean-square diffusion distance x is related to the lifetime t and the diffusion coefficient D by the Einstein equation in three-dimensional space: $x=(6Dt)^{1/2}$ (the average path length can be alternatively calculated as $(16Dt/\pi)^{1/2}$ (Roots and Okada, 1975), numerically similar). Diffusion coefficients are dependent on viscosity and will generally be smaller in a biological matrix than in water, but for a small radical a value of D around 1×10^{-9} m^2 s^{-1} is reasonable (Ford *et al.*, 2002a). With this value, lifetimes of ~1 μs, 1 ms and 1 s correspond to diffusion distances of ~0.1, 2 and 80 μm, respectively.

In calculating lifetimes of species, *all* the reactions that it undergoes must be included. In most biological reactions, the concentrations of biological reactants are usually in large excess over that of the reacting radical, and the kinetics are closely exponential or first order in the concentration of the reagent in excess. The half-life then equals the natural log 2 divided by the sum of the products of rate constant and concentrations for *all* species the radical is reacting with. In this way, it was suggested that the lifetime of an $^{\bullet}$OH radical in a cell was a few nanoseconds and its diffusion distance therefore a few nanometers (Roots and Okada, 1975).

CELL SIGNALING BY REACTIVE OXYGEN/NITROGEN SPECIES: A TWO-STEP HYPOTHESIS

The parallels drawn in the two-step nitration mechanism for nitrotyrosine formation, and the description above of problems with interpretation of experiments involving chemical probes for ROS/RNS, are intentional and instructive. They offer the foundation for a general, mechanism-led hypothesis of cell signaling involving radicals. In both scenarios, the first step was a nonspecific oxidative reaction, often involving highly reactive radicals with short diffusion distances, to produce a radical that in proteins is quite long-lived. (The diffusion distance of NO$_2^{\bullet}$ in the cytosol was recently estimated to be ~0.2 μm (Ford et al., 2002a), and similar calculations for $^\bullet CO_3^-$ indicate about half this value because of its ~ four-fold higher reactivity toward thiols compared with $NO_2^\bullet$.) The second step involves reaction either with radicals that are not very reactive (e.g. $^\bullet O_2^-$ or $NO^\bullet$) or with reactive radicals that are derived from species that act as carriers and so confer a much larger diffusion distance. Thus, S-nitrosothiols are storage forms of $NO^\bullet$, transporting $NO^\bullet$ by trans-nitrosation via thiols (Hogg, 2000). Transfer of $NO^\bullet$ from extracellular S-nitrosothiols to the cytosol may involve basically similar chemistry (Ramachandran et al., 2001), with N_2O_3 as a nitrosating agent at the membrane/cytosol interface. S-nitrosothiols have low reactivity toward $^\bullet O_2^-$ (Ford et al., 2002b), but can release $NO^\bullet$ in tens of seconds in the presence of micromolar levels of Cu(II) under conditions where oxidized GSH (GSSG) does not stabilize the redox metal (Noble and Williams, 2000). It is important to note that there is less than one free copper ion per cell (Rae et al., 1999). Interaction between S-nitrosothiols and thiols has been claimed to generate nitroxyl (HNO/NO⁻) (Wong et al., 1998), since N_2O is formed. The latter product is not an unequivocal indicator for nitroxyl, since it might also be produced by metal-catalyzed formation of $NO^\bullet$ from nitrosothiols and by the reaction of $NO^\bullet$ with thiols releasing N_2O via hyponitrous acid (Hogg et al., 1996).

The formation of S-nitrosothiols such as S-nitrosoglutathione (GSNO) may well be linked into radical damage from ROS and $NO^\bullet$, because a credible route to GSNO formation is the (probably) diffusion-controlled reaction of thiyl radicals ($GS^\bullet$) with $NO^\bullet$. Nonspecific 'repair' of radical damage from ROS is likely to largely involve thiols and the production of thiyl radicals; it is easily calculated from the known equilibrium constants of conjugation of $GS^\bullet$ with thiolate (GS⁻) and oxygen such that an appreciable fraction of $GS^\bullet$ remains unconjugated at equilibrium at physiological oxygen tensions, thiol concentrations and pH (Wardman, 1995; 1999; Wardman and von Sonntag, 1995).

Figure 4 illustrates two aspects of this scenario. Firstly, nonspecific radical damage, for example, from radicals such as $^\bullet OH$ with extremely short diffusion distances, can in effect be 'transported' to other sites via regeneration of $^\bullet O_2^-$, involving the thiyl/disulfide radical ion 'redox switch'. (Some carbon-centered radicals, such as those from carbohydrates, can generate $^\bullet O_2^-$ more directly by addition of oxygen and elimination of $^\bullet O_2^-$ (von Sonntag and Schuchmann, 1997)). Superoxide is rather unreactive and its diffusion distance is controlled largely by SOD and steady-state $NO^\bullet$ levels. While the regeneration of $^\bullet O_2^-$ at a site distant from the original radical damage is not strictly an amplification reaction, it effectively increases the diffusion distance or reaction radius and facilitates possible reaction with a greater range of targets. However, it is difficult to quantitate the increased diffusion distances that might be associated with this mechanism. The lifetime of the disulfide radical-anion at physiological oxygen levels is about 30–40 μs, corresponding to a diffusion distance of around 0.5 μm. The diffusion distance of $^\bullet O_2^-$ will be of the same order in the presence of several micromolar SOD, but that of H_2O_2 could be much greater, since estimates of the turnover time of H_2O_2 in cells range from about 3 to 200 ms (Antunes and Cadenas, 2000). These authors have also noted that the steady-state concentrations of H_2O_2 and $^\bullet O_2^-$ are estimated to be in the ratio ~ 100 : 1 in the mitochondrial matrix but around 1000 : 1 in the cytosol.

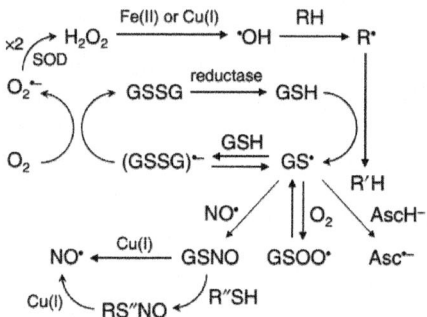

Figure 4: Outline of pathways involved in 'transporting' radical damage via regeneration of $^\bullet O_2^-$ or formation of GSNO, both involving thiyl radicals. 'Repair' of radical damage in the biomolecule RH may not be restituted (i.e. forming R'H) if stereochemical identity is lost in the radical. Radical 'sinks' such as ascorbate (AscH$^-$) or urate may modify pathways.

A second mechanism for radical-driven damage 'migration' is observed under pathological conditions or where local NO$^\bullet$ concentrations can exceed micromolar levels. A fraction of thiyl radicals can conceivably be trapped by NO$^\bullet$ and thus generate the much longer-lived 'transporter' GSNO, provided that the rate constant for reaction between GS$^\bullet$ and NO$^\bullet$ is about 10-fold higher than reaction between GS$^\bullet$ and GS$^-$. The extent of GSNO formation from 'repair' of nonspecific radical damage by GSH in the presence of NO$^\bullet$ will, however, be very dependent on other scavengers for thiyl radicals, particularly urate and ascorbate (Wardman, 1999); the latter is generally absent in cultured cells but not, of course, *in vivo*. An illustration of the expected modifying effects of ascorbate is provided in the case of interaction of the tyrosyl radical with GSH (Sturgeon *et al.*, 1998). A third mechanism for radiation-induced, radical-driven signal propagation, permissive for longer distances including *inter*cellular signaling, is described in more detail below.

This too entails a two-step process with an initial nonspecific oxidative event triggering the second but specific redox sensitive step involving propagation of a reversible mitochondrial permeability transition from one mitochondrion to adjacent mitochondria (e.g. Romashko *et al.*, 1998; Zorov *et al.*, 2000; Leach *et al.*, 2001, 2002). This propagation ultimately results in the Ca^{2+}-dependent activation of constitutive NOS and $NO^{\bullet}$ signaling (Leach *et al.*, 2002).

While much work is needed to establish these concepts on a firmer quantitative basis, the main chemical pathways by which radical damage can be relocated are evident. However, signaling mechanisms are equally likely to involve control via damage amplification or by modulating a specific site. These possibilities are considered below.

POSSIBLE MECHANISMS FOR DAMAGE AMPLIFICATION OR CONTROL INVOLVING RADICALS

The molecular mechanisms for true damage *amplification* are less well founded than for damage migration, except for the well-known lipid peroxidation chain reaction (Halliwell and Gutteridge, 1999) and the analogous reactions with protein (e.g. histone) hydroperoxides (Luxford *et al.*, 2000), and short chain reactions observed in thiol oxidation (Lal *et al.*, 1997) or on eliminating $^{\bullet}O_2^-$ from peroxyl radicals formed from oxygen adding to some carbon-centered radicals (von Sonntag and Schuchmann, 1997). The total damage initially produced in a cell by radiation, calculated on the basis of the ionization yields and by considering track recombination and high scavenging capacity, is around 1 μm/Gy. For comparison, mitochondrial respiration alone generates hundreds of millimoles of $^{\bullet}O_2^-$ per day in a human, that is, of the order of 50 nm/s (Cadenas and Davies, 2000). Thus, a radiation dose of 1 Gy initially generates, at most, as much $^{\bullet}O_2^-$ as humans produce in around 20 s in the same volume, averaged over the cell. The comparison is even more marked considering the rate of production of $^{\bullet}O_2^-$ in mitochondria, which has been estimated to be ~ 0.6 μm/s averaged over the mitochondrial volume (Cadenas and Davies, 2000).

In order to assess the possibility of which sites for free radical damage might function in a controlling/signaling pathway, it is useful to consider the possible general targets for radical reactions. The same logic is used to evaluate whether ONOO⁻ (or its conjugate acid) reacts directly with a target or whether the effects occur via decomposition of ONOO⁻ into radicals by either of the pathways shown in Figure 1. The decomposition of ONOO⁻ at pH 7.4, 37 °C has a first-order rate constant (=ln 2/(half-life)) of about 1 s⁻¹ in the absence of CO_2. At physiological levels of CO_2 *in vivo* (e.g., in plasma), its half-life is ~ 9 ms (rate constant ~75 s⁻¹; Denicola *et al.*, 1996). In Table 1, second-order rate constants are listed for reactions of some radicals and ONOO⁻, with a range of potential physiological targets. The likelihood of significant reaction with a target can be assessed by comparing the product of rate constant and concentration with the first-order rate constants for decomposition and/ or reaction with CO_2, as appropriate. Such calculations have been reported (e.g. Alvarez *et al.*, 1999; Arteel *et al.*, 1999). Obviously, the conclusions are subject to considerable uncertainty because it is difficult to extrapolate measurements of average concentrations to likely values in cellular organelles. Some recent studies specifically consider ONOO⁻ reactions in mitochondria (e.g. Radi *et al.*, 2002).

Table 1: Approximate rate constants (m^{-1} s^{-1}) for reactions of some species involved in oxidative and nitrosative stress, pH 7.4, ~25°C

GSH	9×10^0	2×10^3	2×10^7	5×10^6	6×10^3
Cysteine	2×10^{10}		5×10^7	5×10^7	6×10^3
Ascorbate	1×10^8	5×10^4	4×10^7	1×10^{9b}	5×10^1
Urate	7×10^9		2×10^5		5×10^{2c}
Tyrosine	1×10^{10}	$<1 \times 10^4$	3×10^5	5×10^7	
Tryptophan	1×10^{10}	$<2 \times 10^4$	$<5 \times 10^5$	7×10^6	4×10^{1c}
Methionine	9×10^6	<1		1×10^{9b}	2×10^3
NADH	2×10^{10}	1×10^3	4×10^5	1×10^9	
Tetrahydrobiopterin	9×10^9		1×10^6	5×10^9	
Aconitase		3×10^7			1×10^5
Cytochrome *c* (Fe(II))	$>1 \times 10^{10}$	1×10^6			3×10^4
Mn-SOD	2×10^9				1×10^5
CO_2					3×10^4

ᵃMost rate constants are from the NIST database (Ross *et al.*, 1998) or references mentioned in the text (Radi *et al.*, 2000; Ford *et al.*, 2002a; Kirsch *et al.*, 2002; Patel *et al.*, 2002; Radi *et al.*, 2002). ᵇAt pH 11.4. ᶜAt 37°C

Particularly noteworthy are the high rate constants for reaction of ONOO⁻ with some proteins containing metal-thiolate clusters, compared with reaction with low-molecular-weight thiols. Complexing thiols with metals, particularly iron and zinc, appears to change the redox properties of the thiol moiety, making it more susceptible to oxidation. Thus, the Fe^{2+}/cysteine complex reduces oxygen to •O_2^- rapidly; zinc competes effectively for binding, but reduces •O_2^- formation (Willson, 1977). Typical of iron/sulfur proteins is aconitase (Castro *et al.*, 1994), which is also highly reactive toward •O_2^- (Hausladen and Fridovich, 1994), releasing iron, and also toward ONOO⁻ (Castro *et al.*, 1994). Zinc-thiolate centers are also very reactive toward ONOO⁻, with a rate constant of ~3×10^5 m^{-1} s^{-1} measured in yeast alcohol dehydrogenase (Crow *et al.*, 1995). The low-molecular-weight complex of zinc with 2,3-dimercaptopropanol reacted with ONOO⁻ with a rate constant of ~4×10^4 m^{-1} s^{-1} (Crow, 2000). The importance of zinc-thiolate complexes as targets is that such a center stabilizes dimeric NOS, and ONOO⁻ treatment releases zinc from the endothelial isoform (NOS-3), disrupting NO• synthesis in favor of •O_2^- generation (Zou *et al.*, 2002). Tetrahydrobiopterin (H_4B) is an essential cofactor for all NOS isoforms and is extremely reactive toward NO_2• and •CO_3^- radicals (Patel *et al.*, 2002)

(Table 1). The biopterin radical formed on one-electron oxidation is likely to be repaired by ascorbate *in vivo* (Table 1). While both GSH and ascorbate are protective toward oxidation of H_4B by $NO_2^\bullet$, because the high concentrations of these antioxidants outweigh the lower reactivity toward $NO_2^\bullet$ compared with H_4B, thiol scavenging may not be protective overall because the thiyl radical $GS^\bullet$ oxidizes H_4B very rapidly (Patel *et al.*, 2002). As with the zinc-thiolate complexes, H_4B is important in controlling the relative efficiencies of NOS as either $NO^\bullet$ or $ONOO^-$ generators (Stuehr *et al.*, 2001). Depleting H_4B switches NOS to a 'peroxynitrite synthase' by stimulating more $^\bullet O_2^-$ formation. In turn, this is a true amplification mechanism because the more powerful oxidants thus produced ($NO_2^\bullet$, $^\bullet CO_3^-$, $^\bullet OH$) further deplete H_4B and drive a self-amplifying cascade.

In addition to zinc-thiolate complexes and H_4B as important targets in nitrosative stress, the regulatory role of $NO^\bullet$ and cytochrome *c* oxidase in extending the perfusion distance of oxygen (Brookes *et al.*, 2002) is of note. Peroxynitrite reactions in mitochondria (Radi *et al.*, 2002) and activation of stress response pathways (Klotz *et al.*, 2002) have been recently reviewed. It was concluded (Radi *et al.*, 2002) that cytochrome *c* oxidase (complex IV)-dependent respiration is not affected by $ONOO^-$, but release of mitochondrial cytochrome *c* following exposure to elevated ROS/RNS has been reported and linked to nitration of a critical tyrosine residue. One possible mechanism suggested by Radi *et al.* (2002) is $ONOO^-$-stimulated nitration of the tyrosine-34 site of mitochondrial Mn-SOD. Inactivating SOD by nitration increases the lifetime of $^\bullet O_2^-$ radicals and, hence, further $ONOO^-$ formation from $NO^\bullet$ in a 'positive feedback cycle'. Nitration of Mn-SOD is observed following exposure of cells to low doses of radiation as well, and this target may be of central importance in amplifying the radiation response as discussed below (Leach *et al.*, 2002).

In conclusion, the involvement of nitrosative stress in modulating the function of NOS can be linked to reactions between reactive oxidants and both zinc-thiolate complexes and H_4B. In the wider context, other damage amplification reactions include the release of Fenton-catalytic iron from aconitase by $^\bullet O_2^-$. As discussed below, zinc-thiolate centers are also important in a number of redox-sensitive transcription factors, including those activated by ionizing radiation and that are involved in generating and amplifying cytoprotective responses.

CELLULAR MECHANISMS OF ROS HOMEOSTASIS

ROS free radicals and their reactive nonradical derivatives, such as H_2O_2, are present in all cells at very low concentrations established by a balance of generating and neutralizing processes. Disruption of this balance by adding oxidants (e.g. ionizing radiation) or antioxidants triggers redox signaling and induction of cellular protective mechanisms that re-establish the initial redox/free radical balance (Dröge, 2001).

As discussed above, most cellular ROS in nonphagocytic cells is produced by endogenous mitochondrial electron transport. Another important source of $^{\bullet}O_2^-$ is the plasma membrane NADPH oxidase, which has been extensively studied in phagocytic cells where it plays a prominent role in antimicrobial and tumoricidal oxidative burst. NADPH oxidase isoforms have also been described in nonphagocytic cells. These latter isoforms produce considerably less $^{\bullet}O_2^-$ than the enzyme found in macrophages and neutrophils and that generated by mitochondrial respiration. The roles of the nonphagocytic NADPH oxidase are not completely understood, but appear to be tissue dependent (Dröge, 2001). For example, vascular smooth muscle cells appear to generate intracellular $^{\bullet}O_2^-$ in contrast to extracellular generation in endothelial cells, fibroblasts and neutrophils. NADPH oxidase is an FAD-requiring enzyme. Inhibition of $^{\bullet}O_2^-$ production by the FAD analog, diphenyliodinium, is often taken as evidence for NADPH oxidase-dependent $^{\bullet}O_2^-$ generation. However, other cellular sources of ROS/RNS are also FAD-dependent and sensitive to diphenyliodinium, for example, mitochondrial electron transport (Majander et al., 1994).

Cellular SOD converts the $^{\bullet}O_2^-$ anion to H_2O_2 which is then further reduced to water by catalases and peroxidases in the cytoplasm and in the mitochondria by an NADPH-dependent glutathione peroxidase. As discussed above, NO$^{\bullet}$ also modulates cellular ROS levels. When produced at sufficiently high levels by NOS, NO$^{\bullet}$ can react with $^{\bullet}O_2^-$ to form ONOO$^-$. ONOO$^-$ mostly rearranges to relatively innocuous nitrite/nitrate, converted to such by cytochrome c oxidase, or reacts with GSH to form GSNO. The latter represents a potential storage form of NO$^{\bullet}$. Thus, at modest NO$^{\bullet}$ levels, its reaction with and neutralization of $^{\bullet}O_2^-$ represents a cytoprotective mechanism (Wink et al., 1995; Beckman and Koppenol, 1996).

Other studies have suggested that endogenously produced NO$^{\bullet}$ protects cells by inhibiting cytochrome c oxidase, mitochondrial electron transport and thus endogenous $^{\bullet}O_2^-$ generation (e.g. Kanai et al., 2001; Paxinou et al., 2001;Sarkela et al., 2001; Elfering et al., 2002). However, recent careful measurements of concentration-response curves for mitochondrial respiration in brain slices bring into question the role of cytochrome c oxidase. At least in this tissue the amount of NO$^{\bullet}$ generated under physiological conditions with maximal stimulation of NOS isoform 1 (NOS-1) activity was found to be too low to inhibit mitochondrial respiration but sufficient to fully activate soluble guanylate cyclase (Bellamy *et al.*, 2002). Although the exact mechanism remains to be defined, other evidence also supports a mitochondria-dependent mechanism. Thus, endogenously generated NO$^\bullet$ protects cells against H_2O_2 exposure only when cells are actively respiring (Paxinou *et al.*, 2001). The importance of NO$^\bullet$ in regulating cellular ROS levels is underlined by the recent identification of a Ca^{2+}-activated NOS isoform in the mitochondria of cells (Kanai *et al.*, 2001; Elfering *et al.*, 2002). Pharmacological inhibition of cellular NOS stimulates $^\bullet O_2^-$ production, and NO$^\bullet$ scavengers induce cellular oxidative stress (Niu *et al.*, 1996; Goda *et al.*, 1997; Janssen *et al.*, 1998a, 1998b).

When NO$^\bullet$ levels approach those of SOD, higher, cytotoxic levels of ONOO$^-$ are reached (Beckman and Koppenol, 1996). Cell-damaging effects are seen after treatment of cells with very high concentrations of NO$^\bullet$ donors or by expressing inducible isoforms of NOS that generate large amounts of NO$^\bullet$ (Szabo and Ohshima, 1997; Phoa and Epe, 2002).

CELLULAR MECHANISMS OF NO$^\bullet$ HOMEOSTASIS

In contrast to $^\bullet O_2^-$, which is mostly a by-product of inefficiency in mitochondrial electron transport, NO$^\bullet$ is generated by specific enzymes, the NO$^\bullet$ synthases (Stuehr, 1999). The activities of these enzymes are regulated at multiple transcriptional and post-transcriptional levels. NO$^\bullet$ is released during the catalytic conversion of arginine to citrulline in a complicated mechanism involving multiple cofactors, including H_4B, NADPH, FMN and an Fe–S center. Three types of NOS have been described with the basic catalytic mechanism conserved among the three (Stuehr, 1999). Both the neuronal synthase (NOS-1) and endothelial synthase (NOS-3) enzymes are named from the tissues in which they were first discovered but are constitutively expressed in many cell types. Although considered to be constitutively expressed, NOS-1 mRNA transcription is under a high level of control. The NOS-1 gene is made up of 27 exons and encodes for several different isoforms with molecular sizes ranging from 150 to 165 kDa. A specific isoform of NOS-1 has been identified in the mitochondria of liver and cardiomyocytes (Kanai *et al.*, 2001; Elfering *et al.*, 2002), and its mRNA transcript has been identified in most if not all tissues examined (Elfering *et al.*, 2002). The mitochondrial NOS-1 is myristoylated in contrast to other NOS-1 isoforms.

The NOS isoforms first identified in endothelial cells, NOS-3, are also myristoylated proteins of 130–135 kDa that, like NOS-1, require dimer formation for catalytic activity. NOS-3 is primarily located in the plasma membrane of cells. The requirement for dimer formation has been used to develop, for both NOS-1 and NOS-3, mutants that when expressed inhibit endogenous NOS activity (Lee *et al.*, 1995; Phung and Black, 1999). NOS-1 and NOS-3 produce relatively low amounts of NO$^\bullet$ compared with the third major isoform, inducible NOS (iNOS or NOS-2). NOS-2 activity is normally associated with macrophages but can be transcriptionally activated by different cytokines (e.g. interferon-γ) in a number of cell types (Janssen *et al.*, 1998a;Freeman and MacNaughton, 2000; Yoo *et al.*, 2000). High levels of NO$^\bullet$ and OONO$^-$ formation at inflammatory sites are associated with activation of NOS-2.

The expression of NOS-2 is the predominant factor regulating its cellular activity. This contrasts with the number of post-translational regulatory mechanisms for NOS-1 and NOS-3. The catalytic activities of both of these isoforms are tuned to changes in cellular [Ca^{2+}] by a Ca^{2+}/calmodulin-dependent mechanism. Since NOS-2 tightly binds calmodulin, its activity is largely independent of [Ca^{2+}]. As will be discussed below, a change in cytoplasmic [Ca^{2+}] is one process by which a signal initiated by ROS formation can be converted/translated into NO$^\bullet$-dependent signaling. Several recent studies have shown that NOS-1 and NOS-3 are also regulated by phosphorylation (Michel *et al.*, 1993; Cordelier *et al.*, 1999;Elfering *et al.*, 2002; Leach *et al.*, 2002). In some tissues, NOS-1 associates with adaptor proteins and small monomeric G proteins specific for the tissue. Thus, Dexras1, a G protein found predominantly in the brain, forms a ternary complex with NOS-1 and an adaptor protein (CAPON) and is activated when NOS-1 is activated (Fang *et al.*, 2000).

Breakdown of NO$^\bullet$ has been until recently thought to be due to the reaction of NO$^\bullet$ with oxyhemoglobin or oxymyoglobin producing nitrate and methemoglobin and metmyoglobin. The former reaction has been brought into question (Gow *et al.*, 1999), and it appears unlikely that the oxymyoglobin reaction is relevant at NO$^\bullet$ concentrations obtained under normal noninflammatory conditions. Recent studies (Pearce *et al.*, 2002) suggest that a reaction catalyzed by cytochrome *c* oxidase to produce nitrite is a potential catabolic pathway for NO$^\bullet$ at physiological levels of NO$^\bullet$ in mitochondria-rich tissues. In cells of low mitochondrial content, such as endothelial cells (and tumor cells?), a net result would be to facilitate diffusion of NO$^\bullet$ into the vasculature, minimizing its consumption in the tissue and thus facilitating activation of sGC and other NO$^\bullet$-activatable targets (Pearce *et al.*, 2002).

NO$^\bullet$ can also be produced by routes not involving NOS. In addition to the metal-catalyzed breakdown of *S*-nitrosothiols described above, production of NO$^\bullet$ in tissues subject to ischemia/reperfusion cycles can be demonstrated, involving reduction of nitrate/nitrite catalyzed by xanthine oxidase (Millar *et al.*, 1998; Li*et al.*, 2001).

GSH, SUBCELLULAR COMPARTMENTALIZATION AND THE REDOX STATUS OF CELLS

A key feature of eucaryotic cells in considering redox sensors and homeostasis is cellular compartmentalization. A measure of the redox status of different cellular compartments is the GSH : GSSG ratio. Under normal physiological conditions, the GSH : GSSG ratios of the different cellular compartments are: cytoplasm (>100), endoplasmic reticulum ($\approx$1–2), mitochondria ($\approx$5–10) (Hwang et al., 1992; Morales et al., 1998). There is evidence that the nuclear GSH : GSSG ratio can be different from that of the cytoplasm (Voehringer et al., 1998). Thus, in considering the response of cells to oxidative or nitrosative events, such as produced by radiation, the localization of sensing mechanisms in cellular compartments of differing reductive environments requires consideration.

A number of studies have shown that cells are radiosensitized when total cellular GSH levels drop below 20% (e.g. Meister, 1994 and references therein). Cells depleted of GSH continue to divide at normal rates and only when challenged by an additional oxidative stress such as exposure to ionizing radiation do they commence dying. One interpretation is that radiosensitization is only observed when GSH levels are reduced sufficiently to be unable to counter the amount of ROS produced after radiation. However, even with this extensive depletion, total cellular GSH concentration remains about 1 mm and only small amounts of ROS relative to that produced by metabolism are produced as a direct consequence of the radiation-induced ionization events (Ward, 1994). As a result, this appears to be an unlikely explanation. An alternative possibility follows from an observation by Meister (1994) that mitochondria do not synthesize their own GSH but depend on high-affinity transport and cytoplasmic GSH synthesis. Only when cytoplasmic GSH levels drop below 20%, do mitochondrial GSH levels decrease. The net effect of this decrease is to reduce mitochondrial GSH peroxidase activity, the primary mechanism of H_2O_2 detoxification in mitochondria, and thus increase endogenous levels of metabolically produced H_2O_2. Since we have shown that radiation also stimulates mitochondrial ROS/RNS generation (Leach et al., 2001), the combined increase due to activation of ROS/RNS generation and the elevated ROS generation due to endogenous metabolism may be sufficient to overwhelm the mitochondrial redox homeostasis properties of GSH-depleted cells. Additional evidence for this proposal will come from experiments where mitochondrial GSH is specifically targeted and in studies using ρ^o cells that are deficient in mitochondrial electron transport (King and Attardi, 1996; Leach et al., 2001). An important aspect of GSH metabolism that has not been considered in radiation biology is the GSNO pool. Tissue concentrations of GSNO are in the μm range compared to the nm levels for NO● (Gaston, 1999) and thus GSNO (and other S-nitrosothiols) represents a potential reservoir and buffer of cellular NO. As discussed above, GSNO formation may be linked to radical damage from ROS/RNS, since the repair of this damage involves thiols, the production of thiyl radicals and the probable diffusion-controlled reaction of thiyl radicals with NO●. Intracellular levels of GSNO are also tightly coupled with NOS activity and various GSNO-metabolizing enzymes. Thus, breakdown of GSNO by thioredoxin reductase or glutathione peroxidase results in the release of NO● (Gaston, 1999). Increases of 10-fold in cell S-nitrosothiols are indicative of nitrosative stress and are associated with induction of apoptosis (Eu *et al.*, 2000).

IONIZING RADIATION AND REDOX-SENSITIVE SIGNALING

Ward (1994) originally raised the question of how cells sense and react to the relatively low amounts of ROS produced by clinically relevant radiation doses with background ROS levels 100–1000-fold higher. This differential may be even greater with tumor tissues that in general show much higher rates of ROS generation than normal cells (e.g. Szatrowski and Nathan, 1991). Recent studies measuring ROS/RNS generation in irradiated cells have provided possible answers.

ROS/RNS production induced by radiation has been measured with fluorescent dyes (Clutton *et al.*, 1996; Narayanan *et al.*, 1997; Morales *et al.*, 1998). These studies (keeping in mind the caveats about fluorescent dyes discussed above) demonstrated a relatively early response above baseline metabolic levels at both high and low linear energy transfer radiations and within 30 min of radiation exposure. Measurements of ROS/RNS generated after irradiating cells reveal that much more ROS/RNS is produced than can be accounted for by both primary and secondary products of the ionization events, suggesting an amplification mechanism. Inhibition of the enhanced dye fluorescence with an FAD analog binding inhibitor suggested that alpha-particle radiation stimulated NADPH oxidase activity (Narayanan *et al.*, 1997).

More recent studies used digitized fluorescence microscopy to measure changes in ROS/RNS-sensitive dye fluorescence at the single cell level and a combined pharmacological and genetic approach to establish more definitively the source of the ROS/RNS generated after a radiation exposure (Leach *et al.*, 2001, 2002). A ^{90}Sr eye applicator was mounted on an inverted microscope to permit fluorescence image collection within seconds of commencing radiation exposure. Radiation-stimulated ROS/RNS generation occurred within seconds of starting radiation treatment (1–10 Gy) and persisted for 2–5 min postirradiation in several epithelial cell lines (Leach et al., 2001). The amount of ROS/RNS generated per cell was relatively constant over this radiation dose range, but increasing numbers of cells responded with enhanced ROS/RNS generation with dose escalation. These results are consistent with a threshold, possibly all-or-nothing, response. When plotted as a traditional semilog plot of responding cells versus radiation dose, a straight line extrapolating to 1 was obtained – consistent with a single target.

Cells depleted of mitochondrial DNA and thus deficient in mitochondrial electron transport did not demonstrate a radiation-stimulated increase in ROS/RNS. Furthermore, chelation of intracellular Ca^{2+} or incubating cells with cyclosporin A or bongkrekic acid also inhibited radiation-induced ROS/RNS. The latter two drugs target components of the pore composing the mitochondrial ion channel responsible for the Ca^{2+}-sensitive mitochondrial permeability transition (Ichas and Mazat, 1998). Based on these studies, it was proposed that radiation activated a *reversible* form of the permeability transition that propagated from one mitochondrion to adjacent mitochondria by a membrane depolarization Ca^{2+}-release mechanism with consequential release of ROS. Additional evidence for this proposal came from measuring radiation-induced mitochondrial membrane depolarization and enhanced permeability of the mitochondrial membrane to the fluorescent small molecule, calcein. This model fits with previous findings from other investigators demonstrating such propagating signals within a cell's mitochondrial pool and as a mechanism involved in Ca^{2+} homeostasis and cellular redox modulation (Bernardi and Petronelli, 1996; Vercesi et al., 1997; Ichas and Mazat, 1998; Romashko et al., 1998; Zorov et al., 2000). Specifically, Zorov et al. (2000) proposed that ROS-induced ROS release accompanies the mitochondrial permeability transition and provides a 'self-amplifying' mechanism by which redox signals can be transmitted throughout the cell.

At present, the mitochondrial sensor of the oxidative/ionization event that triggers the mitochondrial permeability transition is not known. A number of studies using sulfhydryl-reacting agents have emphasized the importance of mitochondrial protein thiols in regulating the permeability transition (Bernardi and Petronelli, 1996; Vercesi et al., 1997; Crompton, 1999). Zorov et al. (2000) have evidence for a critical role for mitochondrial thiols in the initiation and propagation of the mitochondrial permeability transition during ROS-induced ROS release in cardiac myocytes. A recent study (Pearce et al., 2001) demonstrated that radiation at relatively high doses inhibited respiratory complexes I (NADH dehoydrogenase) and III (cytochrome *c* reductase) but not complexes II (succinate dehydrogenase), IV (cytochrome *c* oxidase) and V

(ATP synthase). This is not, however, evidence for their roles as sensors, since inhibition was only consistently observed hours after irradiation.

Two features of this model for a redox response mechanism to radiation warrant further comment. Firstly, the model follows traditional radiobiological theory that emphasizes target volume. After the nucleus, the largest cellular target volume is the mitochondria volume representing between 4 and 30% of total cell volume depending on cell type. Since the signal can propagate from one mitochondrion to another, the permeability transition of only one mitochondrion is required for propagation and thus amplification of the initial oxidative event. Release of Ca^{2+} initiates signal propagation to adjacent mitochondria and potentially throughout the mitochondrial pool of a cell. This fits with the apparent all-or-nothing response that is observed. The mitochondrion is also the major source of cellular ROS. A number of past studies have also provided indirect evidence for a mitochondrial role in the cellular response to ionizing radiation (e.g. Gudz et al., 1994 and references therein).

Secondly, the radiation-induced burst of ROS/RNS generation is transient. This has important implications in terms of the final cellular response. If the radiation-stimulated ROS/RNS generation is not transient, one would predict on the basis of a continuous elevated production of ROS/RNS that the final cellular response would be cell death. On the other hand, transient responses suggest that regulatory, possibly cytoprotective, components are being engaged. A linkage between the radiation-stimulated ROS/RNS generation and an important cytoprotective response has been established as described below.

Subsequent studies demonstrated that a major contributor to the radiation-induced oxidation of $DCFH_2$ fluorescence signal was probably $ONOO^-$, observed after activation of NOS (Leach et al., 2002). Molecular analyses with expression of wild-type and a dominant negative mutant of NOS-1 and measurements of NOS catalytic activity with the arginine–citrulline assay confirmed that radiation transiently activated a Ca^{2+}-dependent NOS-1 in Chinese hamster ovary cells within the time frame observed for ROS/RNS generation. Radiation stimulated tyrosine nitration of a number of proteins, including MnSOD, by a mechanism inhibited by expression of the dominant negative mutant of NOS-1, consistent with formation of $ONOO^-$. These findings are also consistent with those from an earlier study with cardiac myocytes, in which a photochemically induced mitochondrial ROS release was measured with $DCFH_2$ (Zorov et al., 2000). $NO^{\bullet}$ was generated as measured with DAF-2, but apparently at significantly lower rates relative to the ROS burst. The radiation-induced MAPK1/2 activation thought to be a cytoprotective response to ionizing radiation and observed by several investigators (reviewed inSchmidt-Ullrich *et al.*, 2000) was found to be a consequence of the radiation-stimulated ROS/RNS generation (Leach *et al.*, 2001, 2002). The radiation-induced MAPK1/2 activation was dependent on NOS-1 activity, as shown by both genetic and pharmacological methods, but the actual mechanism by which this occurs remains undefined. MAPK activity is regulated positively by tyrosine phosphorylation via downstream signaling from growth factor receptors to RAS (a GTP/GDP exchange protein), RAF kinase and MEK1/2 dual function kinases that phosphorylate MAPK1/2 on threonine-202 and tyrosine-204. MAPK is negatively regulated by the activity of different protein tyrosine phosphatases that not only can act on MAPK but also on upstream components including growth factor receptors and RAF-1 kinase. The activities of all these proteins are potentially modulated by ROS/RNS, but preliminary studies suggest that S-nitrosylation and inhibition of a protein tyrosine phosphatase may be critical (see below).

BIOLOGICAL ROS/RNS SENSORS: TECHNICAL ISSUES

ROS/RNS concentrations and reaction rates were discussed in evaluating the relative significance of the different pathways in forming reactive oxidants from $^\bullet O_2^-$ and $NO^\bullet$ radicals (Figure 1). These same considerations apply in identifying physiologically relevant ROS/RNS sensors and signal transduction pathways activated by ROS/RNS. Of most importance for reasons discussed above, measurements of intracellular ROS/RNS concentrations and identification of the specific ROS/RNS involved are inherently difficult, if not impossible, with intact cells and presently available experimental techniques. The problem may be compounded by difficult-to-measure localized subcellular changes in ROS/RNS of sufficient magnitude to be of biological significance.

Much of the work on defining the roles of ROS or RNS as redox second messengers has relied on the use of relatively high concentrations of oxidants, usually H_2O_2, or $NO^\bullet$ donors, as initiators. The difficulties in using such an approach can be seen in studies on protein tyrosine phosphatases (e.g. Lee *et al.*, 1998; Meng *et al.*, 2002). Addition of high concentrations of H_2O_2 to cells clearly oxidizes the active site cysteine of the protein tyrosine phosphatase to sulfenic acid and inhibits catalytic activity. Whether this has anything to do with the *in vivo* situation, where much less H_2O_2 is metabolically generated, is less clear. In some experimental settings, catalase has been added as proof that H_2O_2 is the oxidizing agent. The significance of this control is uncertain since the added catalase is extracellular, whereas the action of H_2O_2 is presumably intracellular, although H_2O_2 gradients across biomembranes can be estimated (Antunes and Cadenas, 2000). Even elevation of intracellular catalase activity by genetic means is not definitive proof. Overexpression of catalase may simply shift the overall redox balance of the cell. One can also explain the H_2O_2-induced inhibition of protein tyrosine phosphatases by an RNS-based mechanism. In this scenario, H_2O_2 treatment or other oxidative stress stimulates the release of Ca^{2+} from intracellular stores with subsequent activation of Ca^{2+}-dependent NOS activity (Todd and Mikkelsen, 1994; Srivastava et al., 1999; Droge, 2001; Leach et al., 2002). $NO^\bullet$ or $ONOO^-$ also oxidizes protein cysteine to sulfenic acid (e.g. Stamler and Hausladen, 1998). S-thiolation is another mechanism that needs consideration (e.g., Eaton et al., 2002).

Besides the potential confusion over what ROS or RNS is involved, the biological response can vary with ROS and RNS. This is to be expected in signal transduction pathways with multiple inputs of variable oxidative or nitrosative sensitivities. The biphasic response of the transcription factor NF-$^\kappa$B to different $NO^\bullet$ concentrations exemplifies this (Sheffler et al., 1995).

TRANSCRIPTION FACTORS AS ROS/RNS SENSORS

ROS sensors in eucaryotic cells remain mostly undefined, but clues to their molecular nature come from what is known about bacterial ROS sensors. The best described are OxyR and SoxR as sensors for H_2O_2 and $^\bullet O_2^-$, respectively. The SoxR protein regulates the synthesis of several proteins at the sox locus involved in the protective response to $^\bullet O_2^-$. SoxR is a homodimer with two 2Fe–2S clusters, one in each subunit, that under normal conditions exist in the reduced state (Bauer et al., 1999). Fe oxidation appears critical for transcription initiation, but the underlying mechanisms remain mostly undefined (Hidalgo et al., 1997). Thus, $^\bullet O_2^-$ can oxidize SoxR but only at supraphysiological levels. Although H_2O_2 can oxidize SoxR Fe, it does so without stimulating biological activity. $NO^\bullet$ does not oxidize SoxR but $NO^\bullet$ modulates biological activity in vivo (Nunoshiba et al., 1995). As pointed out by Marshall et al. (2000), no eucaryotic homologs of SoxR have been described nor have eucaryotic transcription factors with redox-sensitive transition metals been identified.

H_2O_2 activates OxyR by oxidation of cysteine-199 to sulfenic acids which subsequently forms an intramolecular disulfide bond with cysteine-208 (Zheng et al., 1998; Zheng and Storz, 2000). This conversion of reduced to oxidized OxyR can also be achieved by changing the intracellular thiol : disulfide ratio. S-nitrosothiols activate the OxyR protein possibly by S-nitrosylation of cysteine-199 (Marshall et al., 2000; Zheng and Storz, 2000).

Eucaryotic cells also have redox-sensitive transcription factors that potentially act as ROS sensors. Probably, the best studied are AP-1 and NF-$^\kappa$B. The DNA-binding activities of both are responsive to the cellular thiol–

disulfide ratio due to single cysteines in the DNA-binding domains of these transcription factors (e.g.Abate *et al.*, 1990; Matthews *et al.*, 1992). Oxidation of these cysteines blocks binding of the transcription factors to their respective consensus DNA sequences. Not unexpectedly, the sequence-specific binding to DNA is enhanced by reducing conditions. Interestingly, a dual-function protein (REF-1) functions as a reducing agent for AP-1, but it also has another domain with endonuclease activity involved in DNA repair (Xanthoudakis *et al.*, 1992). Thioredoxin restores transcriptional activity of AP-1 by binding to AP-1 and by complexing with and maintaining REF-1 in its reduced state.

The redox- and radio-responsive transcription factor NF-$^\kappa$B is characterized by the redox-sensitive cysteine-62 in the DNA-binding domain of the P50 subunit. As with AP-1, REF-1, chemical thiol-reducing agents and thioredoxin can restore DNA-binding activity of NF-$^\kappa$B (Matthews *et al.*, 1992; Marshall *et al.*, 2000).

Although DNA binding of these transcription factors is favored by reducing conditions, oxidative stress, including ionizing radiation, usually activates both AP-1 and NF-$^\kappa$B. This apparent dichotomy is the result of differential redox regulation in the cytoplasm and nucleus, combined with other signaling pathways promoted by oxidative conditions in the cytoplasm that initiate the activation of NF-$^\kappa$B and AP-1 prior to translocation into the nucleus. Using an elegant approach of protein thiol modification and liquid chromatography–mass spectroscopy analysis, Nishi *et al.* (2002) determined the redox state of different cysteines of the P65 and P50 subunits of NF-$^\kappa$B. The cysteines of both proteins were equivalently reduced in either the cytoplasm or nucleus with one exception. Cysteine-62 in the DNA-binding domain of P50 was highly oxidized in the cytoplasm and reduced in the nucleus. The differential redox state of cysteine-62 was maintained in part by nuclear REF-1. The reduced state of P50 cysteine-62 is critical for transcriptional activity (Matthews *et al.*, 1992; Marshall *et al.*, 2000).

NO$^\bullet$ donors in a similar fashion can sensitize cells to apoptosis by inhibiting NF-$^\kappa$B by different mechanisms depending on cell type (Marshall and Stamler, 2002). In A549 cells, NO$^\bullet$ inhibits NF-$^\kappa$B DNA binding by S-nitrosylation of the P50 subunit. In Jurkat T cells, inhibition occurs in the cytoplasm prior to degradation of I$^\kappa$B.

Examples of signaling pathways activated by oxidative and nitrosative stresses include the release of intracellular Ca^{2+} involved in activation of both NF-$^\kappa$B and AP-1 and observed after treating cells with either H$_2$O$_2$ or with ionizing radiation (Todd and Mikkelsen, 1994; Dröge, 2001). The proteolytic degradation of I$^\kappa$B, the endogenous inhibitor of NF-$^\kappa$B, is also triggered by oxidative stress, although the mechanism by which this occurs may differ depending on the type of cell and stress (Karin, 1999). An alternative mechanism for NF-$^\kappa$B activation involves the phosphorylation of tyrosine-42 of IκB but without its proteolytic degradation. Cells that express IκB mutated at tyrosine-42 are more radiosensitive compared with parental and vector-transfected cells (Miyakoshi and Yagi, 2000). Phosphorylation of tyrosine-42 may be sensitive to oxidative stress since protein tyrosine phosphatase inhibitors enhance tyrosine phosphorylation of IκB with resultant NF-$^\kappa$B activation (Imbert et al., 1996), and protein tyrosine phosphatases are characterized by a redox-sensitive active site cysteine, responsive to oxidative and nitrosative stresses (e.g. Lee et al., 1998; Meng et al., 2002). Other stress-activated serine/threonine protein kinases, depending on cell type and type of stress, may also modulate NF-$^\kappa$B and AP-1 transcriptional activities (e.g. Stephenson et al., 1994; Wilhelm et al., 1997; Norris and Baldwin, 1999; Marshall et al., 2000; Miyakoshi and Yagi, 2000; Park et al., 2000; Droge, 2001; Howe et al., 2002).

Other transcription factors of radiobiological interest whose activities are sensitive to ROS/RNS include those with a zinc-finger (zinc-thiolate cluster) in the DNA-binding domain (e.g. EGR-1 and SP-1; Ahmed et al., 1997; Wang et al., 1999; Raju et al., 2000). Both ROS and RNS inhibit DNA binding of zinc-finger transcription factors by stimulating the release of the metal ion. However, there appears to be an important difference between the ROS and RNS mechanisms. Using the vitamin D receptor (VDR) and retinoid X receptor (RXR) as model zinc-finger transcription factors, Kroncke et al. (2002) showed that the zinc-fingers of VDR and RXR could be repaired after nitrosative but not oxidative stresses. This suggests that RNSs are not just inhibitors but regulators of zinc-finger transcription factor activities.

The tumor suppressor protein, TP53, is activated not only by radiation but also by treatment of cells with NO$^\bullet$ donors (Wang et al., 2002; Brune et al., 2001;Schonhoff et al., 2002). NO$^\bullet$ donors appear to act by two different mechanisms.In vitro studies show that in the short term, S-nitrosylation of the TP53 negative regulator Hdm2, inhibits its interaction with TP53. In vivo this disrupts TP53 ubiquitination and proteolysis, thereby increasing cellular TP53 levels (Schonhoff et al., 2002). NO$^\bullet$ treatment also downregulates Mdm2 (mouse equivalent to Hdm2) expression, resulting in a corresponding drop in TP53 ubiquitination (Wang et al., 2002). It is not known whether NO$^\bullet$ has a significant role in the radiation-stimulated TP53 expression.

EFFECTS OF ROS/RNS ON OTHER ZINC AND IRON PROTEINS

Except for atypical isoforms, protein kinase C is activated either by binding the second messenger, diacylglycerol, or by oxidation facilitated by retinols that convert the protein to an active form (Korichneva *et al.*, 2002 and references therein). The diacylglycerol/retinal-binding domain is contained within the twin cysteine-rich domains of six conserved cysteines and two histidines that form a tetrahedral coordinate complex with two zinc atoms. Treatment with either diacylglycerol or hydrogen peroxide induces the release of zinc, in turn inducing a reversible conformational change activating the kinase. The authors describe zinc as being a 'linchpin' that coordinates protein conformational changes in response to specific signals. These results have important implications for understanding the mechanism of oxidative/nitrosative modulation of other zinc-finger regulatory proteins, for example, RAF-1 kinase.

Calcineurin (protein phosphatase 2B) is activated by changes in cytosolic Ca^{2+} mediated by calmodulin. ROS, including hydrogen peroxide and $^\bullet O_2^-$ as well as certain sulfhydryl-reacting agents, inhibit calcineurin activity. Calcineurin – like some other serine/threonine protein phosphatases, PP1, PP2A, and purple acid phosphatase – contains a binuclear (Fe and Zn) metal center necessary for catalysis. Recent studies indicate that the native and redox-sensitive enzyme is the ferrous ($Fe^{2+}-Zn^{2+}$) form (Namgaladze *et al.*, 2002). Superoxide radical at nanomolar concentrations orders of magnitude lower than hydrogen peroxide inactivates calcineurin in a Ca^{2+}/calmodulin-dependent mechanism that involves oxidation to a redox-insensitive $Fe^{3+}-Zn^{2+}$ form. $NO^\bullet$ blocks the effect of $^\bullet O_2^-$. Thus, the activity of this key enzyme involved in diverse aspects of cell growth (as well as other protein serine/threonine phosphatases) responds to changes in cellular Ca^{2+}, $NO^\bullet$ and $^\bullet O_2^-$ concentrations.

Appropriate regulation of cellular iron levels is essential for providing cells with the necessary iron for metabolic needs (e.g. heme and nonheme iron proteins), while at the same time minimizing iron-catalyzed toxic-free radical formation. Iron regulatory proteins (IRPs) 1 and 2 are key components of iron homeostasis. When cellular iron levels are high, IRP-1 forms a (4Fe–4S) cluster, exhibits aconitase enzymatic activity and is no longer a regulator of ferritin or transferrin receptor expression. $NO^\bullet$ and, at much lower concentrations, $ONOO^-$ activate IRP-1 by inhibiting aconitase activity and removing iron from the Fe–S cluster (Cairo *et al.*, 2002). SOD blunts the effects of $NO^\bullet$, suggesting that $ONOO^-$ is the critical species for disassembling the Fe–S cluster. IRP-2 does not contain an Fe–S cluster and, at low cellular iron levels, acts, as does IRP-1, as a post-transcriptional regulator of ferritin and transferrin receptor expression. IRP-2 is consistently inactivated by ONOO⁻, probably by oxidizing protein cysteine and proteolysis (Kim and Ponka, 2000). The apparent dichotomy in the effects of ONOO⁻ on IRP-1 and -2 activities is not understood but may provide clues as to the role of NO• in regulating cellular iron levels.

NO• also modulates zinc-dependent signal transduction pathways in its role as a regulator of zinc homeostasis. Essentially, all cellular zinc is protein bound. As a consequence, labile Zn^{2+} levels are dynamically maintained by the apposing pathways of NO•-stimulated release from metallothionein and sequestration by the apoprotein, thionein (Maret *et al.*, 1999; St Croix *et al.*, 2002). Thionein is a cysteine-rich protein that binds up to seven zinc atoms per protein molecule as zinc-thiolate clusters. Zinc is released from metallothionein by •NO-catalyzed S-nitrosylation. Thionein has been shown to extract zinc from glyceraldehyde 3-phosphodehydrogenase with a corresponding increase in enzyme activity and from the transcription factors, SP-1 and TF-IIIA, with resulting changes in their transcriptional activities (Zeng *et al.*, 1991a, 1991b; Maret *et al.*, 1999). Thus, the ratio of metallothionein to thionein regulated by •NO can determine cellular zinc levels and the activities of a number of zinc-dependent signal transduction pathways (St Croix *et al.*, 2002).

S-GLUTATHIOLATION MODULATION OF PROTEIN FUNCTION

S-glutathiolation of proteins links redox modulation of protein function directly with the GSH : GSSG ratio of a cell. S-glutathiolation is not that well studied but provides a mechanism for readily reversible oxidative modification of protein SH and represents an intermediate in reduction of protein sulfenic acid or S-nitrosylated Cys to SH. Besides several house-keeping proteins, a number of transcription factors and regulatory proteins have been shown to be S-glutathiolated under conditions of oxidative or nitrosative stress (Padgett and Whorton, 1998; Klatt and Lamas, 2000). Both c-JUN and NF-$^\kappa$B transcription factor DNA binding is redox modulated by specific S-glutathiolation (Pineda-Molina and Lamas, 2001). Recent technical strides in labeling cellular GSH and isolating proteins potentially S-glutathiolated should accelerate research in this area (e.g. Padgett and Whorton, 1998).

NO• ACTIVATION OF SOLUBLE GUANYLATE CYCLASE AND ACTIVATION OF PROTEIN KINASE G

At the low [NO•] produced by the NOS-1 and NOS-3 isoforms, activation of the heme-containing soluble guanylate cyclase and consequently PKG represents a major cellular function of NO$^\bullet$ (Francis and Corbin, 1999). Other heme proteins also bind NO$^\bullet$ and, in the case of cytochrome *c* oxidase, may be of importance in regulating cellular respiration as discussed above.

PKG are classified into two types encoded by different genes (Francis and Corbin, 1999). Two type I isoforms (α and β) arise by alternative splicing from a single gene and differ by 100 amino acids at their amino-terminal autoinhibitory domains but retain the same cGMP binding and catalytic domains. PKG-I isoforms are expressed in diverse cell types. PKG-II is highly expressed in intestinal microvilli and, in contrast to soluble PKG-I, is associated with cellular membranes.

The *in vitro* study of PKG is hampered by its rapid downregulation in most cell types when grown in culture (Chiche *et al.*, 1998; Komalavilas *et al.*, 1999). Thus, the role of PKG in cellular growth regulation has relied on the use of primary cultures, transfection to overexpress PKG or the use of pharmacological activators and inhibitors of PKG or sGC. Modulation of PKG activity by these methods has revealed significant roles for the NO$^\bullet$-activated PKG-dependent signal transduction pathway in cell growth regulation (Cordelier *et al.*, 1997;Chiche *et al.*, 1998; Kim *et al.*, 1999; Komalavilas *et al.*, 1999; Gu *et al.*, 2000). In many cases, this appears to be due to the intersection of PKG signaling with the MAPK and stress-activated kinase-dependent signal transduction pathways (Lander *et al.*, 1996; Suhasini *et al.*, 1998; Go *et al.*, 1999; Komalavilas *et al.*, 1999; Browning *et al.*, 2000) and expression of the cyclin-dependent kinase inhibitor P21$^{Waf1/Cip1}$ (e.g. Gu *et al.*, 2000). PKG has also been shown to modulate the activity of radio-responsive transcription factors (AP-1, CREB, TP53) and the expression of several genes whose expression is also stimulated by radiation, including JUN-B, c-FOS, MKP-1, cyclooxygenase-2, tumor necrosis factor-alpha and P21$^{Waf1/Cip1}$ (Haby *et al.*, 1994; Gudi *et al.*, 1996; Sciorati *et al.*, 1997; Begum *et al.*, 1998; Collins and Uhler, 1999; Gertzberg *et al.*, 2000;Gu *et al.*, 2000; Brune *et al.*, 2001; Pfeilschifter *et al.*, 2001).

PROTEIN TYROSINE-NITRATION

A footprint of ONOO$^-$ formation is tyrosine nitration of proteins. Formation of protein 3-nitrotyrosine has been detected in a number of disease states, but the functions of this protein modification remain unclear. Although there is some suggestion of reversibility, there is no convincing molecular evidence for a eucaryotic denitrase (Gow *et al.*, 1996; Kamisaki *et al.*, 1998). There appears to be some selectivity in the nitration, since not all proteins in a tissue under analysis are nitrated. Souza *et al.* (1999) carefully analysed nitration of three purified proteins under defined conditions and have determined some of the factors governing selectivity of nitration. Their evidence suggested that exposure of tyrosine at the surface of the protein, its location on a loop structure and an adjacent negative charge were some of the determining factors.

Not considered in this *in vitro* analysis but also of importance is subcellular location of the protein relative to the source of the nitrating agent. Thus, a prominent tyrosine-nitrated protein both in rejected human kidney allografts and in tissue culture cells after radiation is the mitochondrial protein Mn-SOD (MacMillan-Crow *et al.*, 1996; Leach *et al.*, 2002). In the case of rejected human allograft tissue, this nitration inhibits mitochondrial Mn-SOD enzymatic activity. As discussed above, inactivation of Mn-SOD increases the lifetime of $^{\bullet}O_2^-$ and hence formation of $ONOO^-$ from $NO^{\bullet}$ by a 'positive feedback cycle' (Radi *et al.*, 2002). Tyrosine nitration of Mn-SOD was also observed after exposing cells to ionizing radiation. A similar significant inhibition in enzymatic activity was not detected, possibly because only a small fraction of a very abundant cellular protein was nitrated (Leach *et al.*, 2002). Nonetheless, if enzyme inhibition is limited to a single mitochondrion and thus not detectable by standard assay, significant biological effects can still accrue. This is because of potential intermitochondrial propagation of an oxidative event from one mitochondrion to adjacent mitochondria via a reversible permeability transition (e.g. Ichas and Mazat, 1998; Zorov *et al.*, 2000; Leach *et al.*, 2001, 2002). Thus, a potentially lethal oxidative event can be limited to a single mitochondrion among hundreds within a cell while at the same time initiating reversible cellular signaling pathways.

Several other biological effects have been attributed to tyrosine nitration. Tyrosine nitration has been shown to inhibit tyrosine hydroxylase activity under conditions of oxidative stress induced with 1-methyl-4-phenyl-1,2,3,6-tetrahydropyridine, a model for Parkinson's disease (Ara *et al.*, 1998). Addition of $ONOO^-$ has also been reported to induce tyrosine nitration and activation of P38, JNK and MAPK and block the interaction of the p85 regulatory subunit of phosphatidylinositol 3-kinase with the catalytic subunit (Hellberg *et al.*, 1998; Schieke *et al.*, 1999). There is recent evidence that $NO^{\bullet}$ donors promote translocation and activation of PKC ε via a mechanism of $ONOO^-$-mediated tyrosine nitration of PKC ε, disrupting its interaction with RACK2 (Balafanova *et al.*, 2002). The biological significance of many of these findings awaits the demonstration of reversibility (and the mechanism thereof) and quantitative evaluation of the degree of inhibition under physiological and pathophysiological endogenous conditions of $ONOO^-$ generation. It may be, as previously suggested, that tyrosine nitration marks the protein for proteolytic degradation (Gow *et al.*, 1996; Ischiropoulos, 1998; Souza *et al.*, 2000).

PROTEIN S-NITROSYLATION

Recent studies have also established S-nitrosylation of protein cysteine as a major mechanism of NO● regulatory signaling (reviewed in Stamler et al., 1997;Broillet, 1999; Hess et al., 2001). Protein S-nitrosylation is linked to the activity of the different NOS isoforms depending on the cell type (Gow et al., 2002). In macrophages, induction of NOS-2 expression with cytokines enhances overall protein S-nitrosylation, as does nerve growth factor induction of NOS-1 in PC12 cells and Ca^{2+}-stimulated NOS-3 in endothelial cells. The MAPK signaling pathway in T-lymphocytes is activated by NO● donors via S-nitrosylation of the GDP–GTP exchange protein RAS on cysteine-118 with subsequent recruitment and activation of the RAF-1 kinase (Deora et al., 2000). On the other hand, NO● inhibits the stress-activated kinase JNK by S-nitrosylation of cysteine-116. This appears to be the primary mechanism by which interferon-γ blocks JNK activation in macrophages (Park et al., 2000). Other proteins whose activities are modulated by S-nitrosylation include the P50 subunit of NF-κB, as discussed above (Marshall and Stamler, 2001), ryanodine, the Ca^{2+} release channel of sarcoplasmic reticulum (Eu et al., 2000), and mitochondrial caspases (Kim et al., 1997; Mannick et al., 1999; Mannick et al., 2001). An essential feature in activation of the FAS-dependent apoptosis pathway is the FAS-induced denitrosylation of caspase 3, freeing the active site cysteine (Mannick et al., 1999). The ROS scavenging properties and antiapoptotic functions of thioredoxin require S-nitrosylation of a specific cysteine (Haendeler et al., 2002).

The cysteine content of a protein is not the sole determining factor for S-nitrosylation (Hess et al., 2001). Specificity is seen in proteins with multiple cysteines, of which only one is S-nitrosylated by NO●. Analysis of S-nitrosylated proteins and the realization that acid–base catalysis alternatively promotes nitrosylation and denitrosylation led to a degenerate consensus sequence, (G,S,T,C,Y,N,Q,)(K,R,H,D,E,)C(D,E) (Stamler et al., 1997; Hess et al., 2001). This acid–base motif does not have to reside in the linear primary sequence, but, as has been shown for methionine adenosyl-transferase, can be generated from the tertiary structure of the protein (Perez-Mato et al., 1999).

This consensus sequence is found in hundreds of proteins beyond those mentioned above, including cyclin D1, Rb, BRCA1 and -2, and protein tyrosine phosphatases. Not all these proteins are S-nitrosylated and other factors that regulate nitrosylation include location of cysteines in hydrophobic compartments (Nedospasov et al., 2000; Hess et al., 2001). As noted above, a hydrophobic environment concentrates the lipophilic reactants, oxygen and NO●, thereby enhancing their rate of reaction and generating NO_x, including N_2O_3 that drives S-nitrosylation.

A potential role for S-nitrosylation in signal transduction pathway activation by ionizing radiation has been tested in some preliminary studies with the protein tyrosine phosphatases SHP-1 and SHP-2 (Figure 5, Barrett et al., in preparation). For this, Chinese hamster ovary cells were transfected with plasmids expressing either wild-type or mutants of these phosphatases in which the active site cysteine was mutated to a serine. After irradiation (4 Gy), S-nitrosylated proteins were purified according to the method of Jaffrey et al. (2001) and fractionated by gel electrophoresis. Western blots were probed with antibodies to either SHP-1 or SHP-2. The results in Figure 5 show a radiation-stimulated transient S-nitrosylation primarily of the active site cysteine of both phosphatases. Since previous investigations demonstrated NO• donor-stimulated S-nitrosylation of RAS, similar experiments were performed with cells transfected with a plasmid encoding RAS. Although basal S-nitrosylation of RAS was observed, no significant increase was observed after radiation.

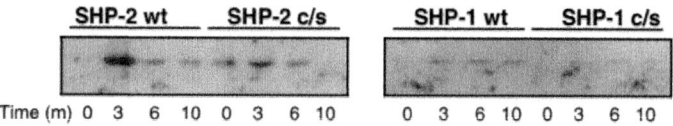

Figure 5: *S*-nitrosylation of protein tyrosine phosphatases. Chinese hamster ovary cells were transiently transfected with expression plasmids, encoding wild-type and dominant negative mutants of the protein tyrosine phosphatases SHP-1 and SHP2. At 48 h after transfection, cells were irradiated (4 Gy), and cell lysates were prepared at the given times postirradiation. *S*-nitrosylated proteins were purified according to Jaffrey*et al.* (2001) and fractionated by gel electrophoresis. Western blots were probed with appropriate antibodies.

We estimate that approximately 5% of each phosphatase was S-nitrosylated after radiation. At least part of the low yield may reflect the inefficiency of the purification procedure. However, the catalytic activities of protein tyrosine phosphatases are in general 100-fold greater than protein tyrosine kinase activities (Tonks and Neel, 1996). Thus, only a very small change in phosphatase activity can translate into a substantial change in net tyrosine phosphorylation. Such transient S-nitrosylation and corresponding inhibition of protein tyrosine phosphatase may account for the enhanced levels of tyrosine-phosphorylated proteins after radiation and activation of signal transduction pathways that require protein tyrosine phosphorylation (e.g. Tuttle *et al.*, 1998; Schmidt-Ullrich*et al.*, 2000).

Of additional radiobiological interest are the mixed function protein tyrosine phosphatases that regulate the different cell cycle checkpoints. These phosphatases are also characterized by an active site cysteine. A recent study demonstrated that treatment of cells with relatively high concentrations of H_2O_2 (>250 μm) stimulated degradation of CDC25C but not CDC25A. *In vitro* studies revealed that hydrogen peroxide stimulated disulfide bond formation between the active site cysteine-377 and an invariant cysteine at position 330 and suggested to the authors that oxidative stress independent of CHK1 activity may induce cell cycle arrest by inducing degradation of CDC25C (Savitsky and Finkel, 2002). In keeping with the homologies between protein tyrosine phosphatases, it would not be surprising to find that NO• *S*-nitrosylates and inhibits CDC25 family members, as found with SHP-1 and SHP-2. Such a mechanism may provide an additional layer of cell cycle regulation important in cellular responses to oxidative and nitrosative stresses and one that is initiated within the cytoplasm.

CONCLUSIONS: RNS AS EFFECTORS OF CELLULAR OXIDATIVE EVENTS

Earlier in this review, we suggested that a two-step nitration mechanism for nitro-tyrosine formation also offers a hypothesis for cell signaling involving radicals. The first step entails a nonspecific oxidative reaction of highly reactive radicals of limited diffusion distance. This in turn produces a more stable radical with chemical reactivity properties of high specificity consistent with a signaling molecule. In developing this hypothesis for biological signaling, we emphasized the importance of reactivity and relative concentrations of reactants. Highly reactive OH• has a low target selectivity and diffusion distance. In contrast, NO• and nitrite/nitrate are chemically inert with most biological molecules and exhibit a high degree of reaction specificity. We have also emphasized with several examples the importance of intracellular location for radical-driven cell signaling. Thus, the hydrophobic domains of membranes concentrate lipophilic reactants, such as NO• and NO_2•, to generate the potent nitrosating agent, N_2O_3. Finally, the importance of RNS, such as NO•, in oxidative signaling can also be seen in the primary goals of cellular redox homeostasis mechanisms. Cells display an array of mechanisms for maintaining nonspecific highly reactive ROS at low manageable levels. On the other hand, cells precisely regulate RNS levels and thus RNS-stimulated signal transduction pathways through anabolic, catabolic and reversible storage mechanisms. For all the above reasons, NO• can be considered the prototypic redox second messenger (Marshall *et al.*, 2000; Hess *et al.*, 2001).

Several studies have shown that cells in responding to a mild oxidative stress, for example, low concentrations of exogenous hydrogen peroxide or ionizing radiation, activate NOS activity. In analogy to the two-step mechanism for nitrotyrosine formation, we propose that cells convert a nonspecific oxidative event to a nitrosative signal. At present, the sensors for mild oxidative stress, including radiation-induced ionization events, have not been definitively defined. We suspect that mitochondria have a key sensing role because mitochondria generate most of a cell's ROS and thus one would expect mitochondria to be part of any homeostatic redox loop. There is also significant experimental evidence for mitochondria as oxidative sensors (e.g. Zorov *et al.*, 2000; Leach *et al.*, 2001). Conversion of the oxidative event to a nitrosative signal provides the amplification mechanisms that activate signal transduction pathways that are cytoprotective or, alternatively, in collusion with proapoptotic pathways induce cell death.

Acknowledgements

The authors' work was supported by the United States Public Health Service Grants CA65896, CA72955 and CA89055 (RBM), and Cancer Research UK (PW).

REFERENCES

1. Abate C, Patel L, Rauscher FJ and Curran T. (1990). *Science*, **249**, 1157–1161.
2. Ahmed MM, Sells SF, Venkatsubbarao K, Fruitwala SM, Muthukkumar S, Harp C, Mohiuddin M and Rangnekar VM. (1997). *J. Biol. Chem.*, **272**, 33056–33061.
3. Alvarez B, Ferrer-Sueta G, Freeman BA and Radi R. (1999). *J. Biol. Chem.*, **274**, 842–848.
4. Antunes F and Cadenas E. (2000). *FEBS Lett.*, **475**, 121–126.
5. Ara J, Przedborski S, Maini AB, Jackson-Lewis V, Trifiletti RR, Horowitz J and Ischiropoulos H. (1998). *Proc. Natl. Acad. Sci. USA*, **95**, 7659–7663.
6. Arnold WP, Mittal CK, Katsuki S and Murad F. (1977). *Proc. Natl. Acad. Sci. USA*, **74**, 3203–3207.
7. Arteel GE, Briviba K and Sies H. (1999). *FEBS Lett.*, **445**, 226–230.
8. Augusto O, Bonini MG, Amanso AM, Linares E, Santos CCX and De Menezes SL. (2002). *Free Radical Biol. Med.*, **32**, 841–859.
9. Balafanova Z, Bolli R, Zhang J, Zheng Y, Pass JM, Bhatnagar A, Tang X-L, Wang O, Cardwell E and Ping P. (2002). *J. Biol. Chem.*, **277**, 15021–15027.
10. Baldus S, Eiserich JP, Brennan M-L, Jackson RM, Alexander CB and Freeman B A. (2002). *Free Radical Biol. Med.*, **33**, 1010–1019.

11. Bauer CE, Elsen S and Bird TH. (1999). *Ann. Rev. Microbiol.*, **53**, 495–523.
12. Beckman JS, Beckman TW, Chen J, Marshall PA and Freeman BA. (1990).*Proc. Natl. Acad. Sci. USA*, **87**, 1620–1624.
13. Beckman JS and Koppenol WH. (1996). *Am. J. Physiol.*, **271**, C1424–C1437. | PubMed | ISI | ChemPort |
14. Begum N, Ragolia L, Rienzie J, McCarthy M and Duddy N. (1998). *J. Biol. Chem.*, **273**, 25164–25170.
15. Bellamy TC, Griffiths C and Garthwaite J. (2002). *J. Biol. Chem.*, **277**, 31801–31807.
16. Bernardi P and Petronelli V. (1996). *J. Bioenerg. Biomembr.*, **28**, 131–138.
17. Bielski BHJ, Cabelli DE and Arudi RL. (1985). *J. Phys. Chem. Ref. Data*, **14**, 1041–1100. | ChemPort |
18. Broillet M-C. (1999). *Cell. Mol. Life Sci.*, **55**, 1036–1042.
19. Brookes PS, Levonen A-L, Shiva S, Sarti P and Darley-Usmar VM. (2002).*Free Radical Biol. Med.*, **33**, 755–764.
20. Browning D, McShane MP, Marty C and Ye RD. (2000). *J. Biol. Chem.*, **275**, 2811–2816.
21. Brune B, von Knethen A and Sandau KB. (2001). *Cell. Signaling*, **13**, 525–533.
22. Burkitt MJ and Wardman P. (2001). *Biochem. Biophys. Res. Commun.*,**282**, 329–333. | PubMed |
23. Burner U, Furtmüller PG, Kettle AJ, Koppenol WH and Obinger C. (2000). *J. Biol. Chem.*, **275**, 20597–20601.
24. Cadenas E and Davies KJA. (2000). *Free Radical Biol. Med.*, **29**, 222–230.
25. Cairo G, Ronchi R, Recalcati S, Camanella A and Minotti G. (2002). *Biochemistry*, **41**, 7435–7442.
26. Candeias LP, Patel KB, Stratford MRL and Wardman P. (1993). *FEBS Lett.*,**333**, 151–153.
27. Castro L, Rodriguez M and Radi R. (1994). *J. Biol. Chem.*, **269**, 29409–29415.
28. Chiche JD, Schlutsmeyer SM, Block DB, de la Monte SM, Roberts JD, Filippov G, Jannssens SP, Rosenzweig A and Block KD. (1998). *J. Biol. Chem.*, **273**, 34263–34271.
29. Clutton S, Townsend K, Walker C, Ansell J and Wright E. (1996).

Carcinogenesis, **17**, 1633–1639.
30. Collins SP and Uhler MD. (1999). *J. Biol. Chem.*, **274**, 8391–8404.
31. Cordelier P, Esteve J-P, Bousquet C, O'Carroll A-M, Schally AV, Vaysse N, Susini S and Buscail L. (1997). *Proc. Natl. Acad. Sci. USA*, **94**, 9343–9348.
32. Cordelier P, Esteve JP, Rivard M, Marletta M, Vaysse N, Susini C and Buscail L. (1999). *FASEB J.*, **13**, 2037–2050.
33. Crompton M. (1999). *Biochem. J.*, **341**, 233–249.
34. Crow JP. (2000). *Free Radical Biol. Med.*, **28**, 1487–1494.
35. Crow JP, Beckman JS and McCord JM. (1995). *Biochemistry*, **34**, 3544–3452. | Article | ISI |
36. Denicola A, Freeman BA, Trujillo M and Radi R. (1996). *Arch. Biochem. Biophys.*, **333**, 49–58.
37. Deora AA, Hajjar DP and Lander HM. (2000). *Biochemistry*, **39**, 9901–9908.
38. Droge W. (2002). *Physiol. Rev.*, **82**, 47–95. ISI | ChemPort |
39. Durocher D and Jackson SP. (2001). *Curr. Opin. Cell Biol.*, **13**, 225–231.
40. Eaton P, Byers HL, Leeds N, Ward MA and Shattock MJ. (2002). *J. Biol. Chem.*, **277**, 9806–2002.
41. Eiserich JP, Butler J, Van der Vliet A, Cross CE and Halliwell B. (1995). *Biochem. J.*, **310**, 745–749.
42. Eiserich JP, Hristova M, Cross CE, Jones AD, Freeman BA, Halliwell B and van der Vliet A. (1988). *Nature*, **391**, 393–397.
43. Elfering SL, Sarkela TM and Giulivi C. (2002). *J. Biol. Chem.*, **277**, 38079–38086.
44. Espey MG, Thomas DD, Miranda KM and Wink DA. (2002). *Proc. Natl. Acad. Sci. USA*, **99**, 11127–11132.
45. Eu JP, Liu L, Zeng M and Stamler JS. (2000). *Biochemistry*, **39**, 1040–1047.
46. Fang M, Jaffrey SR, Sawa A, Ye K, Luo X and Snyder SH. (2000). *Neuron*, **28**, 183–193.
47. Fitzhugh AL and Keefer LK. (2000). *Free Radical Biol. Med.*, **28**, 1463–1469.
48. Ford E, Hughes MN and Wardman P. (2002a). *Free Radical Biol. Med.*, **32**, 1314–1323.
49. Ford E, Hughes MN and Wardman P. (2002b). *J. Biol. Chem.*, **277**, 2430–

2436.
50. Francis SH and Corbin JD. (1999). *Crit. Rev. Clin. Lab. Sci.*, **36**, 275–328.
51. Freeman SL and MacNaughton WK. (2000). *Am. J. Physiol.*, **278**, G243–G250.
52. Fridovich I. (1995). *Annu. Rev. Biochem.*, **64**, 97–112.
53. Gaston B. (1999). *Biochim. Biophys. Acta*, **1411**, 323–333.
54. Gertzberg N, Clements R, Jaspers H, Ferro TJ, Neumann P, Flescher E and Johnson A. (2000). *Am. J. Respir. Cell. Mol. Biol.*, **22**, 105–115.
55. Go YM, Patel RP, Maland MC, Park H, Beckman JS, Darley-Usmar VM and Jo H. (1999). *Am. J. Physiol.*, **277**, H1647–H1653.
56. Goda N, Suematsu M, Mukai M, Kiyokawa N, Natori M, Nozawa S and Ishimura Y. (1997). *Am. J. Physiol.*, **271**, H1893–H1899.
57. Goldstein S, Czapski G, Lind J and Merényi G. (2000). *J. Biol. Chem.*, **275**, 3031–3036.
58. Gow AJ, Chen Q, Hess DT, Day BJ, Ischiropoulos H and Stamler JS. (2002).*J. Biol. Chem.*, **277**, 9637–9640.
59. Gow AJ, Duran D, Malcolm S and Ischiropoulos H. (1996). *FEBS Lett.*, **385**, 63–66.
60. Gow AJ, Luchsinger BP, Pawlowski JR, Singel DJ and Stamler JS. (1999).*Proc. Natl. Acad. Sci. USA*, **96**, 9027–9032.
61. Gu J, Lynch P and Brecher. (2000). *J. Biol. Chem.*, **275**, 11389–11396.
62. Gudi T, Hunar I, Meinecke M, Lohmann SM, Boss GR and Pilz RB. (1996). *J. Biol. Chem.*, **271**, 4597–4600.
63. Gudz TI, Pandelova IG and Novgordov SA. (1994). *Radiat. Res*, **138**, 219–228.
64. Gunther MR, Sturgeon BE and Mason RP. (2002). *Toxicology*, **177**, 1–9.
65. Haby C, Lisovoski F, Aunis D and Zwiller J. (1994). *J. Neurochem.*, **62**, 496–501.
66. Haendeler J, Hoffman J, Tischler V, Berk BC, Zeiher AM and Dimmiler S. (2002). *Nat. Cell Biol.*, **4**, 743–749.
67. Haimovitz-Friedman A, Kan C-C, Ehleiter D, Persaud R, McLaughlin M, Fuks Z and Kolesnick R. (1994). *J. Exp. Med.*, **180**, 525–535.
68. Hallahan DE, Sukhatme VP, Sherman ML, Virudachalam S, Kufe D and Weichselbaum RR. (1991). *Proc. Natl. Acad. Sci. USA*, **88**, 2156–2160.
69. Halliwell B and Gutteridge JMC. (1999). *Free Radicals in Biology and Medicine*. 3rd edn. Oxford University Press: Oxford.

70. Hashimoto S, Kira A, Imamura M and Masuda T. (1982). *Int. J. Radiat. Biol.*, **41**, 303–314.
71. Hausladen A and Fridovich I. (1994). *J. Biol. Chem.*, **269**, 29405–29408.
72. Hellberg CB, Boggs SE and Lapetina EG. (1998). *Biochem. Biophys. Res. Commun.*, **252**, 313–317.
73. Hess DT, Matsumoto A, Nudelman R and Stamler JS. (2001). *Nat. Cell. Biol.*, **3**, E46–E49.
74. Hidalgo E, Ding H and Demple B. (1997). *Cell*, **88**, 121–129.
75. Hodges GR, Marwaha J, Paul T and Ingold KU. (2000). *Chem. Res. Toxicol.*, **13**, 1287–1293.
76. Hoganson CW, Sahlin M, Sjöberg B-M and Babcock GT. (1996). *J. Am. Chem. Soc.*, **118**, 4672–4679.
77. Hogg N. (2000). *Free Radical Biol. Med.*, **28**, 1478–1486.
78. Hogg N, Singh RJ, Joseph J, Neese F and Kalyanaraman B. (1995). *Free Radical Res.*, **22**, 47–56.
79. Hogg N, Singh RJ and Kalyanaraman B. (1996). *FEBS Lett.*, **382**, 223–228.
80. Howe Cj, LaHair MM, Maxwell JA, Lee JT, Robinson PJ, Rodriguez-Mora O, McCubrey JA and Franklin RA. (2002). *J. Biol. Chem.*, **277**, 30469–30476.
81. Hwang C, Sinskey AJ and Lodish HF. (1992). *Science*, **257**, 1496–1501.
82. Ichas F and Mazat JP. (1998). *Biochim. Biophys. Acta.*, **1366**, 33–50.
83. Ignarro LJ (ed.). (2000). *Nitric Oxide. Biology and Pathobiology.* Academic Press: San Diego.
84. Imbert V, Rupec RA, Livolsi A, Pahl HL, Traenchner E, Mueller-Dieckmann C, Farahifar D, Rossi B, Auberger P, Baeurele PA and Peyron J-F. (1996).*Cell*, **86**, 787–798.
85. Ischiropoulos H. (1998). *Arch. Biochem. Biophys.*, **356**, 1–11.
86. Ischiropoulos H, Gow A, Thom SR, Kooy NE, Royall JA and Crow JP. (1999).*Methods Enzymol.*, **301** (Part C), 367–373.
87. Jaffrey S, Erdjument-Bromage H, Ferris CD, Tempst P and Snyder SH. (2001). *Nat. Cell Biol.*, **3**, 193–197.
88. Janssen YMW, Soultanakis R, Steece K, Heerdt E, Singh RJ, Joseph J and Kalyanaraman B. (1998a). *Am. J. Physiol.*, **275**, L1100–L1109.
89. Janssen YMW, Van den Berge DL, Verovski VN, Monsaert C and Storme GA. (1998b). *Cancer Res.*, **58**, 5646–5648.

90. Jin F, Leitich J and von Sonntag C. (1993). *J. Chem. Soc., Perkin Trans.*, **2**, 1583–1588.
91. Jourd'heuil D. (2002). *Free Radical Biol. Med.*, **33**, 676–684.
92. Jourd'heuil D, Jourd'heuil F L, Kutchukian PS, Musah RA, Wink DA and Grisham MB. (2001). *J. Biol. Chem.*, **276**, 28799–28805.
93. Kamisaki Y, Wada K, Kian K, Balabanli B, Davis K, Martin E, Behhod F, Lee Y-C and Murad F. (1998). *Proc. Natl. Acad. Sci. USA*, **95**, 11584–11589.
94. Kanai AJ, Pearce LL, Clemens PR, Birder LA, Van Bibber MM, Choi S-Y, de Groat WC and Peterson J. (2001). *Proc. Natl. Acad. Sci. USA*, **98**, 14126–14131.
95. Karin M. (1999). *Oncogene*, **18**, 6867–6874.
96. Kasid U, Suy S, Dent P, Ray S, Whiteside TL and Sturgill TW. (1996). *Nature*, **382**, 813–816.
97. Kavanagh B, Todd D, Chen P, Schmidt-Ullrich RK and Mikkelsen RB. (1998).*Radiat. Res.*, **149**, 579–587.
98. Kettle AJ, van Dalen CJ and Winterbourn CC. (1997). *Redox Rep.*, **3**, 257–258.
99. Khanna KK, Lavin MF, Jackson SP and Mulhern TD. (2001). *Cell Death Differ.*, **8**, 1052–1065.
100. Kim S and Ponka P. (2000). *J. Biol. Chem.*, **275**, 6220–6226.
101. Kim YM, Talanian RV and Billiar TR. (1997). *J. Biol. Chem.*, **272**, 31138–31148.
102. Kim Y-M, Bombeck CA and Billiar TR. (1999). *Circ. Res.*, **84**, 253–256.
103. King MP and Attardi G. (1996). *Methods Enzymol.*, **264**, 304–313.
104. Kietzmann T, Fandrey J and Acker H. (2000). *News Physiol. Sci.*, **15**, 202–208.
105. Kirsch M, Korth H-G, Sustmann R and de Groot H. (2002). *Biol. Chem.*,**383**, 389–399.
106. Klatt P and Lamas S. (2000). *Eur. J. Biochem.*, **267**, 4928–4944.
107. Klotz L-O, Schroeder P and Sies H. (2002). *Free Radical Biol. Med.*, **33**, 737–743.
108. Kojima H, Hirotani M, Nakatsubo N, Kikuchi K, Urano Y, Higuchi T, Hirata Y and Nagano T. (2001). *Anal. Chem.*, **73**, 1967–1973.
109. Kojima H, Nakatsubo N, Kibuchi K, Kawahara S, Kirino Y, Nagoshi H, Hirata Y and Nagano T. (1998). *Anal. Chem.*, **70**, 2446–2453.

110. Komalavilas P, Shah PK, Jo H and Lincoln TM. (1999). *J. Biol. Chem.*, **274**, 34301–34309.
111. Koppenol WH, Moreno JJ, Pryor WA, Ischiropoulos H and Beckman JS. (1992). *Chem. Res. Toxicol.*, **5**, 834–842.
112. Korichneva I, Hoyos B, Chua R, Levi E and Hammerling U. (2002). *J. Biol. Chem.*, **277**, 44327–44331.
113. Kroncke K-D, Klotz L-O, Suschek CV and Sies H. (2002). *J. Biol. Chem.*, **277**, 13294–13301.
114. Lal M, Rao R, Fang X, Schuchmann H-P and von Sonntag C. (1997). *J. Am. Chem. Soc.*, **119**, 5735–5739.
115. Lander HM, Jacovina AT, Davis RJ and Tauras JM. (1996). *J. Biol. Chem.*, **271**, 19705–19709.
116. Leach JK, Black SM, Schmidt-Ullrich RK and Mikkelsen RB. (2002). *J. Biol. Chem.*, **277**, 15400–15406.
117. Leach JK, Van Tuyle G, Lin P-S, Schmidt-Ullrich RK and Mikkelsen RB. (2001). *Cancer Res.*, **61**, 3894–3901.
118. Lee SR, Kwon KS, Kim SR and Rhee SG. (1998). *J. Biol. Chem.*, **273**, 15366–15372.
119. Lee CM, Robinson LJ and Michel T. (1995). *J. Biol. Chem.*, **270**, 27403–27406.
120. Li H, Samouilov A, Liu X and Zweier JL. (2001). *J. Biol. Chem.*, **276**, 24482–24489.
121. Liu X, Miller MJS, Joshi MS, Thomas DD and Lancaster Jr JR. (1998). *Proc. Natl. Acad. Sci. USA*, **95**, 2175–2179.
122. Long CA and Bielski BHJ. (1980). *J. Phys. Chem.*, **84**, 555–557.
123. Luxford C, Dean RT and Davies MJ. (2000). *Chem. Res. Toxicol.*, **13**, 665–672.
124. Lymar SV. (2001). *McGraw-Hill Yearbook of Science & Technology.* Staff M-H (ed). McGraw-Hill: New York, pp 263–266.
125. Lymar SV and Hurst JK. (1995). *J. Am. Chem. Soc.*, **117**, 8867–8868.
126. Lymar SV, Jiang Q and Hurst JK. (1996). *Biochemistry*, **35**, 7855–7861.
127. MacMillan-Crow LA, Crow JP, Kerby JD, Beckman JS and Thompson JA. (1996). *Proc. Natl. Acad. Sci. USA*, **93**, 11853–11858.
128. Majander A, Finel A and Wikstrom M. (1994). *J. Biol. Chem.*, **269**, 21037–21042.
129. Mannick JB, Hausladen A, Liu L, Hess DT, Zeng M, Miao QX, Kane LS, Gow AJ and Stamler JS. (1999). *Science*, **284**, 651–654.

130. Mannick JB, Schonhoff C, Papeta N, Ghafourifar P, Szibor M, Fang K and Gaston B. (2001). *J. Cell Biol.*, **154**, 1111–1116.
131. Marchesi E, Rota C, Fann YC, Chignell CF and Mason R P. (1999). *Free Radical Biol. Med.*, **26**, 148–161.
132. Maret W, Jacob C, Vallee BL and Fischer EH. (1999). *Proc. Natl. Acad. Sci. USA*, **96**, 1936–1940.
133. Marshall HE, Merchant K and Stamler JS. (2000). *FASEB J.*, **14**, 1889–1900.
134. Marshall HE and Stamler JS. (2002). *J. Biol. Chem.*, **277**, 34223–34228.
135. Matsumoto H, Hayashi S, Hatashita M, Ohnishi K, Shioura H, Ohtsubo T, Kitai R, Ohnishi T and Kano E. (2001). *Radiat. Res.*, **155**, 387–396.
136. Matthews JR, Wakasugi N, Virelizier J-L, Yodoi J and Hay RT. (1992). *Nucleic Acids Res.*, **20**, 3821–3830.
137. Meister A. (1994). *Cancer Res.*, **54**, 1969s–1975s.
138. Meng T-C, Fukada T and Tonks NK. (2002). *Mol. Cell*, **9**, 387–399.
139. Michel T, Li GK and Busconi L. (1993). *Proc. Natl. Acad. Sci. USA*, **90**, 6252–6256.
140. Millar TM, Stevens CR, Benjamin N, Eisenthal R, Harrison R and Blake DR. (1998). *FEBS Lett.*, **427**, 225–228.
141. Miyakoshi J and Yagi K. (2000). *Br. J. Cancer*, **82**, 28–33. | Article |
142. Morales A, Miranda M, Sanchez-Reyes A, Biete A and Fernandez-Checa J. (1998). *Int. J. Rad. Oncol.*, **42**, 191–204.
143. Namgaladze D, Hofer HW and Ullrich V. (2002). *J. Biol. Chem.*, **277**, 5962–5969.
144. Narayanan P, Goodwin E and Lehnert B. (1997). *Cancer Res.*, **57**, 3963–3971.
145. Nauser T and Koppenol WH. (2002). *J. Phys. Chem.*, **106**, 4084–4086.
146. Nedospasov A, Rafikov R, Beda N and Nudler E. (2000). *Proc. Natl. Acad. Sci. USA*, **97**, 13543–13548.
147. Nishi T, Shimizu N, Hiramoto M, Sato I, Yamaguchi Y, Hasegawa M, Aizawa S, Tanaka H, Kataoka K, Watanabe H and Handa H. (2002). *J. Biol. Chem.*,**277**, 44548–44556.
148. Niu XF, Ibbotson G and Kubes P. (1996). *Circ. Res.*, **79**, 992–999.
149. Noble DR and Williams DLH. (2000). *Nitric Oxide*, **4**, 392–398.
150. Norris JL and Baldwin AS. (1999). *J. Biol. Chem.*, **274**, 13841–13846.
151. Nunoshiba T, DeRojas-Walker T, Tannenbaum SR and Demple B. (1995).

Infect. Immun., **63**, 794–798.

152. Padgett CM and Whorton AR. (1998). *Arch. Biochem. Biophys.*, **358**, 232–242.
153. Palmer RMJ, Ferrige AG and Moncada S. (1987). *Nature*, **327**, 524–526.
154. Park H-S, Huh S-H, Kim M-S, Lee SH and Choi E-J. (2000). *Proc. Natl. Acad. Sci. USA*, **97**, 14382–14387.
155. Patel KB, Stratford MRL, Wardman P and Everett SA. (2002). *Free Radical Biol. Med.*, **32**, 203–211.
156. Paxinou E, Weisse M, Chen Q, Souza JM, Hertkorn C, Selak M, Daikhin E, Yudkoff M, Sowa G, Sessa WC and Ischiropoulos H. (2001). *Proc. Natl. Acad. Sci. USA*, **98**, 11575–11580.
157. Pearce LL, Epperly MW, Greenberger JS, Pitt BR and Peterson J. (2001). *Nitric Oxide*, **3**, 128–136.
158. Pearce LL, Kanai AJ, Birder LA, Pitt BR and Peterson J. (2002). *J. Biol. Chem.*, **277**, 13556–13562. | Article
159. Perez-Mato I, Castro C, Ruiz FA, Corarales FJ and Mato JM. (1999). *J. Biol. Chem.*, **274**, 17075–17079. | Article ISI |
160. Pfeiffer S, Leopold E, Hemmens B, Schmidt K, Werner ER and Mayer B. (1997). *Free Radical Biol. Med.*, **22**, 787–794.
161. Pfeilschifter J, Eberhardt W and Beck K-F. (2001). *Pflugers Arch-Eur. J. Physiol.*, **442**, 479–486.
162. Phoa N and Epe B. (2002). *Carcinogenesis*, **23**, 469–475.
163. Phung YT and Black SM. (1999). *IUBMB Life*, **48**, 333–338.
164. Pineda-Molina E and Lamas S. (2001). *Biofactors*, **15**, 113–115.
165. Prütz WA, Butler J and Land E J. (1983). *Int. J. Radiat. Biol.*, **44**, 183–196.
166. Prütz WA, Mönig H, Butler J and Land EJ. (1985). *Arch. Biochem. Biophys.*, **243**, 425–434.
167. Radi R, Cassina A, Hodara R, Quijano C and Castro L. (2002). *Free Radical Biol. Med.*, **33**, 1451–1464.
168. Radi R, Denicola A, Alvarez B, Ferrer-Sueta G and Rubbo H. (2000). *Nitric Oxide Biology and Pathobiology*. Ignarro LJ (ed). Academic Press: San Diego, pp. 57–82.
169. Radi R, Peluffo G, Alvarez MN, Naviliat M and Cayota A. (2001). *Free Radical Biol. Med.*, **30**, 463–488.
170. Rae TD, Schmidt PJ, Pufahl RA, Culotta VC and O'Halloran TV. (1999). *Science*, **284**, 805–808.

171. Raju U, Gumin GJ and Tofilon PJ. (2000). *Int. J. Radiat. Biol.*, **76**, 1045–1053.
172. Ramachandran N, Root P, Jiang X-M, Hogg PJ and Mutus B. (2001). *Proc. Natl. Acad. Sci. USA*, **98**, 9539–9544.
173. Romashko DN, Marban E and O'Rourke B. (1998). *Proc. Natl. Acad. Sci. USA*, **95**, 1618–1623.
174. Roots R and Okada S. (1975). *Radiat. Res.*, **64**, 306–320.
175. Ross AB, Mallard WG, Helman WP, Buxton GV, Huie RE and Neta P. (1998).*NDRL-NIST Solution Kinetics Database: Ver 3*. Notre Dame Radiation Laboratory and National Institute of Standards and Technology: Notre Dame, IN and Gaithersburg, MD.
176. Rota C, Chignell CF and Mason RP. (1999a). *Free Radical Biol. Med.*, **27**, 873–881.
177. Rota C, Fann YC and Mason RP. (1999b). *J. Biol. Chem.*, **274**, 28161–28168.
178. Sarkela T, Berthiaume J, Elfering S, Gybina A and Guilivi C. (2001). *J. Biol. Chem.*, **276**, 6945–6949.
179. Savitsky PA and Finkel T. (2002). *J. Biol. Chem.*, **277**, 20535–20540.
180. Schieke SM, Briviba K, Klots LO and Sies H. (1999). *FEBS Lett.*, **448**, 301–303.
181. Schmidt K, Desch W, Klatt P, Kukovetz WR and Mayer B. (1997). *Naunyn-Schmiedeberg's Arch. Pharmacol.*, **355**, 457–462.
182. Schmidt-Ullrich R, Dent P, Grant S, Mikkelsen RB and Valerie K. (2000). *Radiat. Res.*, **153**, 245–257.
183. Schonhoff CM, Daou MC, Jones SN, Schiffer CA and Ross AH (2002). *Biochemistry*, **41**, 13570–13574.
184. Sciorati C, Nistico G, Meldolesi J and Clementi E. (1997). *Br. J. Pharmacol.*,**122**, 687–697. | Article |
185. Sheffler LA, Wink DA, Melillo G and Cox GW. (1995). *J. Immunol.*, **155**, 886–894.
186. Shiloh Y and Kastan MB. (2001). *Adv. Cancer Res.*, **83**, 209–254.
187. Souza JM, Choi I, Chen Q, Weisse M, Daikhin E, Yudkoff M, Obin M, Ara J, Horowitz J and Ischiropoulos H. (2000). *Arch. Biochem. Biophys.*, **380**, 360–366.
188. Souza JM, Daikhin E, Yudkoff M, Raman CS and Ischiropoulos H. (1999).*Arch. Biochem. Biophys.*, **371**, 169–178.
189. Srivastava RK, Sollott SJ, Khan L, Hansford R, Lakatta EG and Longo

DL. (1999). *Mol. Cell. Biol.*, **19**, 5659–5674.
190. St Croix CM, Wasserloos KJ, Dineley KE, Reynolds IJ, Levitan ES and Pitt BR. (2002). *Am. J. Physiol.*, **282**, L185–L192.
191. Stamler JS and Hausladen A. (1998). *Nat. Struct. Biol.*, **5**, 247–249.
192. Stamler JS, Toone EJ, Lipton SA and Sucher NJ. (1997). *Neuron*, **18**, 691–696.
193. Stephenson MA, Pollock S, Coleman NC and Calderwood S. (1994). *Cancer Res.*, **54**, 12–15.
194. Stuehr DJ. (1999). *Biochim. Biophys. Acta*, **1411**, 217–230.
195. Stuehr DJ, Pou S and Rosen GM. (2001). *J. Biol. Chem.*, **276**, 14533–14536.
196. Sturgeon BE, Sipe Jr HJ, Barr DP, Corbett JT, Martinez JG and Mason RP. (1998). *J. Biol. Chem.*, **273**, 30116–30121.
197. Suhasini MH, Li SM, Lohmann GR, Boss RB and Pilz. (1998). *Mol. Cell. Biol.*,**18**, 6983–6994.
198. Szabo C and Ohshima H. (1997). *Nitric Oxide*, **1**, 373–385.
199. Szatrowski TP and Nathan CF. (1991). *Cancer Res.*, **51**, 794–798.
200. Todd D and Mikkelsen RB. (1994). *Cancer Res.*, **54**, 5224–5230.
201. Tonks NK and Neel BG. (1996). *Cell*, **87**, 365–368.
202. Tuttle S, Horan AM, Koch CJ, Held K, Manevich Y and Biaglow J. (1998).*Int. J. Rad. Oncol. Biol. Phys.*, **42**, 833–838.
203. Uppu RM and Pryor WA. (1999). *J. Am. Chem. Soc.*, **121**, 9738–9739.
204. Vercesi AE, Kowaltowski AJ, Grijalba MT, Meinicke AR and Castilho RF. (1997). *Biosci. Rep.*, **17**, 43–52.
205. Voehringer DW, McConkey DJ, McDonnell TJ, Brisbay S and Meyn RE. (1998). *Proc. Natl. Acad. Sci. USA*, **95**, 2956–2960.
206. von Sonntag C and Schuchmann H-P. (1997). *Peroxyl Radicals*. Alfassi ZB (ed). Wiley: New York, pp. 173–234.
207. Wang S, Wang W, Wesley RA and Danner RL. (1999).*J. Biol. Chem.*,**274**, 33190–33193.
208. Wang X, Michael D, de Murcia G and Oren M. (2002).*J. Biol. Chem.*,**277**, 15697–15702.
209. Ward J. (1994). *Radiat. Res.*, **138**, S85–S88.
210. Wardman P. (1995). *Biothiols in Health and Disease*. Packer L and Cadenas E (eds). Marcel Dekker: New York, pp. 1–19.
211. Wardman P. (1998). *The Chemistry of N-Centered Radicals*. Alfassi ZB

(ed). Wiley: New York, pp. 155–179.
212. Wardman P. (1999). *The Chemistry of S-Centered Radicals*. Alfassi ZB (ed). Wiley: New York, pp. 289–309.
213. Wardman P, Burkitt MJ, Patel KB, Lawrence A, Jones CM, Everett SA and Vojnovic B. (2002). *J. Fluoresc.*, **12**, 65–68.
214. Wardman P and von Sonntag C. (1995). *Methods Enzymol.*, **251**, 31–45.
215. Wilhelm D, Bender K, Knebel A and Angel P. (1997). *Mol. Cell. Biol.*, **17**, 4792–4800.
216. Williams DLH. (1997). *Nitric Oxide*, **1**, 522–527.
217. Williams DLH. (1988). *Nitrosation*. Cambridge University Press: Cambridge.
218. Willson RL. (1977). *Iron Metabolism. Ciba Foundation Symposium 51 (new series)*. Elsevier/Exerpta Medica/North-Holland: Amsterdam, pp. 331–354.
219. Wilson I, Wardman P, Cohen GM and d'Arcy Doherty M. (1986). *Biochem. Pharmacol.*, **35**, 21–22.
220. Wink DA, Cook JS, Pacelli R, Liebmann J, Krishna MC and Mitchell JB. (1995). *Toxicol. Lett.*, **82/83**, 221–226.
221. Winterbourn CC and Kettle AJ. (2000). *Free Radical Biol. Med.*, **29**, 403–409.
222. Wong PS-Y, Hyun K, Fukuto JM, Shirota FN, DeMaster EG, Shoeman DW and Nagasawa HT. (1998). *Biochemistry*, **37**, 5362–5371.
223. Wu LJ, Randers-Pehrson, Xu A, Waldren CA, Geard CR, Yu Z and Hei TK. (1999). *Proc. Natl. Acad. Sci. USA*, **96**, 4959–4964.
224. Xanthoudakis S, Miao G, Wang F, Pan Y-C and Curran T. (1992). *EMBO J.*, **11**, 3323–3335.
225. Yang J, Liu X, Bhalla K, Kim CN, Ibrado AM, Cai J, Peng T-I, Jones DP and Wang X. (1997). *Science*, **275**, 1129–1132.
226. Yoo JC, Pae HO, Choi BM, Kim WI, Kim JD, Kim YM and Chung HT. (2000).*Free Radical Biol. Med.*, **28**, 390–396.
227. Zeng J, Heuchel R, Schaffner W and Kagi JH. (1991a). *FEBS Lett.*, **279**, 310–312.
228. Zeng J, Valle BL and Kagi JH. (1991b). *Proc. Natl. Acad. Sci. USA*, **88**, 9984–9988.
229. Zheng M, Azlund F and Storz G. (1998). *Science*, **279**, 1718–1721.
230. Zheng M and Storz G. (2000). *Biochem. Pharmacol.*, **59**, 1–6.

231. Zorov DB, Filburn CR, Klotz L-O, Zweier JL and Sollott SJ. (2000). *J. Exp. Med.*, **192**, 1001–1014.
232. Zou M-H, Shi C and Cohen RA. (2002). *J. Clin. Invest.*, **109**, 817–826.

CITATION

CHAPTER 1
Muhammad Abdul Mojid Mondol, Hee Jae Shin and Mohammad Tofazzal Islam, Diversity of Secondary Metabolites from Marine Bacillus Species: Chemistry and Biological Activity, doi:10.3390/md11082846.

CHAPTER 2
Mijat Božović, Adele Pirolli and Rino Ragno, Mentha suaveolens Ehrh. (Lamiaceae) Essential Oil and Its Main Constituent Piperitenone Oxide: Biological Activities and Chemistry, doi:10.3390/molecules20058605.

CHAPTER 3
Shohei Sakuda, Hiromasa Inoue and Hiromichi Nagasawa, Novel Biological Activities of Allosamidins, doi:10.3390/molecules18066952.

CHAPTER 4
Sevim Rollas and Ş. Güniz Küçükgüzel, Biological Activities of Hydrazone Derivatives, doi:10.3390/12081910.

CHAPTER 5
Yasmeen Gull, Nasir Rasool, Mnaza Noreen, Ataf Ali Altaf, Syed Ghulam Musharraf, Muhammad Zubair, Faiz-Ul-Hassan Nasim, Asma Yaqoob, Vincenzo DeFeo and Muhammad Zia-Ul-Haq, Synthesis of N-(6-Arylbenzo[d]thiazole-2-acetamide Derivatives and Their Biological Activities: An Experimental and Computational Approach, doi:10.3390/molecules21030266.

CHAPTER 6
Bhuwan K. Chhetri, Nasser A. Awadh Ali and William N. Setzer, A Survey of Chemical Compositions and Biological Activities of Yemeni Aromatic Medicinal Plants, doi:10.3390/medicines2020067.

CHAPTER 7
Sidney A. Katz, The Chemistry and Toxicology of Depleted Uranium, doi:10.3390/toxics2010050.

CHAPTER 8
Yoshiko Miura, Shunsuke Onogi and Tomohiro Fukuda, Syntheses of Sulfo-Glycodendrimers Using Click Chemistry and Their Biological Evaluation, doi:10.3390/molecules171011877.

CHAPTER 9
Ross B Mikkelsen and Peter Wardman, Biological chemistry of reactive oxygen and nitrogen and radiation-induced signal transduction mechanisms, doi:10.1038/sj.onc.1206663.

INDEX

A

Acetylcholine esterase (AChE) 169
Alanine amino transferase (ALT) 49
Alkaline phosphatase (ALP) 49
Aminobenzothiazole derivatives 134
Aspartate amino transferase (AST) 49
Atomic force microscopy (AFM) 233

B

Bronchoalveolar lavage (BAL) 82
Butylated hydroxytoluene (BHT) 47

C

Chitin synthesis 71
Circular dichroism (CD) 233
Cyclic lipopeptides (cLPs) 4

D

Depleted uranium (DU) 193
Dimethylformamide (DMF) 135
Dynamic light scattering (DLS) 239

E

Electroencephalographic (EEG) 212
Electrospray ionization mass spectroscopy (ESI-MS) 236

F

Fatty acid synthases (FAS) 13
Fluorescein isothiocyanate (FITC) 238

G

Glycol-clusters containing sulfonated N-acetyl-D-glucosamine (GlcNAc) 231
Glycosaminoglycans (GAGs) 232

H

Herpes simplex virus-1 (HSV-1) 123
Higher Education Commission (HEC) 155
Higher Education Quality Enhancement Project (HEQEP) 23
Human immunodeficiency virus (HIV) 123

I

International Commission on Radiological Protection (ICRP) 207
Isonicotinoylhydrazones (ISNEs) 113

M

Matrix-assisted laser desorption ionization and time of flight mass spectroscopy (MALDI-TOF-MS) 236
Maximal electroshock (MES) 98
Methicillin-resistant Staphylococcus aureus (MRSA) 170
Meticilin-resistant Staphylococcus aureus (MRSA) 55
Minimum bactericidal concentration (MBC) 170
Minimum inhibitory concentrations (MIC) 116
Minimum inhibitory concentrations (MICs) 7

N

Non-nucleoside inhibitors (NNIs) 123
Nonribosomal peptides (NRPs) 7

P

Peptide deformylase (PDF) 16
Piperitenone oxide (PO) 33, 34, 38
PKS (Polyketide synthetase) 4
Productive agricultural region 162
Pyrroloquinoline quinone (PQQ) 198

R

Reactive nitrogen species (RNS) 259
Reactive oxygen species (ROS) 46, 259, 260

Reverse transcriptase (RT) 123

S

Southern Arabian Peninsula 162, 169
Subcutaneous pentylenetetrazole (scPTZ) 98
Substrate-assisted mechanism 73, 86
Superoxide dismutase (SOD) 262
Suzuki cross coupling methodology 134, 135

T

Thiobarbituric acid reactive substances (TBARS) 170

U

United Nations Environment Programme (UNEP) 219
Urease inhibition 134, 138, 141, 152, 155, 158
US Food and Drug Administration (US-FDA) 20

V

Vancomycin-resistant enterococcus (VRE) 8

W

Wheat germ agglutinin (WGA) 233
World Health Organization (WHO) 34